Gold from Newton's Apple Tree

Gold from Newton's Apple Tree

HISTORICAL RECIPES *for* NATURAL INKS, PAINTS & DYES

NABIL ALI

PRINCETON UNIVERSITY PRESS
PRINCETON AND OXFORD

Published by Princeton University Press
41 William Street, Princeton, New Jersey 08540
99 Banbury Road, Oxford OX2 6JX
press.princeton.edu

GPSR Authorized Representative: Easy Access System Europe - Mustamäe tee 50, 10621 Tallinn, Estonia
gpsr.requests@easproject.com

Library of Congress Control Number: 2025935063
ISBN: 978-0-691-27821-6
Ebook ISBN: 978-0-691-27822-3
British Library Cataloging-in-Publication Data is available

This book was conceived, designed, and produced by UniPress Books Limited

Commissioning Editor Jason Hook
Associate Publisher Daniel Mills
Art Director Alex Coco
Project Manager Ruth Patrick
Designer Luke Herriott
Editor Caroline West
Picture Researcher Julia Ruxton
Cover design by Luke Herriott

Printed in Malaysia
10 9 8 7 6 5 4 3 2 1

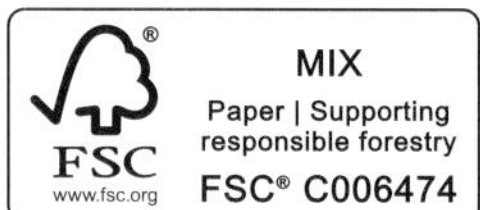

Contents

Introduction

Gold from Newton's Apple Tree is an exploration of the art of extracting colors from plants, and draws its inspiration from sources dating back to antiquity and continuing in the present day through my own work as a visual artist and researcher. The potential for plant material to be transformed into an array of colors is examined through the art of organic alchemy. This process harnesses the diverse pigments and dyes found in plants and develops them into a new and sometimes unpredictable palette of colors.

People often ask how I began using plants to create organic paint systems. My inspiration to make an herb garden developed from a wood engraving in Adam Lonicer's *The Herbal* (1545), which depicted a gardener working in an herb garden surrounded by raised beds and potted plants, pear trees, and grapevines. The first seed I planted was woad, primarily for the second year's seed crop. Most of my herbs were grown from seed, and within two years, I had established a potted grapevine, trees, and bushes, and thirty different herb plants for cooking, scent, brewing tea, and, of course, making colors. This became my herb garden, which was encouraged by a local elderly lady named Ivy, whose little cat sometimes hid in my laurel bushes.

In 2002, I found myself contemplating whether to pursue a fine art degree, so I decided to integrate plants into my academic journey, creating a paint palette using the herbs to produce a collection of artist's colors and installations. The first plant I processed was soldier's woundwort, more commonly known as yarrow (*Achillea millefolium*), which took more than four hours to grind into a fine pigment before gum tragacanth was mixed in as a binder to produce a golden-colored paint. I then continued to process other plants into dry pigments for further exploration.

Soon after, I traveled to Italy to meet a specialist dyer linked to the Spindigo Project (Sustainable Production of Plant-derived Indigo), carrying a silver-colored briefcase that contained eighteen jars of plant pigments segmented and cushioned in foam. The dyers were intrigued by my work, as were the airport security personnel when my pigment briefcase went through the X-ray scanners, prompting them to open and inspect the contents with puzzled expressions.

My research continued following graduation, focusing on plants mentioned in historical technical manuscripts. I discovered that some of the most significant medieval plant colors referenced in the manuscripts of painters, illuminators, and dyers included weld, madder, woad, brazilwood, and buckthorn. This research culminated in the *Artist's Garden* exhibition, which was displayed in Lisbon, Portugal, and organized by the Dyes in History and Archaeology (DHA) committee. DHA is an international annual conference that focuses on discussion of dyes and organic pigments used in the past. The exhibition showcased colors

A gardener in an herb garden. Wood engraving from Adam Lonicer's *The Herbal* (1545).

Von Bauung der Gärten / und Pflantzung der Bäume.

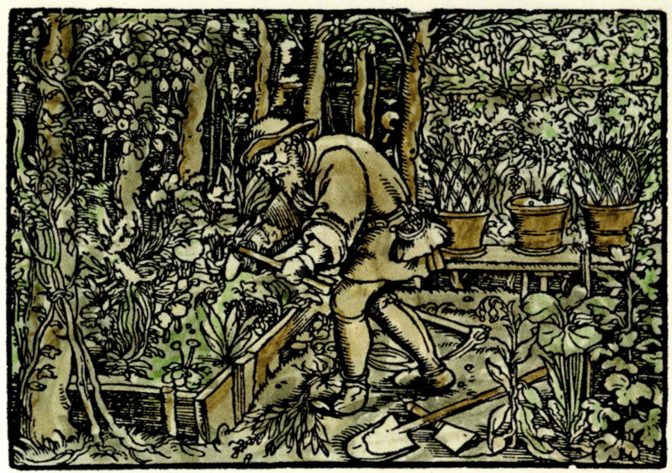

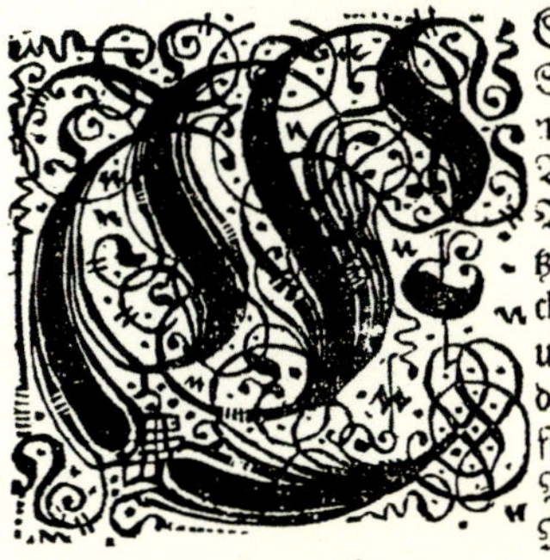

S hat der Allmächtige Gott anfangs der Schöpffung der Welt / das Erdrich zum zierlichsten mit allerhand Edlen Gewächsen / von Kräutern / und Bäumen / in aller vollkommenheit / dem Menschen zum Wollust / und zu seiner Notdurfft zu gebrauchen / besetzet und geschmückt / auch den Menschen in den herrlichen Lustgarten deß Paradeiß / denselbigen zu besitzen / und seinen Lust darinnen zu haben / eingesetzet / und hat der Erdboden damals alles für sich selbst / ungebauet herfür gegeben. Solche Freude und Wollüste / herrliches Wesen / und müssige Leben / hat der leydige Satan dem Menschen mißgönnet / und alle seine Liste angestellet / daß er ihn möchte zur übertrettung deß Göttlichen Gebotts reitzen / und also Gottes Zorn wider ihn erregen / daß er auß solchem Lustgarten verstossen würde / und mit Mühe und Arbeit sich ernehren müste. Da nun der böse Feind den Menschen also betrogen hat / da ist der Fluch über uns ergangen / und dem Erdreich seine Krafft auch genommen worden / Wie dann Genes. am 3. Cap GOtt zum Menschen spricht: **Verflucht sey der Acker um deinet willen / mit Kummer solt du dich darvon nehrē dein Lebenlang / Dorn und Disteln soll er dir tragē / und solt das Kraut auf dem Feld essen. In Schweiß deines Angesichts solt du dein Brodt essen / biß du wieder zur Erden werdest / davon du genommen bist.**

Also haben wir nichts dann Mühe und Arbeit / so lang wir auf dieser Welt leben / müssen wir mit Kummer / Angst / und grosser Mühe das Feld bauen und pflantzen / und wann wir lang grosse Arbeit und Unkosten geführet / so kommt offtmals Reiffe / Hagel / Ungewitter und Ungezieffer / beschädiget und verderbet alles / was gepflantzet ist / an Obs / an Früchten und Wein / daß nichts dann Jammer und Elend zu sehen ist,

Dieweil dann nun unsere Leibs notdürfftige Underhaltung erheischet / die Gärten und das Feld stättigs mit säen und Pflantzen / daß sie Getreyde / Wein und Obs / jährlich uns ertragen / zu erbauen / so haben je und allweg unsere erste Eltern und deren Nachkömlinge mit allem Fleiß deß Feldbauens und Gartenpflantzens sich angenommen / und den grösten Reichthumb und Nahruug an den Feldgütern und Viehezucht gehabt /

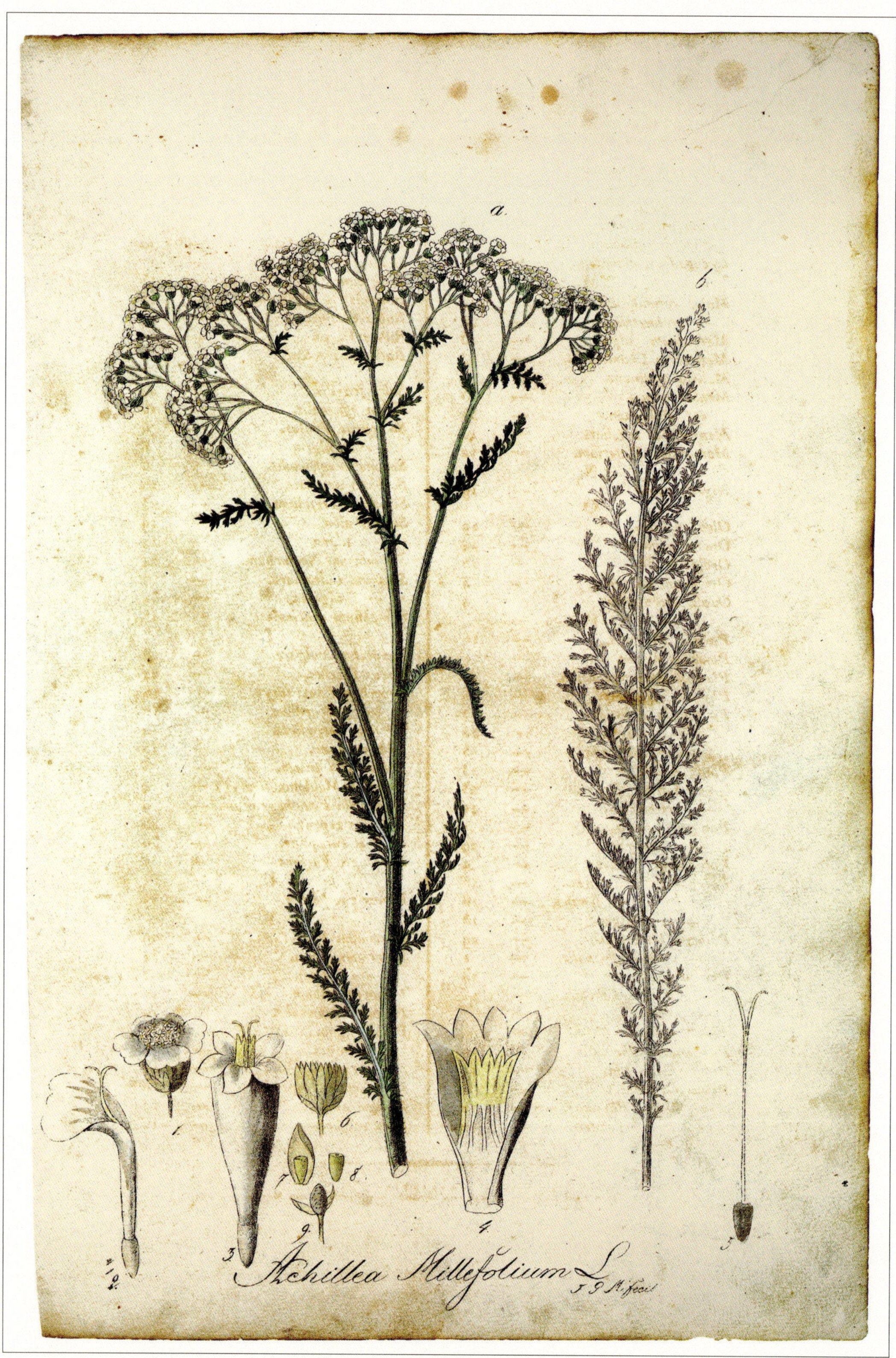
a.
b.
Achillea Millefolium L.

derived from plants and featured a collection of images illustrating those recipes, influenced by historical technical manuals and techniques, and used by craftsmen and artisans throughout the centuries. I also created a hybrid ink for a conceptual art piece, which combined a 15th-century German poppy recipe and a 14th-century Venetian recipe, with additional ingredients such as the bladder of a sturgeon and white wine. The ink was kept in a Victorian inkwell, which was fixed to a military brass gun case dated 1916 to commemorate those affected by the First World War.

Another artwork drew on a 15th-century manuscript called *The Göttingen Model Book*, which illustrated how to draw acanthus leaves in illuminated manuscripts. I transformed the leaf shapes into large abstract sculptures with spheres that represented repetition, rotation, and circular motion. The spheres were painted green according to a French medieval recipe that used purple iris flower dye adhered to powdered chalk and mixed with tree gum. I continued my investigations by creating a collection of abstract paintings using egg tempera. These were painted using green stinging nettle pigment from the crushed dried leaves, black pigment from charred nettle stalks, and crushed eggshell to include a white pigment.

For a SciArt Residency at FrukLab at the University of Cambridge in 2024–25, I developed a permanent painting collection using a range of different plant pigments, such as weld (*Reseda luteola*), madder (*Rubia tinctorum*), and indigo (*Indigofera tinctoria)*, which focuses on color composition. Further colors were processed from the leaves of an on-site staghorn sumac tree (*Rhus typhina*) following a 14th-century recipe to produce a black colorant. Additional cancer treatment plants were used in the work, including the Madagascar periwinkle (*Catharanthus roseus*), which makes a green-black color, and a yellow color from leaves of the mayapple (*Podophyllum peltatum*), both collected from Cambridge University Botanic Garden glasshouse and woodland area. This collection also included 23.5-karat gold leaf and chalk collected from the Cambridge Cherry Hinton chalkpits.

The following pages will guide you through a color-by-color investigation of historical and technical texts, illuminated manuscripts, medical herbals and remedies, physicians and artists, the natural and horticultural worlds, and various cultures. Here, you'll find information on a selection of plants from around the world, including well-known species and those that may be less familiar. As well as long-forgotten stories, people, and recipes, also included are recipes that I developed to create colorants and as a way of engaging with nature. At the end of each chapter you will find a summary of the colors that can be created using the plants, from the medieval recipes to the featured contemporary recipes. Exploring plant dyes, inks, and paints for yourself will give the possibilities for making old and new colors.

Yarrow (*Achillea millefolium*) can be ground into a pigment to create a golden color. Handcolored lithograph from Johann Gottlieb Mann's *Deutschlands Wildwachsende Arzney-Pflanzen* (Germany's Wild Medicinal Plants; ca. 1828).

The physician and painter

There were two main guilds in medieval Europe and these represented the craftsmen and merchants, who fulfilled economic, educational, social, and religious functions. Florentine painters and illuminators belonged to the Arte dei Medici e degli Speziali

(Guild of Physicians and Pharmacists), which emerged during the mid-1300s. This was also a guild for physicians, apothecaries, and spice merchants, along with other overlapping trades, and supplied artist's pigments and a range of raw materials, from organic materials to mineral and earth pigments and other ingredients. Venice established the Arte dei Depentori (Guild of Painters) in 1271, whose members included figure painters, sign painters, furniture painters, mask makers, illuminators, gilders, leatherworkers, textile designers and embroiderers, and playing-card makers.

Each European city had its own guilds. Such guilds provided rules and regulations for skilled artisans, although these were not standardized or universal. Nor would the artisans necessarily share the same patron saint. For example, in the city of York, in England, the medieval painters, stainers, and gold-beaters—whose patron saint was Dunstan and sometimes St. Eligius—all belonged to one guild. The guilds were all distinctly different from each other, especially in London, where the guild for painters and stainers stipulated the use of specific materials; along with the illuminators in Bruges, for instance, stainers were only allowed to use watercolor paints and not oils, which were reserved for painters. Indeed, the Painter-Stainers' Company was established in 1502 in London due to previous disputes between the painters and the stainers. The organization still maintains traditional customs and remains under the guardianship of St. Luke, the patron saint of painters and doctors as well as other tradespeople such as tailors, cheesemakers, and gold-beaters.

Establishing guilds in this way separated the trades into sectors relating to specific skills and artistic purposes: for example, stainers were responsible for staining cloth, flag designs using stencils for pageants and funeral ceremonies, and imitation tapestries; painters undertook the painting of portraits, wooden panels, and barges or murals on walls; and saddlers painted saddles, shields, chariots, and banners for a different purpose. In Spain there were separate guilds for different painting applications—for example, painters of altarpieces, painters of interiors, and fabric painters—with similar trends in Paris and elsewhere in Europe.

Illuminators who used organic color materials were also associated with the monastic life of the scriptorium. They were not part of any craft guild outside their religious orders and demonstrated multiple skills. However, one renowned late 12th-century monk called Walter of Colchester was celebrated as a leading painter and sculptor at the English Benedictine Abbey of St. Albans; he was brought in as a skilled layman to paint the abbey, staying on to become an artistic monk-craftsman.

Organic paint colors, as well as mineral and earth pigments, were generally used by illuminators for book work in monasteries until the 11th–12th century and also outside them in workshops. Vitruvius states in his 1st-century *De Architectura Libri Decem* (Ten Books on Architecture) that some plant colors—for example, mountain pansies (*Viola lutea*; see page 136)—can be used by fresco painters, indicating that different craftsmen, other than dyers, could use plants for color, which is also seen in the 14th-century *Liber Diversarum Arcium* (Book of Various Arts), a collection of paint and dye recipes.

St. Luke, the patron saint of painters and doctors, was a doctor and painter who allegedly painted the Virgin Mary in person. Illuminated manuscript by the Bedford Master, showing St. Luke painting an image of the Virgin Mary (ca. 1440–50).

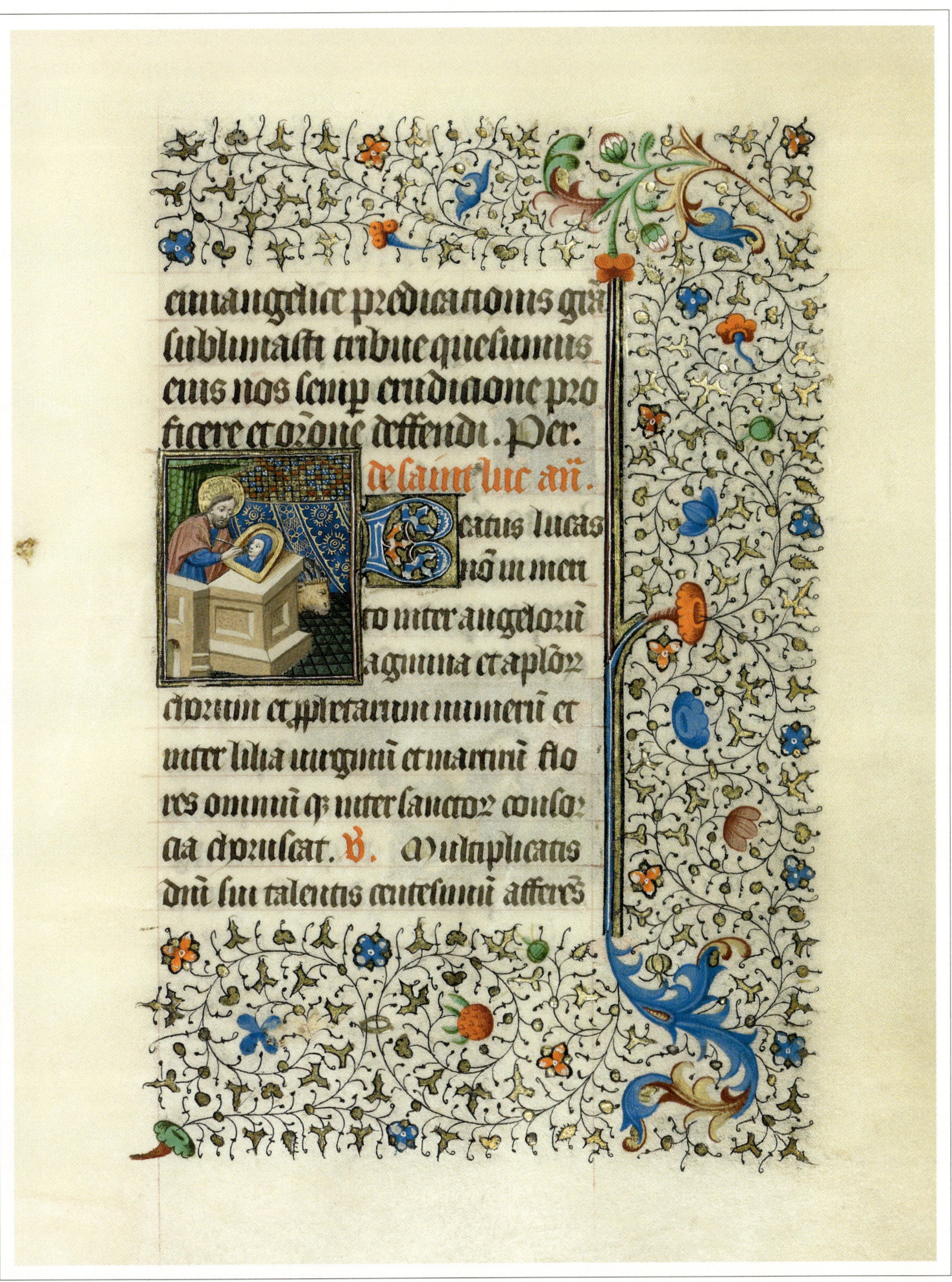

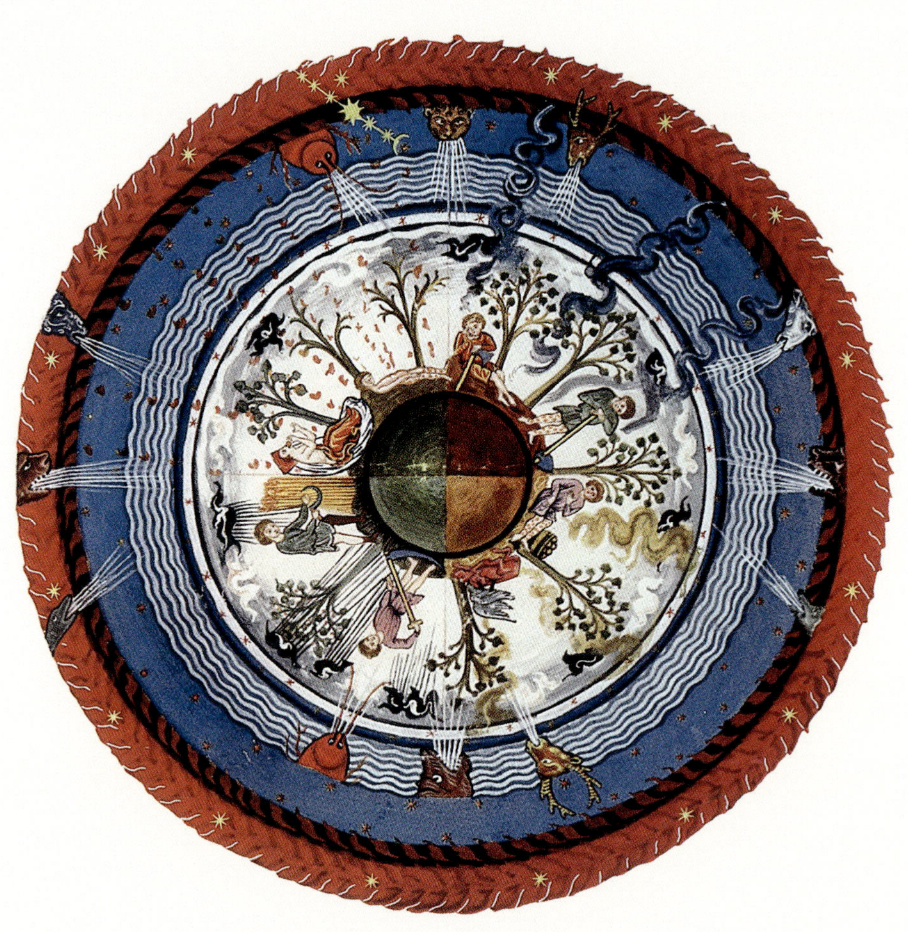

Natural dyes were highly prized by the spinners, weavers, and dyers of the wealthy Guild of Clothiers. They enjoyed a monopoly over the manufacture of the cloth depicted in medieval art by English royal court painters. Some of the guild's members were, indeed, skillful Benedictine monks from Westminster Abbey.

It is believed that St. Luke was a doctor and painter from the 1st century CE who created the only portrait in person of the Virgin Mary. The Guild of Saint Luke catered for both physicians and artists, primarily painters, who used many of the raw materials commonly employed in each field—for example, herbs; pigments; natural materials like cinnabar, sepiolite, and red ocher; and elements frequently found in apothecaries and dispensaries. Consequently, it is perhaps not surprising to find the recipes of painters, illuminators, dyers, and stainers alongside medical texts and manuscripts that mention plants, as the two were closely related.

Numerous technical recipes highlight the use of plants by both physicians and painters, with the recipes originally used by painters for panel paintings, as well as illuminators and dyers, and by textile craftsmen to dye cloth using "colored waters" for wall hangings. Technical manuscripts outlined detailed staining techniques, copied in many 15th-century Middle English manuscripts. One untitled manuscript (MS Ee.i.13, fol.131r–135v.), held in the Cambridge University Library, clearly highlights a collection of stainer's dye recipes, perhaps set out by the master of the workshop for the apprentice.

Ancient Mesopotamian clay tablets mention ingredients such as turmeric, saffron, hellebore, and alum mordants alongside a hair-dye recipe made from leeks and cassia extracts. The *Eber Papyrus*, an Egyptian herbal from around 1550 BCE, contains a wide array of plant remedies, including more than 800 medical plant recipes. These were further documented in early herbal medical texts and technical manuscripts used by medieval illuminators and painters.

Many of the materials were also well known to painters, dyers, and stain-makers. For instance, saffron (*Crocus sativus*) is referred to by the ancient Egyptians as "the Blood of Thoth," while other notable materials include aloe gum (*Aloe* species), gum tragacanth (*Astragalus verus* and *A. gummifer*), indigo (*Indigofera tinctoria*), woad (*Isatis tinctoria*), smoke tree (*Rhus* species), madder (*Rubia tinctorum*), dyer's broom (*Genista tinctoria*), and the weld plant (*Reseda luteola*).

Centuries later, the Worshipful Society of Apothecaries was established by a royal charter from James I in 1617 as the equivalent of the community pharmacy. In 1673 the society founded London's Chelsea Physic Garden, which continues to cultivate a variety of medical, edible, and dye plants known since antiquity.

Understanding the origins of colorants in relation to their historical context, especially organic colors and their ties with the medical field, is important, as this is how the medieval European guilds started to influence and standardize the quality and function of trade: by regulating industries through a range of rules. Ignoring these aspects diminishes the connection between other disciplines and our knowledge of the past, specifically regarding the use of plants. Furthermore,

This illustration shows the **cosmos, body, and soul** as having influences on mankind, animals, and plants through creation, fire, ether, water, the stars, and the winds. Color features strongly: the blue circle represents the water, the red the celestial fire, and the black and red circle represent judgment on Earth. Illustration from Hildegard of Bingen's *Liber Divinorum Operum* (Book of Divine Works; 13th century).

integrating this wisdom enhances the exploration of color and fosters a harmonious relationship with the natural world through creativity and a multidisciplinary perspective from past practice into the modern day.

Plants in technical manuscripts

The earliest surviving technical manuscripts date from around 1700 BCE and originated in Mesopotamia. The recipes highlight processes for making natural dyes, colored glazes, and imitation gems, which were recorded considerably later in the 3rd-century-CE *Leiden Papyrus X* and the *Papyrus Graecus Holmiensis* (often referred to as the *Stockholm Papyrus*) of around 300 CE.

Many centuries later in Europe, a collection of recipes from the early medieval period survives in the *Mappae Clavicula* manuscripts (written in Greek and Latin). The manuscripts include the *Lucca Manuscript* (Biblioteca Capitolare Feliniana, Codex 490); *Sélestat Lectionary* (Bibliothéque Humaniste, MS 17); and the *Phillipps-Corning Manuscript* (Rakow Research Library, Corning Museum of Glass, New York). They date from the 9th to the late 12th century CE, with some recipes from the classical period. They explain processes for using plants and organic matter, such as dragon's blood (see page 92) mixed with orpiment and juniper juice to make a golden sealing wax called *bero inbriome*. Other recipes cover using saffron with vermilion to produce an orange shade; the juice of a "lupine cluster" (probably of the plant's roots) combined with the sprouts of sea leek (*Allium ampeloprasum* var. *babingtonii*) and red natron salt to clean tarnished silver; and a recipe for "lulax," a shade of light blue made using the flowers of parsley (*Petroselinum crispum*), common flax (*Linum usitatissimum*), and violets (*Viola* species) along with blue lily and woad leaves. (In the past, the term *lily* might, in fact, have referred to iris.)

Recipes were often copied into later manuscripts, with a fine Latin example being the 14th-century *Liber Diversarum Arcium* (Book of Various Arts), preserved in manuscript MS H 277, an early-15th-century physical copy of the 14th-century collection held in the Bibliothèque Universitaire Historique de Médecine, Montpellier. It lists over thirty different plants for making paint and ink, as well as other craft skills. The manuscript also includes a handful of dyeing processes using various plants, including the roots of madder, brazilwood, weld, indigo, young fustic, sumac leaves, lichens, and ripe buckthorn berries. Another 14th-century Middle English manuscript called the *Five Medical Texts* (MS O.9.39), held at Trinity College, Cambridge, contains recipes for manufacturing pigments and dye, preparing skins and furs, imitating expensive imported leathers, counterfeiting semiprecious materials, adulterating verdigris, and making soaps and confectionery.

Later in the 16th century, Giovanventura Rosetti's *Plictho de Larte de Tentori* (Instructions in the Art of Dyers; 1548) reveals in-depth processes. The organics are oak galls (see page 204), pomegranate (*Punica granatum*; see page 208), peach stones, walnut husks, turnsole (also known as folium), and common rue (*Ruta graveolens*; see page 172), which causes skin irritation and blisters in sunlight.

Astragalus verus is a source of gum tragacanth, a common natural binder used in pigments. Illustration by J. H. Colen from Joseph Carson's Illustrations of Medical Botany (1847).

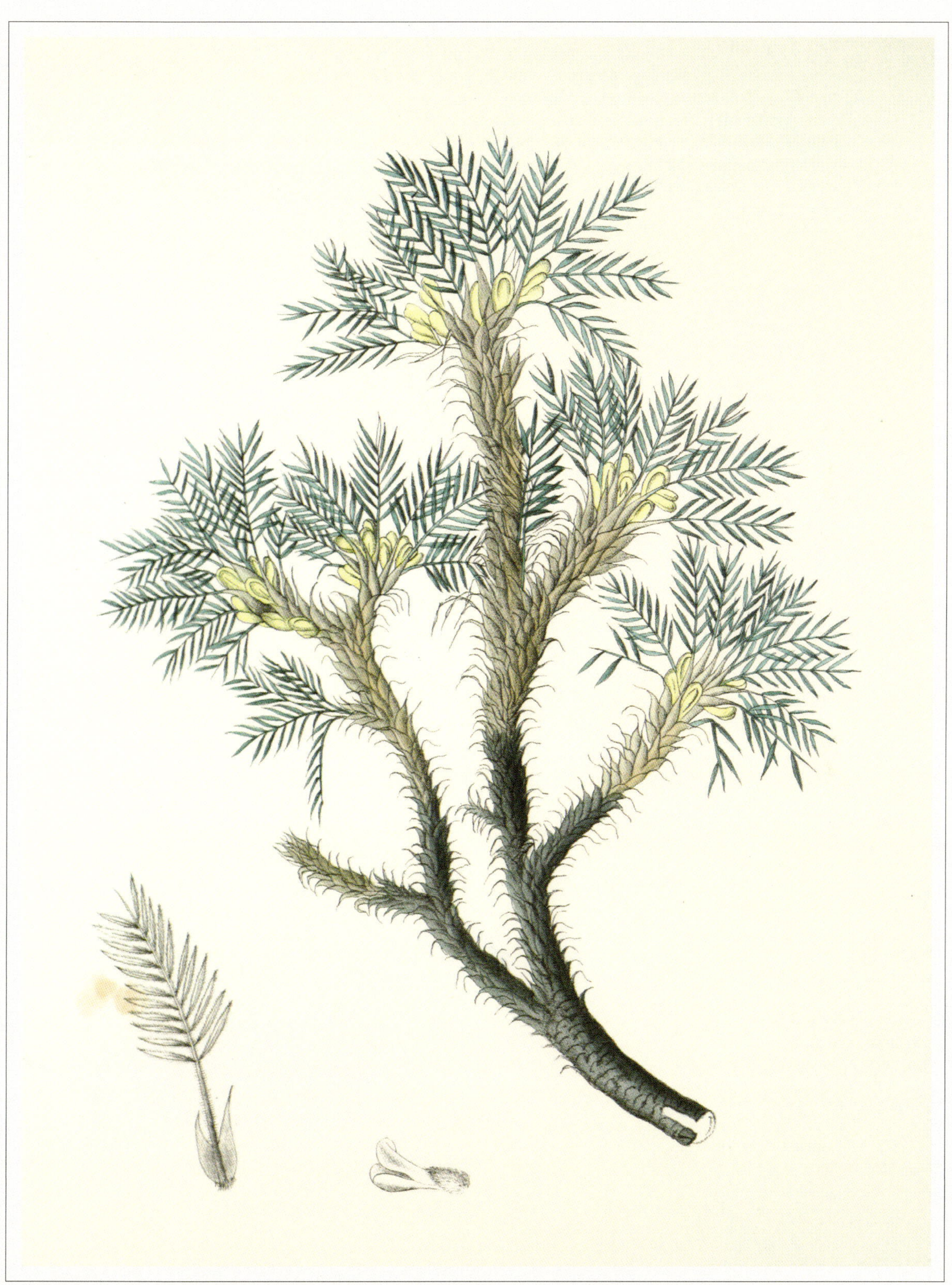

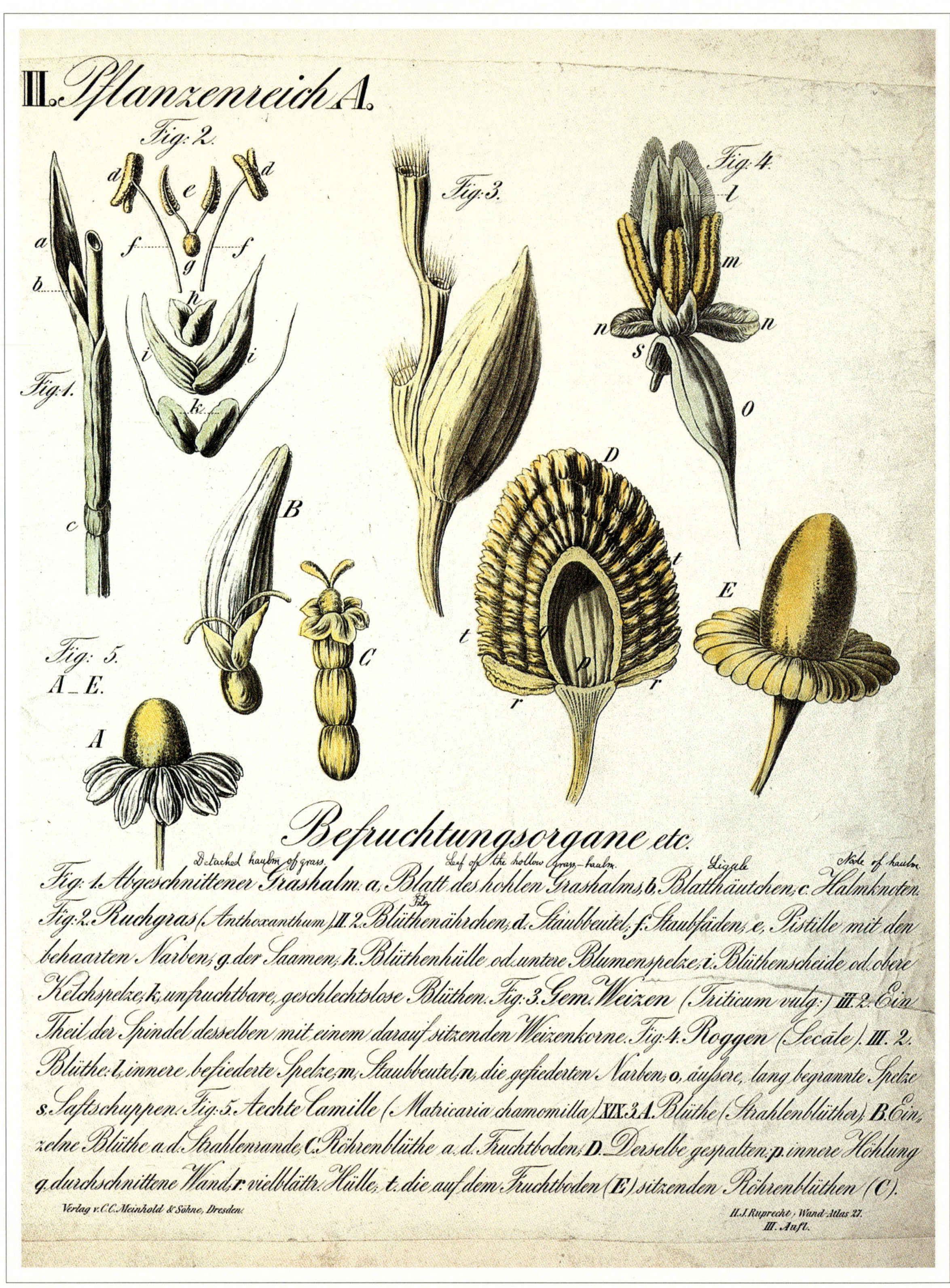
II. Pflanzenreich A.
Fig: 1.
Fig: 2.
Fig: 3.
Fig: 4.
Fig: 5.
A_E.
Befruchtungsorgane etc.
Detached haulm of grass
Leaf of the hollow grass-haulm
Ligule
Node of haulm
Fig: 1. Abgeschnittener Grashalm. a, Blatt des hohlen Grashalms, b. Blatthäutchen, c. Halmknoten.
Fig: 2. Ruchgras (Anthoxanthum) II. 2. Blüthenährchen, d. Staubbeutel, f. Staubfäden, e, Pistille mit den
behaarten Narben, g. der Saamen, h. Blüthenhülle od. untere Blumenspelze, i. Blüthenscheide od. obere
Kelchspelze, k, unfruchtbare, geschlechtslose Blüthen. Fig: 3. Gem. Weizen (Triticum vulg:) III. 2. Ein
Theil der Spindel desselben mit einem darauf sitzenden Weizenkorne. Fig: 4. Roggen (Secāle). III. 2.
Blüthe: l, innere befiederte Spelze, m, Staubbeutel, n, die gefiederten Narben, o, äußere, lang begrannte Spelze
s. Saftschuppen. Fig: 5. Aechte Camille (Matricaria chamomilla) XIX. 3. A. Blüthe (Strahlenblüthe). B. Ein-
zelne Blüthe a. d. Strahlenrande, C. Röhrenblüthe a. d. Fruchtboden, D. Derselbe gespalten: p. innere Höhlung
q. durchschnittene Wand, r. vielblättr. Hülle, t. die auf dem Fruchtboden (E) sitzenden Röhrenblüthen (C).
Verlag v. C.C. Meinhold & Söhne, Dresden.
H. J. Ruprecht, Wand-Atlas 27.
III. Aufl.

Thousands of such recipes appear in technical manuscripts throughout the world's libraries and this book highlights only a small proportion of them. Some originate from much earlier cultures, including those of ancient Egypt, Greece, and Rome, and were then translated over the centuries. Inevitably, such recipes featured in palimpsests (manuscripts written over due to the scarcity of expensive papyrus), appeared in lost or destroyed texts, or were the result of inaccurate interpretations.

The Framework

Working from ancient texts and medieval manuscripts, and even their modern translations, throws up a wealth of artistic and scientific information that can be difficult to navigate, especially for a beginner new to the world of organic colors. I have found that using a structured system to create these colors can help beginners grasp the basic principles of developing a paint palette from plants in their own garden with confidence. I have named this process "The Framework" (see Box, page 19), which serves as a practical working system.

When we observe a plant, we might appreciate its flowers, leaves, foliage, color, the texture of its bark, or the fruits and berries it produces. Yet every part of a plant has the potential to yield a colorant. The Framework consists of six sections, each containing the ingredients you can use to make an organic dye, paint, or ink.

1. Parts of plants

The first section outlines the parts of plants that can be processed. This includes the leaves; roots; berries, fruits, and seeds; flowers, petals, and stamens; and bark. To create organic paint, it is necessary to first extract a dye from one or more of these parts.

2. Liquids

The second section lists suitable liquids for extracting the dye from a plant. These include white wine, spirits (such as vodka, rum, or brandy), clear vinegar, and water. When using water, it is advisable to add a preservative.

3. Preservatives

The third section focuses on preservatives to prevent bacterial or mold growth in water. Grapefruit seed extract is very effective, as are concentrated rosemary extract and cloves. You can also add 3% Preventol® (sodium-2-phenylphenolate), which should be handled with care. Preservatives are less essential when using clear vinegar, spirits, or wine.

4. Mordants/additives

The fourth section lists some mordants or additives. A mordant is a metal salt that can alter the color characteristics of a dye and help it adhere to fabric fibers, stabilizing the color. When creating a dye, which is the midway point when making a paint, I use four little dishes for side tests. Each dish contains a different mordant,

Dissected flowers of a grass, rye, wheat, vernal grass, and chamomile, which can be used to make dye colors. Chromolithograph by H. J. Ruprecht (1877).

allowing me to add a small amount of dye and observe any color changes. Mordants can modify the pH level of the dye from acidic (pH 1–4) to alkaline (pH 9–12), influencing the final color. Suitable mordants include:

Alum (aluminum potassium sulfate) – helps to brighten the dye (pH 2–3)
Tin (stannous chloride) – brightens color and is acidic (pH 2–3)
Copper (copper sulfate) – produces greener hues (pH 3–4)
Iron (iron II sulfate/ferrous sulfate) – darkens color, reacts with tannins (pH 2–3)
Baking soda (bicarbonate of soda) – makes dye more alkaline (pH 9)
Potash (potassium carbonate) – enriches color and is very alkaline (pH 12)
Chrome (potassium dichromate) – brightens color (pH 3–5)

Safety notes: Handle mordants with care, as they can be skin irritants and should not be ingested or inhaled. Wearing gloves and goggles or a mask is advisable. Chrome is highly toxic and should be disposed of appropriately at a waste facility.

5. Fillers

The fifth section covers fillers, those substances to which the dye adheres when creating a colored pigment. The following materials can be used:

- Chalk (calcium carbonate)
- Eggshells (calcium carbonate)
- Marble dust (calcium carbonate)
- Gesso (contains gypsum or calcium sulfate dihydrate)
- Cuttlefish bone (calcium aragonite)
- Sepiolite clay (magnesium silicate)
- Kaolin or China clay (aluminum silicate hydrate)

6. Natural binders

The sixth section focuses on natural binders or glues that are soluble in water and used to fix the pigment to the surface. Other natural materials can serve as a binder, including non-animal substitute products. For instance, onion or garlic juice can act as a mild adhesive, making them ideal for gilding, along with fig tree latex. Which binder you use is a matter of personal preference.

- Gum arabic – *Acacia senegal* (now *Senegalia senegal*) and *Acacia seyal* (now *Vachellia seyal*) trees
- Gum tragacanth – from the *Astragalus gummifer* and *A. verus* shrubs
- Rabbit skin glue – processed animal hide in granule form
- Egg white (glaire), yolk – chicken eggs, which are easily accessible
- Isinglass glue – glue made from the bladder of the sturgeon fish
- Parchment glue – glue made by boiling scraps of parchment (from animal hide)

The Framework is a system for creating organic colors according to established guidelines. By using this system, the processes involved become significantly simpler and more fluent, allowing you to achieve your objectives when collecting plants from your garden and turning them into colorful paints, pigments, dyes, stains, and inks, many of which may surprise you.

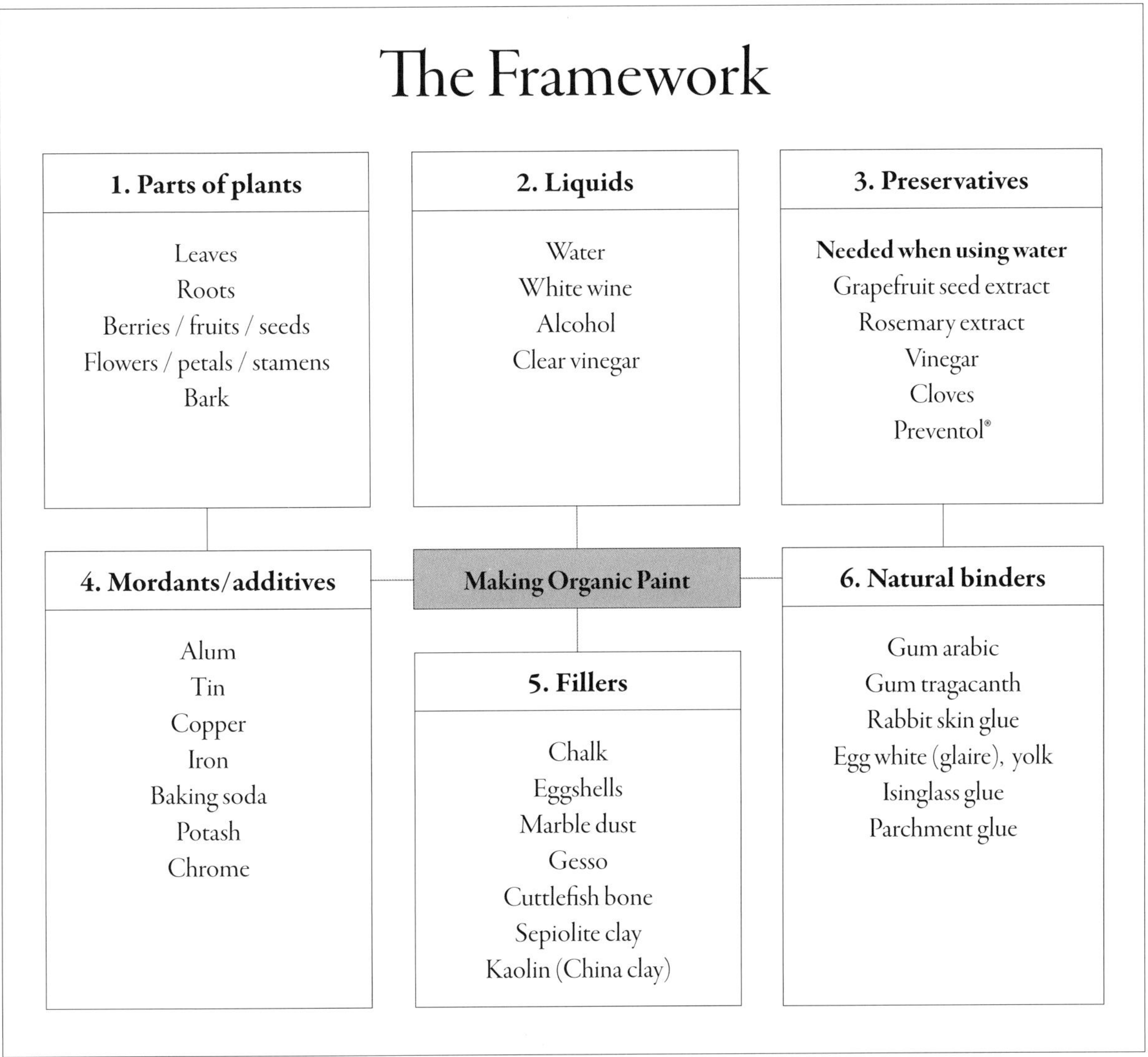

A note on recipe quantities

It is important to note that the quantities of plants used can vary significantly. For instance, leaves typically contain more colorant material than berries, although this is not always the case. The amount of colorant can also change based on factors such as the seasonal crop and whether the leaves are harvested in early spring or late summer. A good starting point is to try a small amount, then through experience you will learn how much is needed for the result you are trying to achieve.

Therefore, the recipes featured in this book should be viewed as guidelines, in terms of both the quantities and methods used, and can be adjusted according to your own experience and working practice. Some of the featured recipes do not include specific measurements so as to encourage a more organic approach.

Cambridge University Botanic Garden

The botanic garden at Cambridge University, in the heart of Cambridge, was founded in 1762 to provide medical students with a knowledge of plants used for treatments. Around 1846, Professor John S. Henslow relocated the garden from its original site, near today's New Museums Site, to its current location. Professor Henslow, a botanist, geologist, and clergyman, created wall charts and diagrams as teaching tools for his students, emphasizing the importance of plant diversity. He enlisted hundreds of collaborators, one of whom was Charles Darwin. Henslow's son, George, owned a 14th-century manuscript (MS Egerton 2852), now held by the British Library, London, which includes twenty-eight recipes for dyeing with plants. He referenced these recipes in his 1899 publication *Medical Works of the Fourteenth Century*.

The Cambridge University Botanic Garden is highly significant for plant research and has been invaluable for organic artists like myself. The garden provided an ideal environment for developing organic paint systems and exploring a diverse range of plants from around the world that could potentially produce dye colors.

I arrived at the gates of the garden as artist-in-residence in 2023, and was given the opportunity to explore the vast world of the plant kingdom. My goal was to create a public index catalog of dyes showcasing colors produced through a specific processing framework, as outlined on pages 17–19. The garden offered thousands of color possibilities, all waiting to be discovered and transformed into shades derived from nature. And so began my alchemical journey into the world of organic colors.

First, I obtained a detailed map of the garden, highlighting specific areas such as the glasshouses, David Rogerson Mellor's circular fountain, Cory Lodge, and the custodian hut, which historically housed the on-site policeman. I converted this hut into a temporary art space for a night installation called *The Hut of Curiosity*. This installation featured large apothecary jars, bottles, and Victorian decanters, artificially lit from below and filled with natural dyes that glowed in the dark.

I spent a considerable amount of time exploring the dye plants growing in the systematic garden beds. This became my starting point for transforming familiar plants, such as the weld plant, madder root, and woad leaves, into dyes. I also ventured into creating a colorant using the endangered Wollemi pine (*Wollemi nobilis*), working with several of its leaves. By adding iron sulfate and employing a heat and reducing method, I was able to produce a black ink. Furthermore, by applying a tin and alum mordant, I created a yellow, as well as achieving an orange color by adding potash.

Map of the Cambridge University Botanic Garden, drawn by the garden's first curator, Andrew Murray, in 1845.

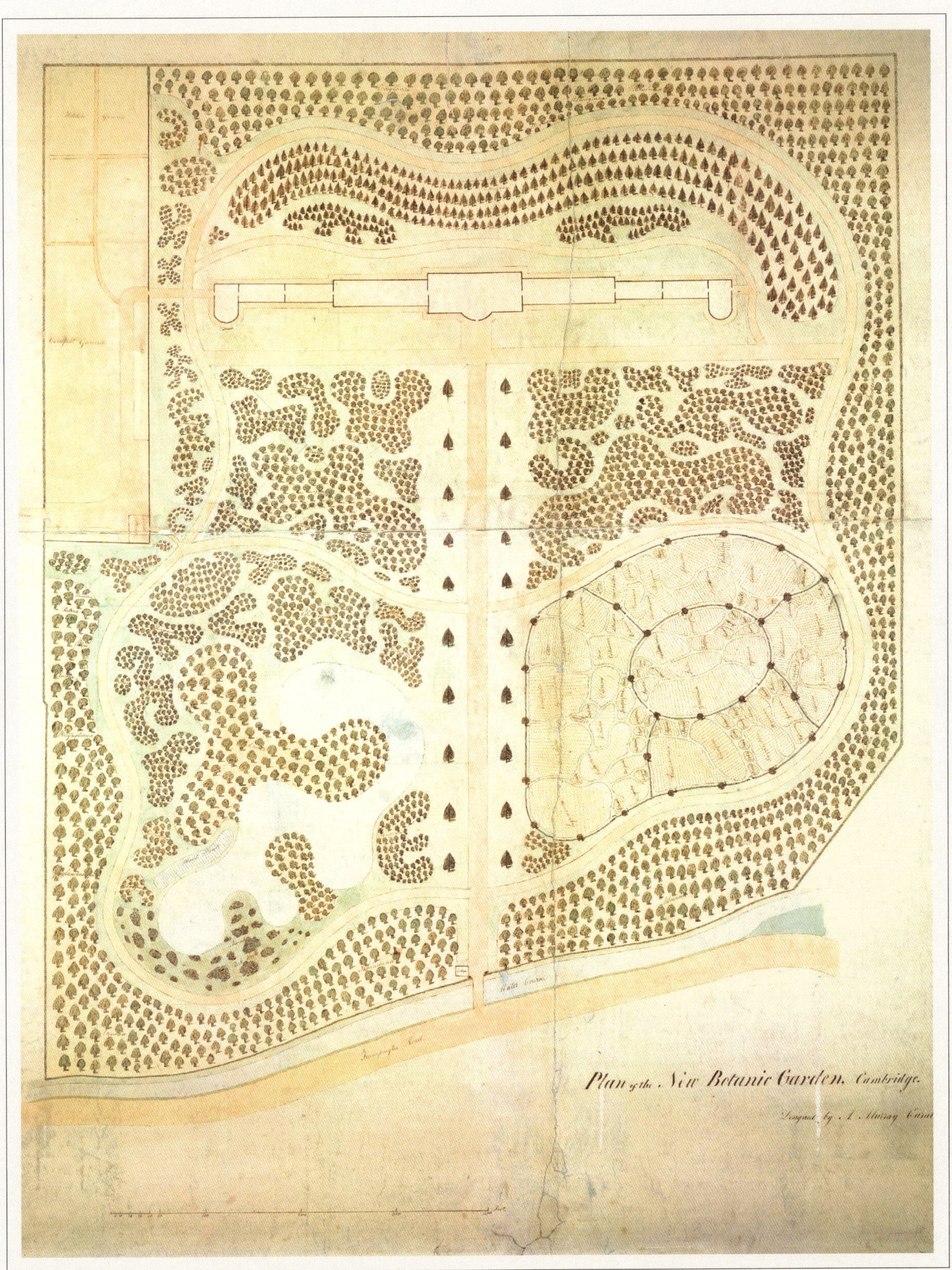
Plan of the New Botanic Garden, Cambridge.
Designed by A. Murray Curat

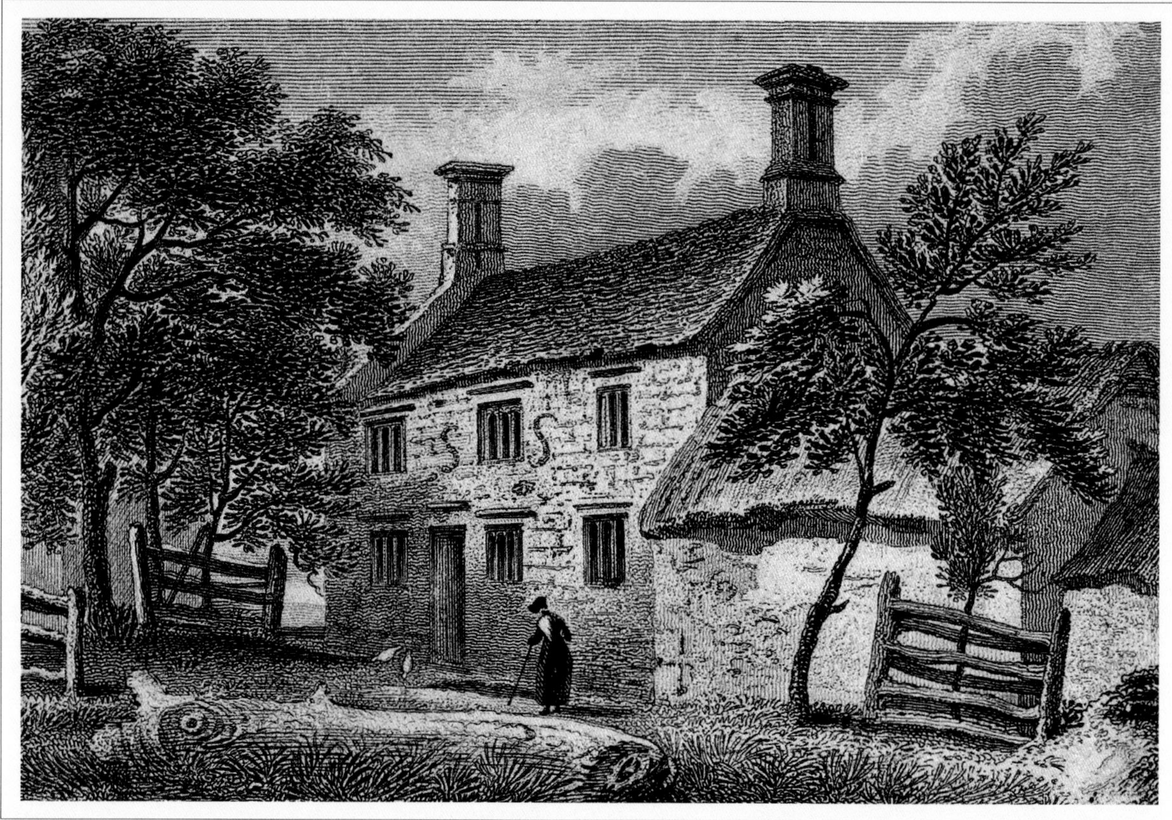

Many sections of the garden introduced new plants for my research. I roamed through the woodland area to the rock gardens up the main walk, then onward through the old pinetum, the trees of which were planted when the garden was first established in the 19th century. As I went deeper into the garden, I found the location of Newton's apple tree, which had been uprooted after dying from disease.

Newton's apple tree

Woolsthorpe Manor, in Lincolnshire, was the birthplace of the famous scientist and alchemist Sir Isaac Newton. It is said that Newton was inspired by a falling apple while walking in his garden. Twenty-six years after graduating from Trinity College, Cambridge, this led him to develop his fundamental work on the theory of gravity, *Philosophiæ Naturalis Principia Mathematica* (The Mathematical Principles of Natural Philosophy), published in 1687, which focuses on the laws of motion, the universal law of gravity, and the concept of absolute time and space.

Scientists at Cambridge University Botanic Garden took a scion from the original tree and replanted the grafted tree in the garden in 1954. Unfortunately, the tree died of honey fungus, and was uprooted in 2022 by Storm Eunice. After sixty-eight years of producing pink and white blossoms and adequate apples,

Newton's tree was logged and stored away from public view. The apple tree (*Malus pumila* 'Flower of Kent') was cloned and grafted by a horticulturist and is currently in the garden's nursery. Once mature, it will be planted again in the garden.

Back in 2016, before the tree died, I walked past it while collecting garden plants to process, and noticed many newly fallen apples on Brookside Lawn. I took one of them to my garden studio to clone artistically. To do this, I made a plaster mold of the apple, from which I produced a stone plaster replica, so I could make it into a singular piece of art. I covered the final shaped apple in thin copper tape and placed it inside a sealed container where it was exposed to the vapors of clear vinegar. Over time, this transformed the copper color of the apple into a vivid, turquoise–greenish blue hue.

This singular sculpture, entitled *Newton's Apple*, is permanently on display in Brookside House at Cambridge University Botanic Garden. The color concept and process were inspired by a technical recipe in the late-12th-century *Phillipps-Corning Manuscript* (in New York's Corning Museum of Glass) for making "Greek Green" or verdigris, which can be widely seen naturally forming on weathered bronze statues and copper pipes. Similar verdigris recipes are described in the 14th-century *Liber Diversarum Arcium* (Book of Various Arts). However, making verdigris dates back to antiquity, with color recipes appearing in the *Papyrus Graecus Holmiensis* (also known as the *Stockholm Papyrus*) and centuries earlier in Pedanius Dioscorides' 1st-century-CE *De Materia Medica* (On Medical Substances).

At Cambridge University Botanic Garden, my aim was to undertake organic color research using the garden's Living Collections and to produce site-specific artwork relating to nature. A few months into my residency, I revisited Newton's apple tree and chiseled off some bark from the logs. My aim was to process the bark into a colorant for an art installation called *The Tree That Once Was*. The bark was made into a golden-yellowish dye in a process inspired by a recipe for apple dye in an early 16th-century German manuscript, *Liber Illuministarum* (Book of the Illuminator), held in the Bavarian State Library in Munich, Germany (MS BSB Cgm. 821). I used the dye to paint sixty-eight duplicated stone plaster apples covered with thin cotton paper, dyed with yellow apple bark dye and then sealed with a beeswax binder.

The yellow replica apples were raised 4 inches (10 centimeters) off the ground and fixed to thin transparent rods, which gave the illusion that they were levitating above the grass, seemingly moving harmoniously together as you walked past them. The site where Newton's apple tree once stood on Brookside Lawn was covered with these yellow apples. They gave the appearance of falling fruit suspended by the absence of the laws of gravity, frozen in time and space, with one central apple placed exactly where the rooted trunk would have grown. This single apple was gilded according to a medieval process of fixing 24-karat gold leaf with fish glue over a ground of red bole (naturally forming red clay) mixed with rabbit skin glue.

The Tree That Once Was references Newton's alchemic work and its relation to the *Principia*, with gold representing alchemical purity and perfection, echoing the gold standard brought in when Newton became Master of the Royal Mint in 1699.

Woolsthorpe Manor, Lincolnshire, where Sir Isaac Newton was born and was inspired by a falling apple. Engraving (ca. 1810).

Chapter 1
Gold

Cornflower

Latin name: *Centaurea cyanus*
Other name: Bachelor's buttons
Family: *Asteraceae*
Genus: *Centaurea*
Native: Southeastern Europe, Western Asia
Type: Annual

The bluish-purple cornflower (*Centaurea cyanus*), a delicate flower with slender petals, often blooms with the vibrant red corn poppy (*Papaver rhoeas*; see page 88) and can be seen painted in ancient Egyptian tombs. Intriguingly, archaeologists in Egypt revealed dried cornflowers in the floral collars of Tutankhamun's embalming cache from the 18th Dynasty (ca. 1550–1295 BCE) and in artifacts in the Greco-Roman era (up to 395 CE). Some artistic representations of the flower, can be dated even earlier, to the 4th millennium BCE, showcasing the cornflower's long-standing cultural significance.

Historical evidence indicates that flowers of Iranian knapweed (*Centaurea depressa*)—which differs from the cornflower in both origin and the color of the flowers—were combined with the fragrant leaves of European olive (*Olea europea*), willow foliage (*Salix*), petals of blue water lily (*Nymphaea nouchali* var. *caerulea*), and wild celery leaves (*Apium graveolens*) for use in sacred rituals.

One of the earliest literary mentions of the cornflower appears in the *Garland*, an anthology of Greek poems woven together by Greek poet Meleager of Gadara in the 1st century BCE. This work includes a vivid image of the cornflower, juxtaposed with "young shoots of Alexander's olive." Pliny the Elder, writing in the 1st century CE, also refers fleetingly in his *Naturalis Historia* (Natural History) to the cornflower, noting its scarcity during the era of Alexander the Great—perhaps the small, blue flower was hidden in plain sight or forgotten in the centuries before. Yet it remains today, growing among the weeds and wildflowers to be enjoyed.

The beauty of the cornflower also inspired medieval and Renaissance artists. Along with other wildflowers, it was often depicted in the borders of illuminated manuscripts, in easel paintings, on altarpieces, and in woven tapestries. More recently, the flower was celebrated in floriography, the Victorian language of flowers, in which it embodied the poignant message of "Hope in Love." The cornflower remains a testament to human emotion, signifying the remembrance of fallen French soldiers in the First World War. It still inspires artists today, in expressive images and through the ancient practice of crushing the fresh flowers to extract the blue juice that is exposed to sulfur to create a coloration that imitates gold.

Cornflower
(*Centaurea cyanus*). Illustration from Johannes Zorn's *Icones Plantarum Medicinalium* (Illustrations of Medicinal Plants; 1779–90).

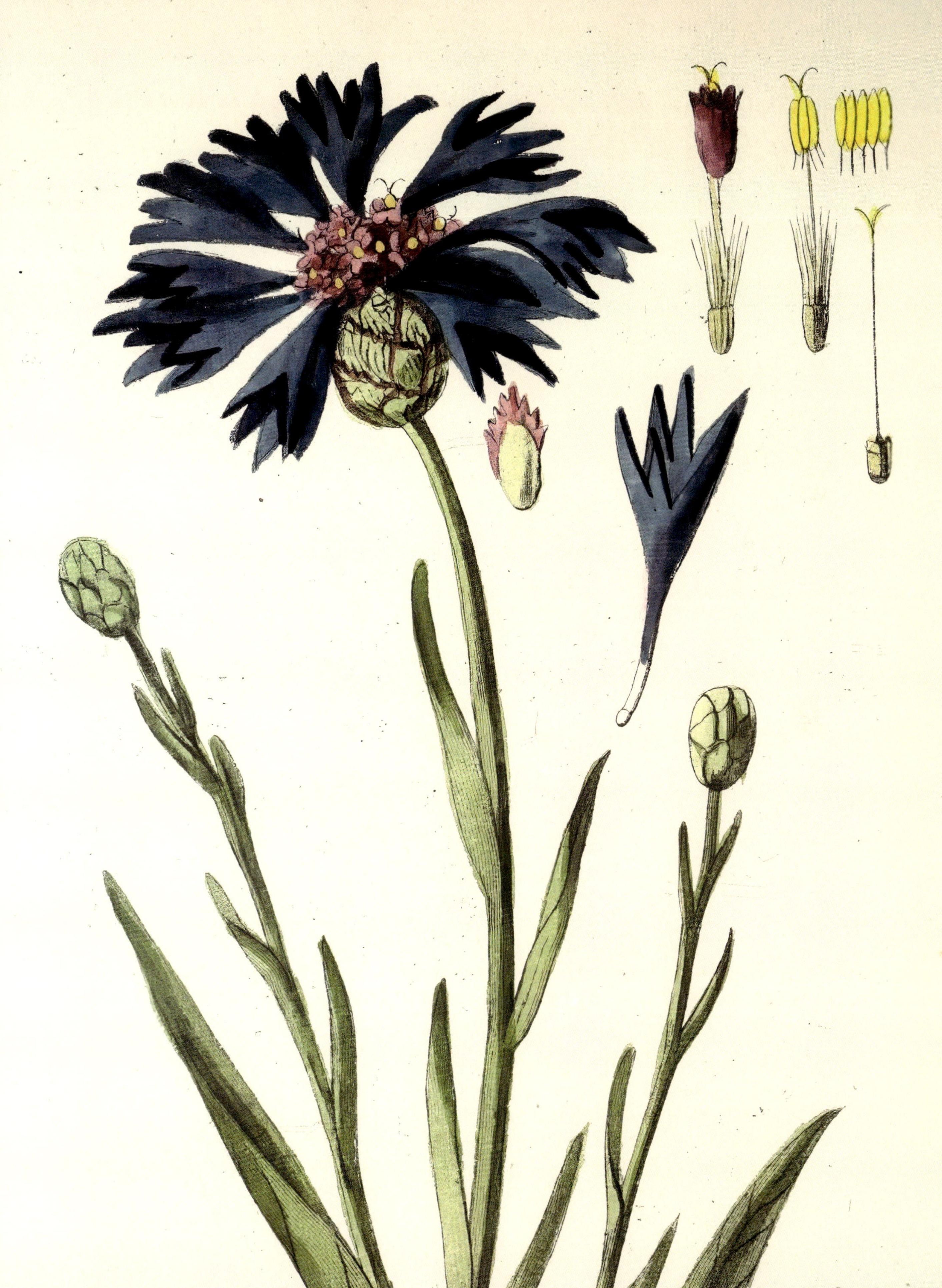

Recipes from Cornflower

The 12th-century *Mappae Clavicula* manuscript contains recipes for metallic gold made by adding juice from greater celandine (*Chelidonium majus*) to mercury and using saffron (*Crocus sativus*) as a glaze to transform tin leaf into imitation gold (see pages 34 and 30, respectively). Fig juice, from the *Ficus carica* leaf, was used to stick silver leaf to paper, and layers of yellow-orange celandine juice could be painted over the top, allowing the silver to shimmer in the light and resemble gold leaf.

Theoretically, therefore, blue cornflower juice may have been applied to silver leaf to create an imitation gold, as suggested in the 15th-century *Strasbourg Manuscript*, a medieval painters' handbook (see *Recipe 1*), and in a 16th-century German recipe for *Dunnckhel blaw* (a dark blue translucent color on silver). Although historical recipes do not imply that this was the intention, in my own experiments I created a golden color by placing a silver leaf sample that was glazed using concentrated cornflower juice in a sealed container. I then exposed the sample to hydrogen sulfide from an egg yolk inside the container. After several days, the transparent glaze changed from dullish blue to a golden color that shimmered in the daylight—all made possible through organic alchemy. Compared with other plants, the cornflower technique is perhaps the most convincing way to make silver, or lesser metals like tin, look like gold, although each organic colorant has its own qualities. Cornflower gold can be made through a unique process using only a handful of flowers and metal leaf stuck down with egg glaire or the sap from fig leaves.

Cornflower juice on its own can produce a blue, as outlined in Recipe 1.3.4A in the *Liber Diversarum Arcium* (Book of Various Arts) and in German recipes held in Heidelberg University Library (Cod.Pal.Germ.489, ff.23–4), but the raw juice can be light-fugitive, fading more quickly than dyer's indigo (*Indigofera tinctoria*) or woad (*Isatis tinctoria*). Several medieval recipes from the 15th century describe similar techniques to that of the *Liber Diversarum Arcium* recipe, briefly explaining how blue flowers can be crushed for the juice to write the blue initials at the beginning of illuminated texts (see *Recipe 2*).

Cornflower (*Centaurea cyanus*). Colored line engraving by C. H. Hemerich (ca. 1759).

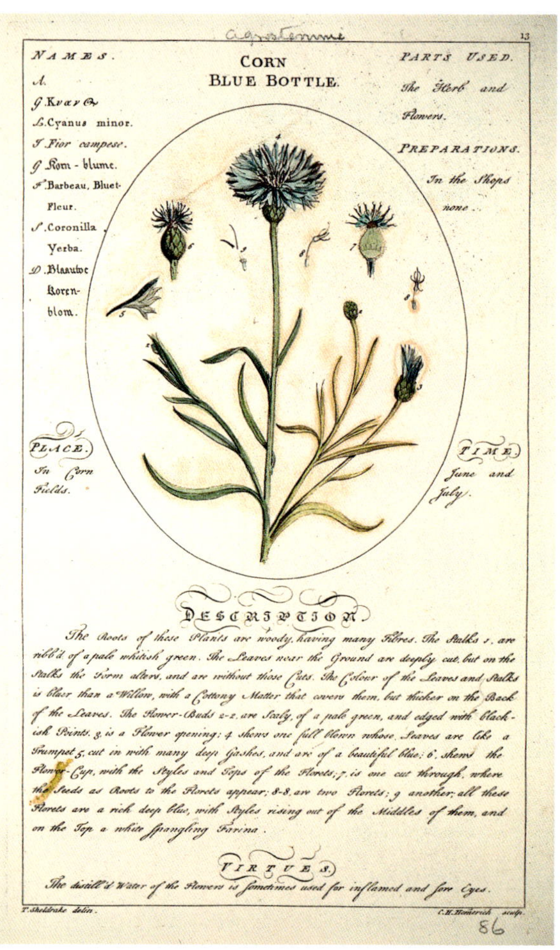

Recipe 1: Making azure-blue or gold

"If you want to make good blue color, take the flowers of cornflower—you know well when—and dry softly and grind them with good wine, and let that dry. Take a little camphor [*Camphora officinarum*] and half as much sal ammoniac [ammonium chloride] and grind it also into it. So you have to apply onto silver or where you want good blue as fine azure. Temper [mix] it with gum or with medium of egg glaire."

—Recipe 108, *Strasbourg Manuscript*, 15th century

Recipe 2: Creating blue letters

"If blue flowers which are found in [corn-] fields are crushed, from them blue letters can be made."

—Recipe 1.3.4A, *Liber Diversarum Arcium*, 14th century

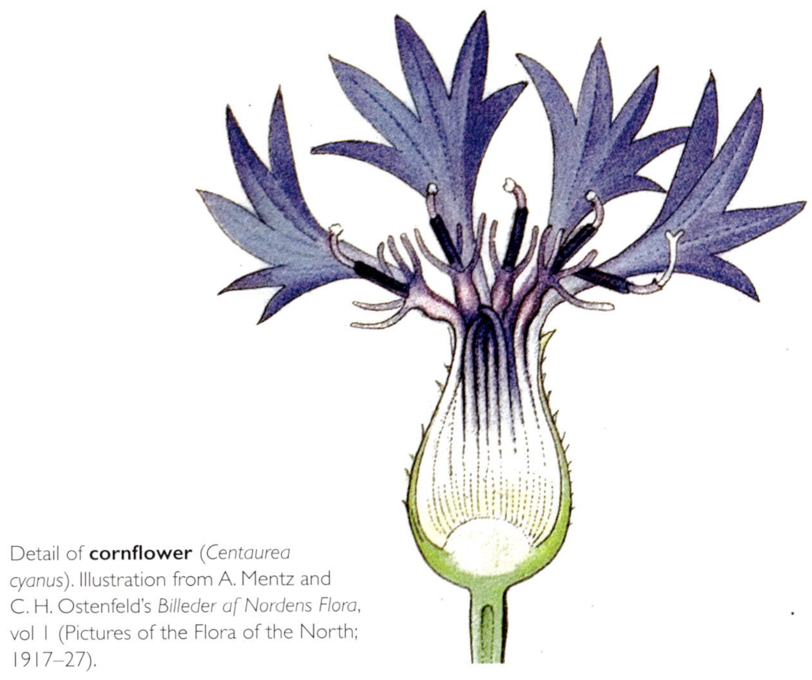

Detail of **cornflower** (*Centaurea cyanus*). Illustration from A. Mentz and C. H. Ostenfeld's *Billeder af Nordens Flora*, vol I (Pictures of the Flora of the North; 1917–27).

Saffron Crocus

Latin name: *Crocus sativus*

Other name: Autumn crocus

Family: *Iridaceae*

Genus: *Crocus*

Native: Greece, Lebanon, Jordan

Type: Perennial corm

Saffron, or *za'faran* in Arabic, is an aromatic spice derived from the delicate stigmas of *Crocus sativus* flowers. It has an illustrious history stretching back thousands of years. Saffron crocus flourishes in the sun-soaked landscapes of the Middle East. Ancient Minoan wall frescoes at Akrotiri on Crete, which survived a volcanic eruption in 1625 BCE, depict women gathering crocuses as an important medicinal and dye crop.

The fragrance of the flowers is strikingly potent and richly aromatic, captivating the senses with every breath. Historical accounts from Pliny the Elder emphasize that this intoxicating scent is best when the flowers are harvested under warm, clear skies. This is especially true in hot climates where the blooms flourish, infusing the air with a tantalizing bouquet that lingers long after they have been gathered.

Saffron's golden color and distinctive flavor have made it a prized possession for culinary and medicinal purposes, as well as for yellow dye and paint for illuminating the pages of books. In fact, saffron was so valuable that it was often referred to as "red gold" due to the high prices it fetched. The ancient Greek physician and botanist Pedanius Dioscorides described saffron in his monumental *De Materia Medica* (1st century CE), a pharmacopoeia of medicinal plants. He highlighted its many benefits and uses as a digestive aid, emollient, mild astringent, and diuretic.

In recent times, Norfolk Saffron, in East Anglia, UK, has excelled in cultivating premium saffron, achieving a quality on a par with the renowned Spanish producers. Interestingly, England has had historical ties with this treasured spice since the Middle Ages. The market town of Saffron Walden derives its name from the saffron that flourished there, serving as a significant cash crop in the 15th and 16th centuries. England had already established itself as a prominent saffron producer in the 14th century, with the spice grown in monasteres across the south and on the premises of such Cambridge colleges as Peterhouse, Pembroke, King's, and St. John's, and later at Clare, Trinity, and Jesus.

Saffron's rich history tells a captivating story through time, from being revered by ancient cultures to receiving modern recognition as a prized spice. Its legacy continues to shine brightly, showcasing how valuable saffron remains today.

Saffron crocus (*Crocus sativus*). Illustration by Dame Ann Hamilton from *162 Drawings of Plants* (1752–66).

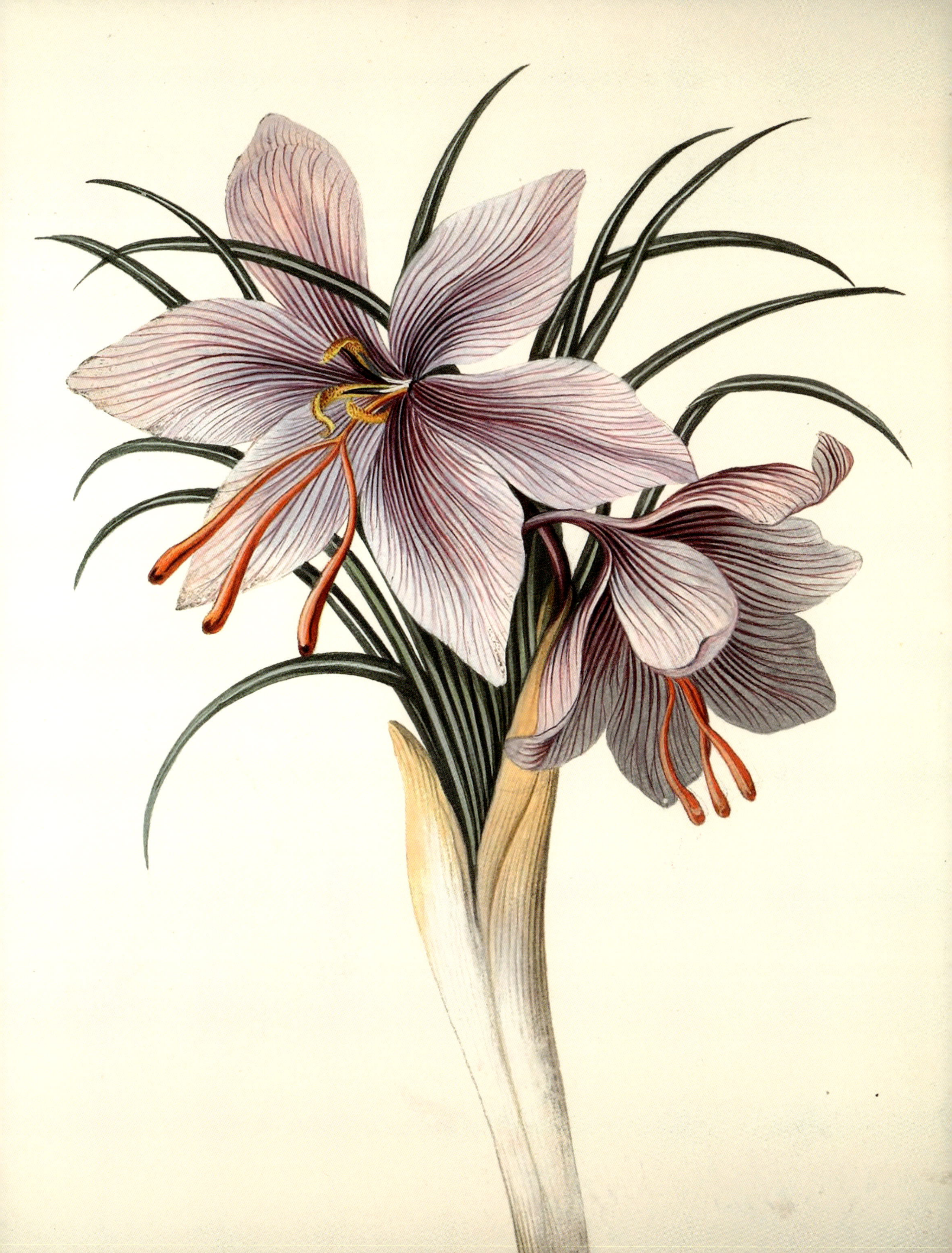

Recipes from Saffron Crocus

Saffron has been used for hundreds of years as a colorant to dye fabrics, produce ink, and make paint. Evidence for these uses can be found in color recipes in technical manuscripts that survived both the English Reformation in the 16th century—when many old recipes were destroyed—and the passage of time. However, learning the practical aspects of using saffron, from its cultivation to application, can be a lengthy process due to the unpredictability of seasonal conditions and variations in crop yields in different parts of the world. Generally, saffron is sourced from multiple countries to meet the high demand for this valuable product that can be used to provide instant color quite easily (see *Recipe 1*).

The Greco-Roman Egyptian *Leiden Papyrus V* (ca. 300 CE) documents the use of saffron to make arrows look golden, while the 3rd-century *Papyrus 121* (P.Lond. I 121), one of the Greek Magical Papyri, held in the British Library, London, explains: "To make an egg become like an apple: Boil the egg and smear it with a mixture of egg-yolk or saffron and [red] wine."

In medieval times, one of the purposes of the dye was to produce an imitation golden color. The 12th-century German Benedictine monk Theophilus Presbyter provided a recipe in his treatise *De Diversis Artibus* (On Diverse Arts) for creating saffron yellow from the dried stigmas, primarily for staining the surface of tin as a glaze to mimic the appearance of gold (see *Recipe 3*). This is a fairly simple process that can be done quickly. The stigmas were sometimes combined with other pigments to create different colors. For example, to make an orange-gold called "glaucus," Recipe 1.27.10 of the 14th-century *Liber Diversarum Arcium* (Book of Various Arts) suggests mixing orpiment (or king's yellow, a toxic color containing arsenic sulfide that is found in large natural deposits but can also be made artificially) with vermiculum (dried bodies of female *Coccus ilicis* insects that produce a red colorant), along with a small amount of saffron. Making a golden saffron glaze with egg glaire is useful when applying the saffron over an azure (any form of blue) color base that has been mixed with chalk to create a yellow-green.

Saffron crocus (*Crocus sativus*). Colored line engraving by C. H. Hemerich (ca. 1759).

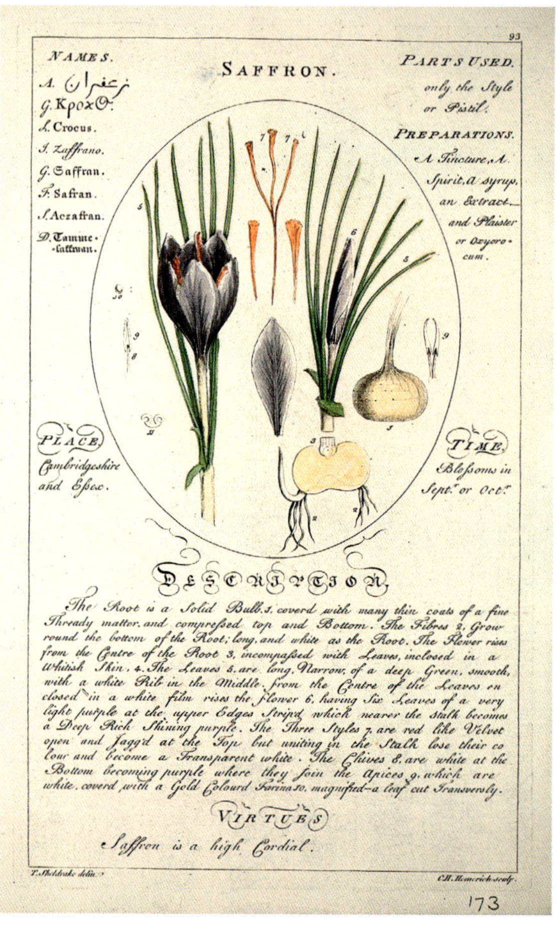

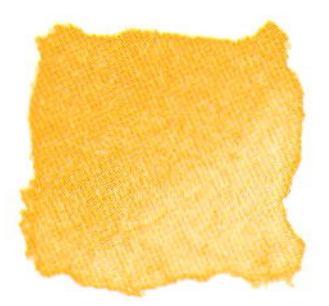

Recipe 1: Tempering crocus to make gold

"Its tempering is of this kind: put it in a container, lay clean glaire on top, and leave until the glaire becomes good and yellow and if it will be too strong, introduce water."

—Recipe 1.19.1B, *Liber Diversarum Arcium*, 14th century (Clarke, 2011, p.116)

Recipe 2: Gilding tin sheets

"Take leaves of tin, dip them in vinegar and alum, and glue them together with glue made from parchment. Then take saffron and pure (i.e., clear and transparent) glue, drench them both in water with vinegar, and cook them with filings over a slow fire. When the glue shines, coat the tin leaves [with it] and they will appear golden to you. But be careful not to add elidrium [flaxseed/linseed oil]. Now, if you have everything already ground, do not add glue, for your work will be hardened. If it has been with glue, for gold writing add elenusia [flaxseed/linseed oil], so that you may extend it."

—Recipe 60, *Mappae Clavicula*, 12th century (Smith and Hawthorne, 1974, p.36)

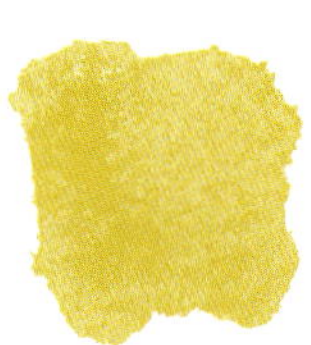

Recipe 3: Making saffron yellow

"Saffron Yellow, from the dried stigma of the saffron plant or crocus. Theophilus uses the pigment mostly for staining the surface of tin to represent gold, but the term *croceum*, is also applied to the color of a clear yellow glass."

—Theophilus Presbyter, *De Diversis Artibus*, early 12th century (Smith and Hawthorne, 1963/1979, Ch.1, p.14)

Greater Celandine

Latin name: *Chelidonium majus*

Other names: Nipplewort, swallow wort, celandine poppy

Family: *Papaveraceae*

Genus: *Chelidonium*

Native: Northwestern Europe, North Africa, Siberia, Asia

Type: Perennial

The word *chelidon* (ancient Greek for "swallow") gave rise to the name for a common weed found growing in gardens and hedges, on roadside verges, and other unexpected places. Greater celandine (*Chelidonium majus*) is part of the poppy family and flowers from late spring to fall. It was used in folk medicine to treat chronic eczemas and, according to Dioscorides, writing in the 1st century CE, it is good for vision when the juice is heated with honey. In the 3rd century BCE, Theophrastus, a student of Plato, explained in his *Historia Plantarum* (Enquiry into Plants) that when celandine blooms, it is a sign that the swifts and swallows will return to the warmer weather and if their fledglings are blind, then the mother bird feeds them with this herb to cure their blindness. Indeed, other ancient writers noted that doves use common vervain (*Verbena officinalis*) for similar purposes, while hawks turn to hawkweed (*Hieracium*) and linnets to eyebright (*Euphrasia*) to restore lost eyesight to their blinded young.

If the little hairy stalks and leaves of greater celandine are broken, yellow-orange latex bleeds from the wounds. This can be slightly irritating to the skin and poisonous in large doses, but the plant has been used by herbalists through the ages. The chemical responsible for the latex's yellow-orange color is chelidonic acid, which is toxic if ingested. In natural folklore it is known as the devil's milk; it was used in herbal medicine and spoken of by Pliny the Elder in his *Naturalis Historia*.

A Russian Orthodox nun called Sister Gabriela (formerly known as Sister Maria), an artist working in the traditional style on religious icons in a monastery in the southeast of England, once spoke to me joyfully of how she would use celandine juice when she was younger to paint her fingernails yellow to create a natural yellow-orange nail varnish in spring and summer. Celandine now grows wild in monastery gardens, staining everything it touches, having become established due to the many seeds in the elongated fruits that follow the small yellow flowers.

Yellow and orange dyes from greater celandine can be stable if the color is kept away from sunlight and used as a colorant on paper as well as fabric. It makes a good natural stain, which was used throughout the Middle Ages for glazes on metals and illuminated pages, yet, as with many organic yellow colors, it can change over time.

Greater celandine (*Chelidonium majus*). Illustration from John Stephenson and James Churchill's *Medical Botany* (1836).

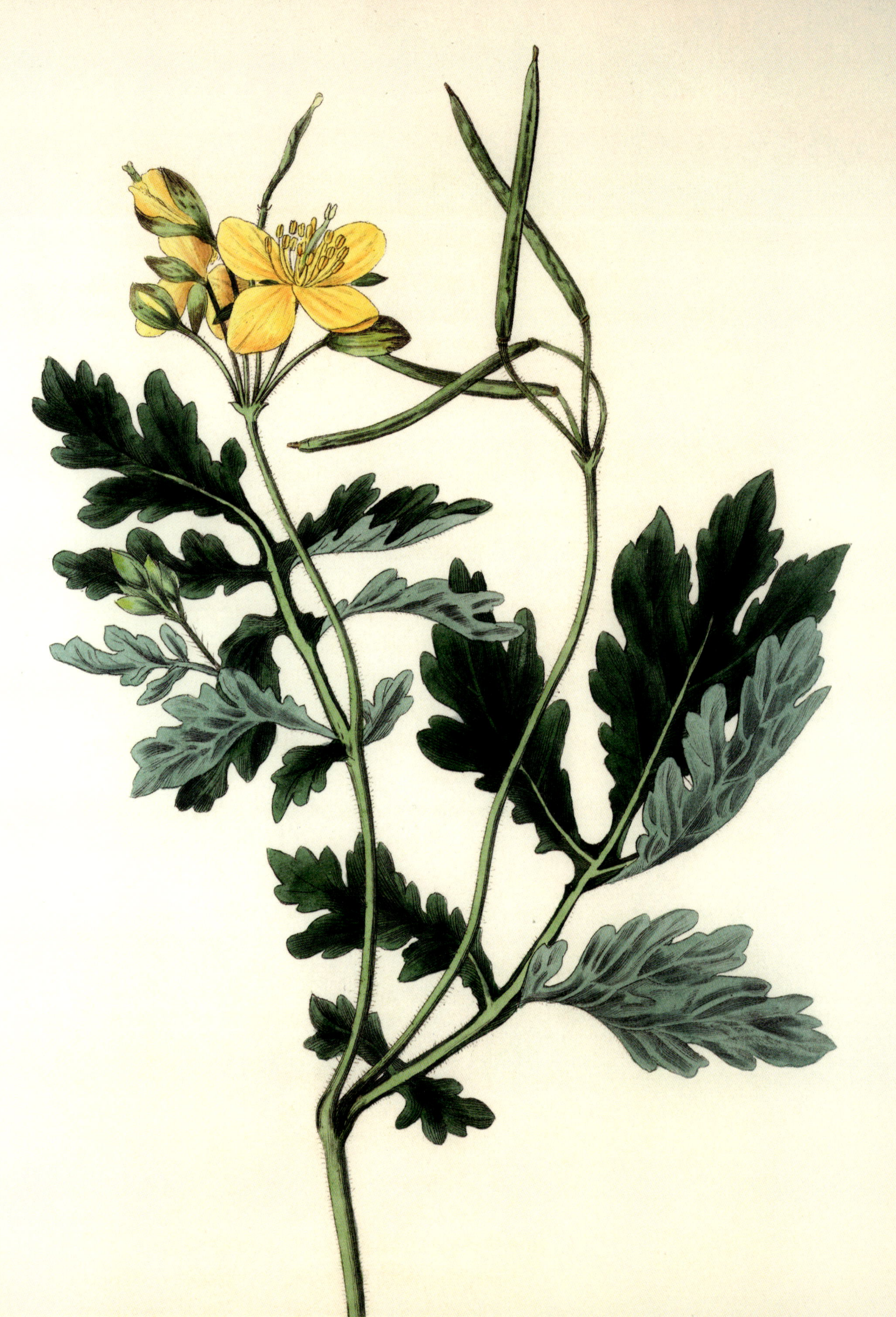

Recipes from Greater Celandine

The medieval *Mappae Clavicula* manuscript features recipes for metallic gold, including adding greater celandine juice to mercury and using saffron as a glaze to transform tin leaf into imitation gold. This highlights the variety of plants that can be used to make an imitation gold colorant.

The 15th-century manuscript *Experimenta de Coloribus* (Experiments upon Colors), compiled by Frenchman Jehan Le Bégue and held in the Bibliothèque Nationale de France, Paris (MS Latin 6741), focuses on illumination, as well as other painting and color techniques. Originally written by Johannes Alcherius in the late 14th century, the manuscript was translated by Mary P. Merrifield in the 19th century and provides several examples using greater celandine.

One recipe focuses on "letters that may seem to be of gold. —Mix sal nitrinum [potassium nitrate] with water and write upon parchment, and illuminate it with juice of celandine and warm the paper, and the letters will appear like gold."

Another recipe suggests using celandine as an additive "to make a good temper for iron utensils." It focuses on collecting a large quantity of plant material early in the morning when the celandine is "wet and full of dew" and pounding its juice with green wolf's milk (*Euphorbia esula*) before heating in water with antimony (a brittle, powdered silver metal), which has been used since antiquity in cosmetics and medicine.

Celandine was used in combination with other dyes to make additional colors. For example, the *Mappae Clavicula* manuscript suggests heating celandine roots until a feather is dyed yellow-orange—as a color indicator—then adding indigo with powdered gum arabic to make a "beautiful blue," though in reality it would make a bluish-green. Other recipes from the manuscript suggest mixing "hairy saffron" with celandine and "myrtle water or pomegranate" to make a violet color.

In my recipe, juice from a fig tree (*Ficus carica*) was used to fix silver leaf to paper, so the yellow-orange juice from the celandine could be applied. Several layers of the juice were painted on the silver sample using part of the plant's broken stem. The bright orange juice is semi-transparent, which allows the silver layer to shimmer in the light and resemble gold leaf. The color will hold for a period of time if kept out of direct sunlight.

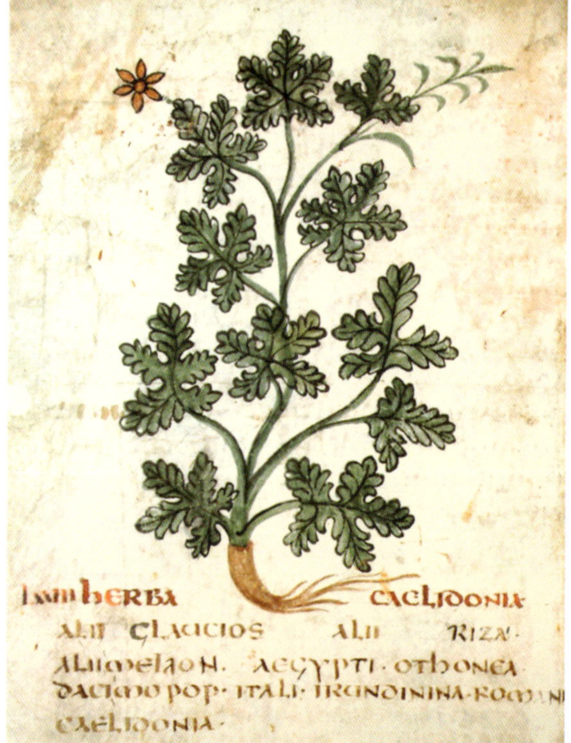

Greater celandine (*Chelidonium majus*). Illustration from a 6th-century copy of the herbal Pseudo-Apuleius's *Herbarius* (5th century).

Recipe 1: Dyeing of fabrics

"By celandine one means a plant root. It dyes a gold color by cold dyeing. Celandine is costly, however. You should accordingly use the root of the pomegranate tree and it will act the same. And if wolf's milk is boiled and dried it produces yellow. If, however, a little verdigris [copper salts of acetic acid] is mixed with it, it produces green; and safflower blossom likewise."

—Recipe 139, *Papyrus Graecus Holmiensis*, also known as the *Stockholm Papyrus*, ca. 300 CE (Caley and Jensen, 2008, p.82)

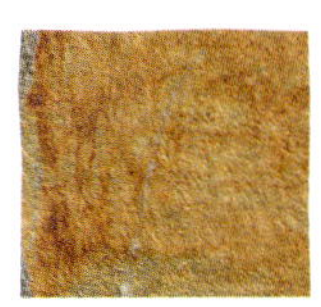

Recipe 2: Creating imitation gold over metal leaf

Use either the root or the broken stems of the greater celandine plant, harvested in late spring, summer, or fall, then apply the yellow-orange juice to silver or tin leaf and this will give the appearance of a golden color. The juice does not need a fixative or mordant (a substance used to set dye in a fabric).

—Nabil's modern recipe

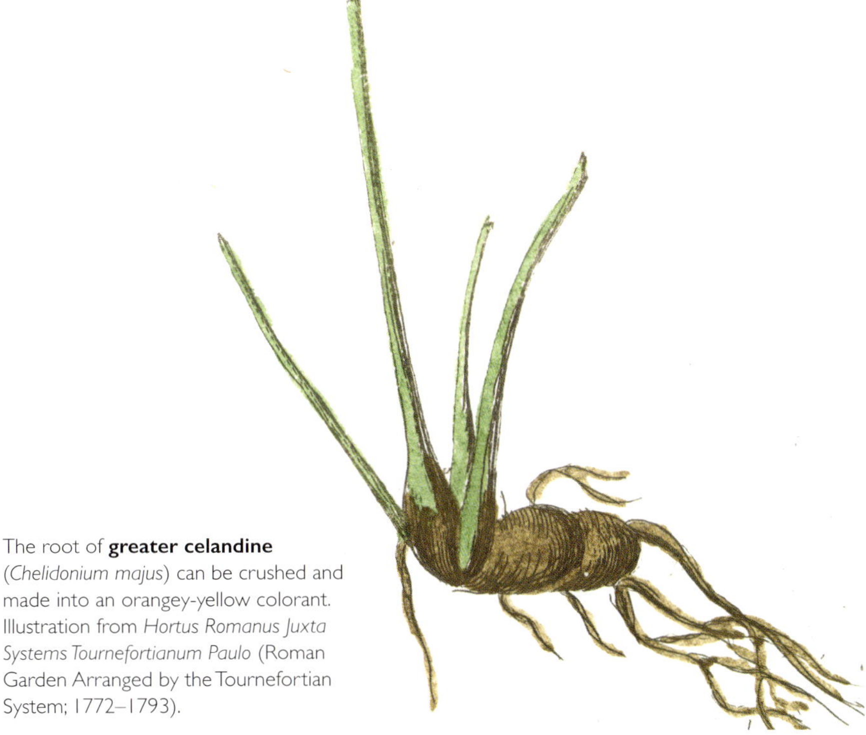

The root of **greater celandine** (*Chelidonium majus*) can be crushed and made into an orangey-yellow colorant. Illustration from *Hortus Romanus Juxta Systems Tournefortianum Paulo* (Roman Garden Arranged by the Tournefortian System; 1772–1793).

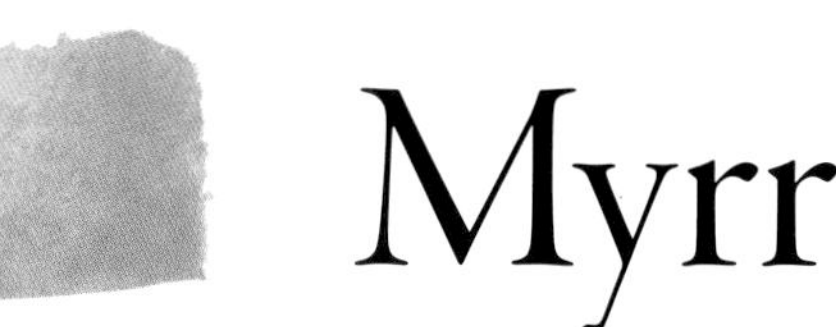

Myrrh

Latin name: *Commiphora myrrha*
Other names: African myrrh, common myrrh
Family: *Burseraceae*
Genus: *Commiphora*
Native: Middle East, northeast Africa
Type: Deciduous shrub or small tree

There are over 190 species of *Commiphora*. Many have different aromas, while the color of the gum resin can be milky white, dark brown, yellowish-amber, or colorless. In a process called tapping, the resin is extracted by cutting the bark and allowing it to drain into bowls tied to the trunk; a similar method is used to extract frankincense from *Boswellia* trees. Myrrh is a scented, translucent yellow gum resin from *Commiphora myrrha*. The ancient Egyptians used it with cassia (a spice made from cinnamon bark) and other aromatics when preserving the dead. Myrrh has reportedly been found in calcite jars in tombs dated to the Middle Kingdom (11th–13th Dynasties, ca. 2124–1640 BCE) and was mixed with oil to create a scented ointment.

Pliny the Elder noted in his *Naturalis Historia* that myrrh was mixed with other resins, such as mastic gum from the shrub *Pistacia lentiscus*, then added to cucumber juice and a natural form of lead called litharge to make the product heavier. He commented that Indian myrrh was inferior to Arabian myrrh from North Africa, which may have come from different species, although Theophrastus, centuries earlier in his *Historia Plantarum* (Enquiry into Plants), implies that Arabian, Syrian, and Indian myrrh are exceptional and distinct from that of other lands.

As well as being used for incense, and in perfumes and cosmetics, myrrh also had other practical applications. In his 1st-century writings, Dioscorides described how an aromatic medicine was made by adding myrrh, balsam, pepper, iris, and myrtle to a wine concoction. In medieval times, myrrh was added to a painting medium, gum arabic, with the 14th-century *Liber Diversarum Arcium* (Book of Various Arts) explaining that painters should "put in myrrh and incense such that flies do not devour it." Myrrh has similar properties to many other natural resins, and these were harnessed to make varnish, which can be included as an ingredient for making a stiff sealant or added to pigments when making medieval paint.

Historically, few recipes highlight myrrh as a main ingredient for making a golden glaze, although several suggest that the making of imitation gold was embedded in working practices and detail the value of myrrh as a minor colorant enriched with aroma.

Myrrh (*Commiphora myrrha*). Illustration from *Köhler's Medizinal-Pflanzen* (Köhler's Medicinal Plants; vol.2, 1890).

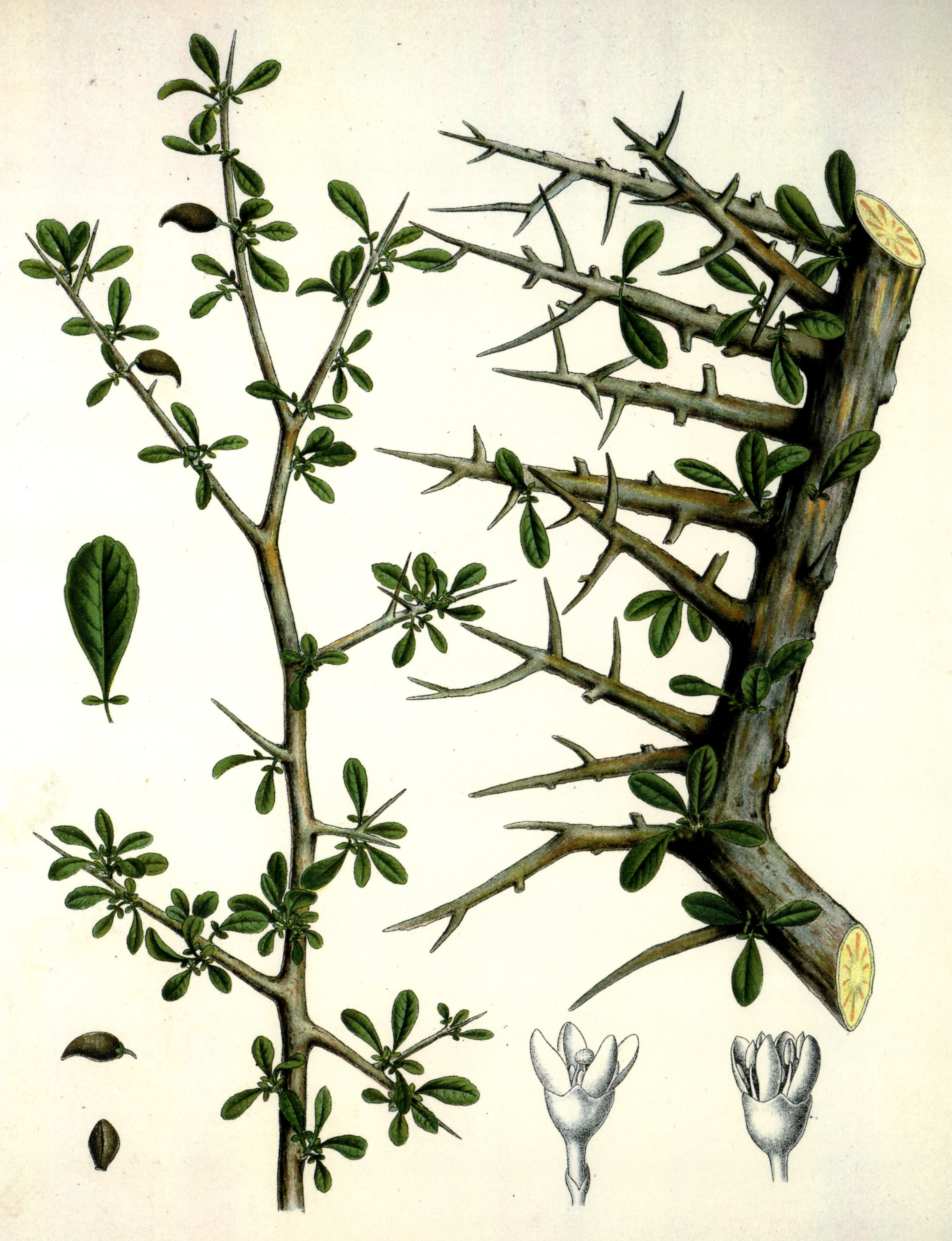

Recipes from Myrrh

The medieval *Mappae Clavicula* manuscript includes a recipe for "a gold-colored transparent varnish" using flaxseed (linseed) oil, Spanish pitch (a sticky resin from pine trees), and ox glue with some oriental saffron, frankincense, mastic gum (a resin from the *Pistacia lentiscus* tree), cherry tree resin, the poplar flower, and, of course, some myrrh resin. This would produce a dense color and, depending on the transparency of the varnish, be quite thick, although the quantity of the ingredients could change the consistency.

Most resins are insoluble in water but will dissolve in oil-based liquids. In a modern recipe, you would grind dried gum resin into a powder using a pestle and mortar, then heat it in an oil, such as flaxseed or walnut oil, and mix well. Next the mixture would be applied over silver or tin leaf to create imitation gold. Medieval accounts included warnings when heating oil-based solutions, with a varnish-based recipe in the 14th-century *Liber Diversarum Arcium* (Book of Various Arts) advising "to beware of fire, because it is very dangerous, and [the oil] is not easily extinguished." The manuscript also notes that the varnish was an additional "ingredient" rather than a final step in the varnishing process.

A method for giving tin leaf a golden color using a mixture of myrrh and aloe gum resin also appears in medieval manuscripts. Indeed, in 1849, M. P. Merrifield published *Original Treatises in the Arts of Painting*, in which she translates a 13th–14th century recipe from the *Liber Magistri Petri de Sancto Audemaro de Coloribus Faciendis* (Book of Master Peter of St. Audemar on Making Colors), held at the Bibliothéque Nationale de France, Paris (MS 6741), for "how to gild leaves or beaten plates of tin":

"Take the herb which is called 'myrrh', and aloes, of equal weights, and having mixed them together, put them in a proper quantity of water. Then boil them well, and after they have been boiled, pour the water into a vessel, and take the leaves of tin, covered on one side with varnish, immerse it in the liquor as long as necessary. Then boil the middle bark of the black plum well in a vessel, and afterward dip the same tin in this water. Then lay it on a table to dry."

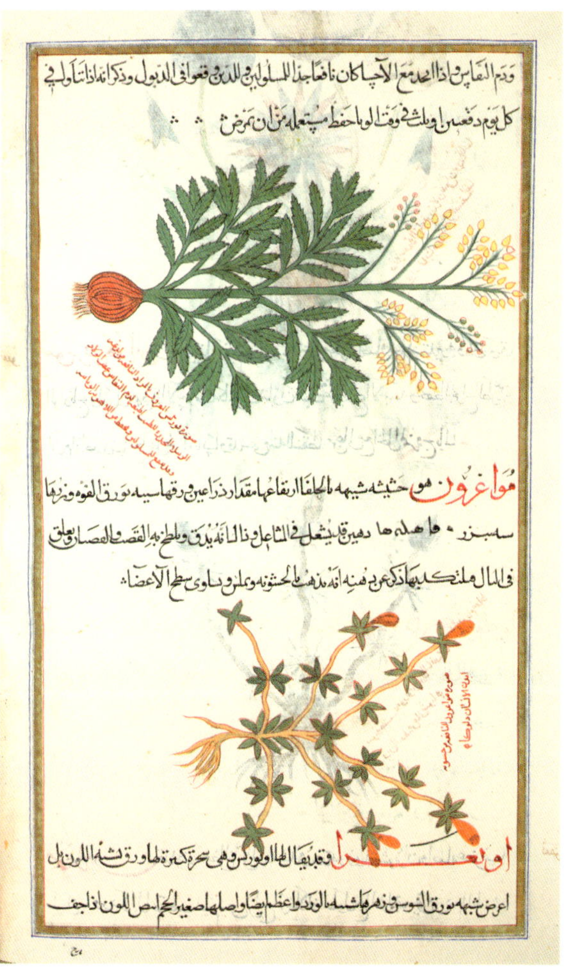

Recipe 1: Writing in gold

"Cook down in a pot the juice of a mulberry or fig tree and [add] to the juice a quarter part of alum, and smear it on the vessel that is to be gilded, and so gild it. Before doing this smear it with a quarter [part of] myrrh, or the juice of a mulberry or fig tree, and so gild."

—*Mappae Clavicula*, 12th century (Smith and Hawthorne, 1974)

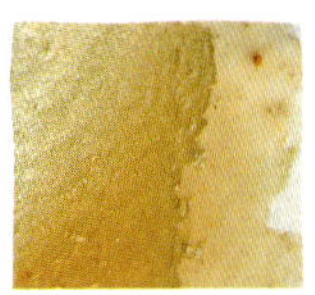

Recipe 2: Preparing gold-colored ink

"3 dried figs, 3 stones of the Nicolaus date, 3 fragments of wormwood, and 3 lumps of myrrh; [mix together] after pulverizing them [the ingredients]."

—*Papyrus 121* (P.Lond. I 121), 3rd century CE, British Museum, London (Betz, 1986, p.144)

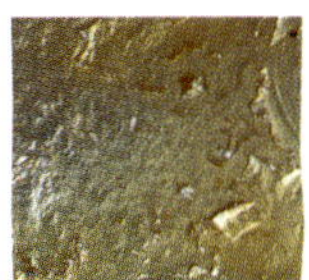

Recipe 3: Making myrrh glaze

Crush some myrrh pieces in a pestle and mortar, then add the crushed myrrh to some hot water in a small glass jar, along with a small amount of isinglass powder (isinglass comes from the swim bladders of sturgeon fish). Put the jar in a pan of very hot water and keep warm while you apply the myrrh to silver leaf. Several layers are needed to create a golden transparent glaze.

—Nabil's modern recipe

Garden myrrh or sweet cicely (*Myrrhis odorata*), which is not related to the resin-producing myrrh (top). Illustration by Persian artist Mirzā Bāqir for an 1889–90 edition of a 9th-century Persian translation of Pedanius Dioscorides' *De Materia Medica* (1st century CE).

Turmeric

Latin name: *Curcuma longa*
Other names: Haldi, turmeric root
Family: *Zingiberaceae*
Genus: *Curcuma*
Native: India
Type: Rhizomatous perennial

Turmeric is used as a culinary spice and for its medicinal qualities. It is made from the dried root of *Curcuma longa*, which requires a warm climate to thrive. Assyrian cuneiform tablets from the 7th century BCE refer to turmeric growing in dust along walls. Turmeric was popular in ancient Greece and used in other cultures to make a yellow dye—to color the robes of monks and priests in India, for example. The dye can be "fixed" to textiles without a mordant (a substance that "binds" dyes to fabric), making it a favored colorant and far cheaper than saffron dye. Turmeric was also used as a supporting colorant, mixed with cochineal beetles in the production of British military redcoats, with Dutch dyers making the dye to enhance the appearance of the coats.

Mixing the yellow dye with indigo produces a parrot-green colorant, which can be seen on Islamic textiles in Egypt's Cairo Museum. Paper dyed with turmeric was used by Islamic scribes (see *Recipe 1*, page 45). It has been suggested that the blank end pages in Arabic manuscripts were dyed with turmeric root or saffron due to their protective qualities. To some extent, these dyes—mainly turmeric—do indeed have an anti-microbiological effect, with dyed pages providing a barrier against bacteria or insects when inserted in the front and back of books. Other plants like rosemary (*Rosmarinus officinalis*) or the seeds of white mustard (*Sinapis alba*) could similarly be used due to their antioxidant and antimicrobial properties.

Analysis of 16th- to 18th-century artifacts in New York's Metropolitan Museum of Art has revealed traces of turmeric in the yellows and light and bluish greens of velvet and silk fragments from India and Persia. Further color examples are found in an 18th-century beaten inner bark cloth collected on Captain James Cook's voyage to the Pacific Islands. It depicts red human handprints made with dye from noni (*Morinda citrifolia*), which were painted on bark cloth dyed with a yellow flavonoid dyestuff from the turmeric plant.

Turmeric, with its earthy aroma, produces a seductive color, but unfortunately the dye can be sensitive when exposed to direct sunlight. Combining it with other plants, such as the weld plant (*Reseda luteola*; see page 144) or safflower (*Carthamus tinctorius*; see page 104), can create a more stable yellow color.

Turmeric (*Curcuma longa*). Illustration from Étienne Denisse's *Flora d'Amérique* (Flora of America; 1843–46).

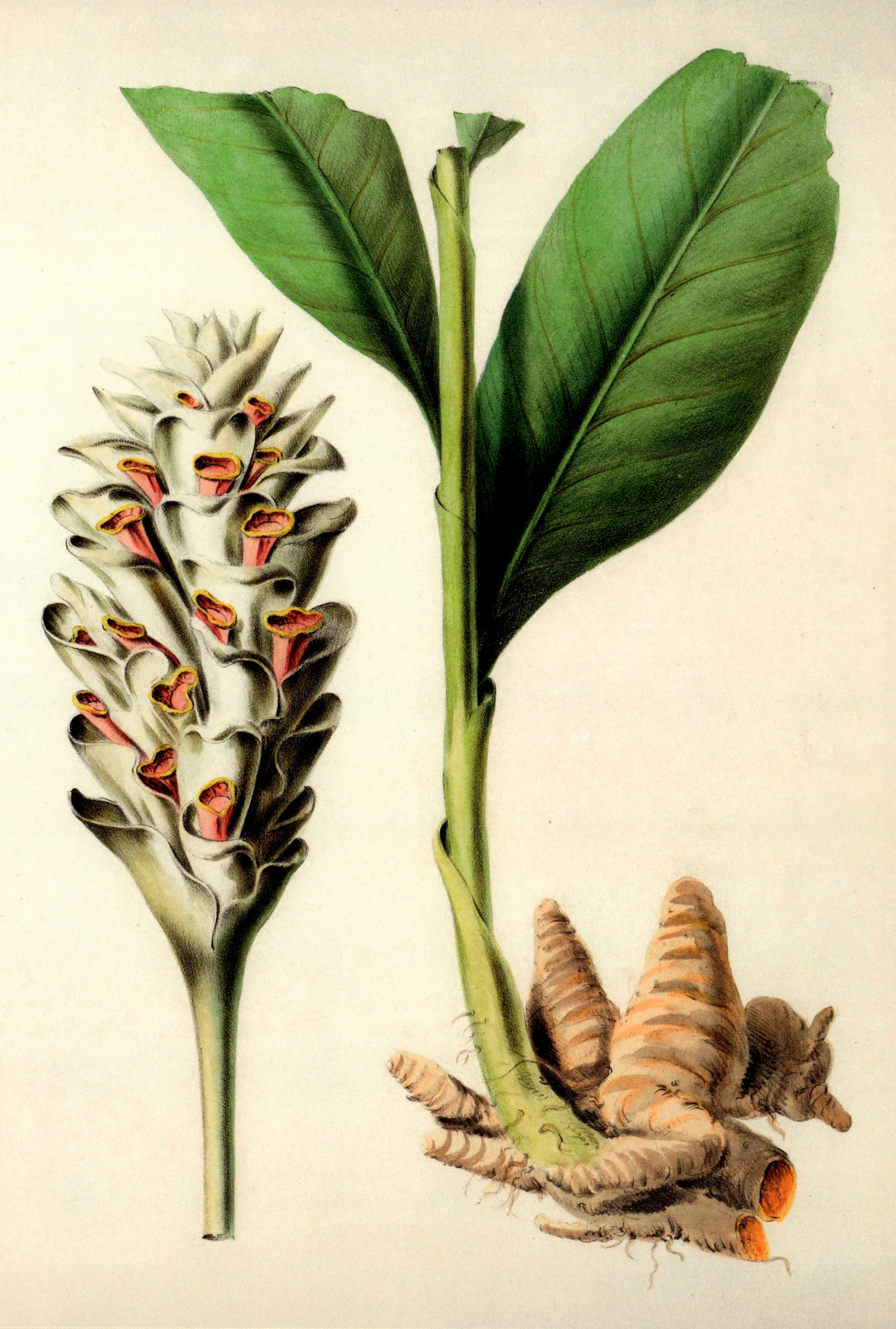

Recipes from Turmeric

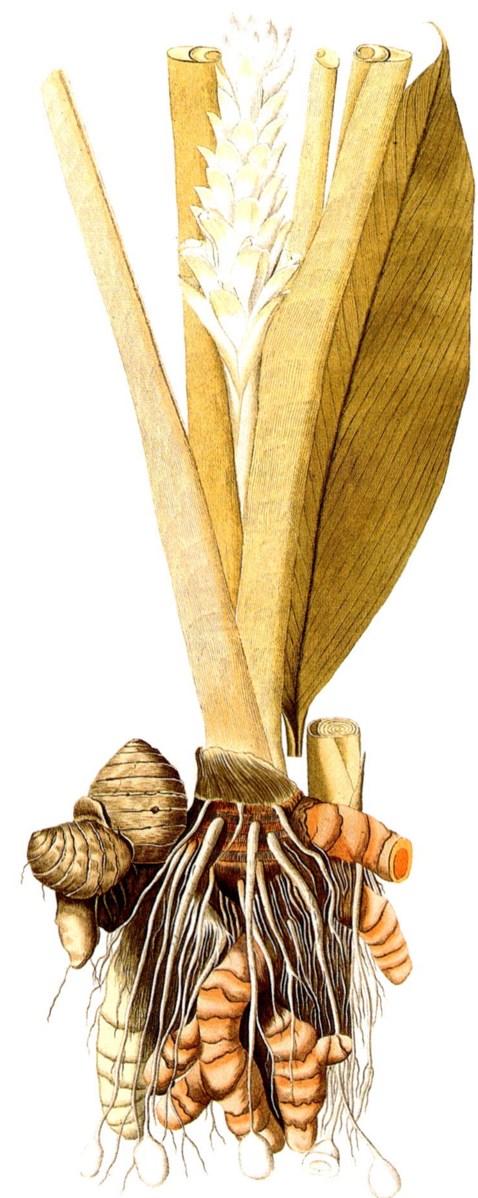

Turmeric *(Curcuma longa)*. Colored engraving after F. von Scheidl (1776).

Turmeric root produces a strong yellow dye when crushed into powder. It colors the world yellow, from the robes of Buddhist monks to painted symbols on elephant skins to repel evil spirits in Malaysia. It is used by multiple cultures across Asia, the Pacific Islands, and beyond.

The earliest Greco-Egyptian alchemic text is the 1st-century-CE *Physika Kai Mystika* (Natural and Mystical Matters) by Pseudo-Democritus, which explains how yellow medicinal drugs can be used to dye metals yellow, turning them gold. A handful of plants and materials are included: saffron crocus, safflower, (greater) celandine, rhubarb, and gum arabic from *Acacia* trees. Although turmeric root is absent, I found that applying a glaze of yellow turmeric dye over silver leaf worked very well to create a golden color (see *Recipe 2*). The yellow dye imparts a gold color to silver metals through the basic processes of organic alchemy.

Although few historical recipes describe using turmeric as a glaze, it is documented in the 14th-century technical manuscript *De Arte Illuminandi* (On the Art of Illumination), held in the Biblioteca Nazionale Vittorio Emanuele II, Naples (MS Latin XII.E.27). This manuscript explains how the dye can be mixed with different pigments rather than just applying it as a yellow organic glaze over silver or tin to produce gold.

Like saffron, turmeric was added to verdigris to make a green-yellow color, as described in a 16th-century Venetian manuscript (MS Sloane 416, fol.138) held in the British Museum, London. This manuscript focuses purely on alchemic and color recipes, which use the same methods as those outlined in numerous medieval manuscripts centuries earlier.

In a modern recipe, if potash (pH 12) is added to ground turmeric root—making the dye more alkaline—the color changes from yellow to a rich red. This can be observed in parts of Africa where leather hides are dyed with a paste made from ground turmeric root and water, then soaked in sodium bicarbonate (pH 9), resulting in the color changing to red. Once the hides are washed several times in lime juice and water, the red leather hides turn yellow again.

Recipe 1: Dyeing paper yellow

"Take half istar [5.8 grams] pure yellow turmeric and coarsely grind it. Then place it in a big pot. Add nearly 5 istar [58 grams] water and bring it to [a] boil. Filter it through a piece of muslin cloth. Then take some alkaline ash (aškār) and grind it. Pass it through a piece of muslin cloth and add it to the filtered turmeric water. Hand-mix the mixture and pour it in an open mouth vessel, so that [the] paper does not tear or fold. Keep two other open mouth vessels and pour filtered water in one and a filtered lemon juice in another.

First dip paper in the turmeric water thoroughly. If deep yellow is required, repeat the process as many times as needed until the desired color is obtained. If light yellow is required, add more water until you get the desired color.

After that, take out the paper from the turmeric dye, dip it in the lemon juice, and shake it well till it fully penetrates the paper. Take out the paper from the lemon juice, dip it in the water vessel, and wash it thoroughly. Take it out and let it dry in the shade."

—Anonymous, *Resāleh dar Bayān-e Kāğad Morakkab va Hall-e Alvān* (a treatise on paper, ink, and dyes), early or mid-9th century AH, Parliament Library copy, No.1 and No.4767, Tehran, Iran, 1100 AH/15th century CE (Barkeshli, M. 2015).

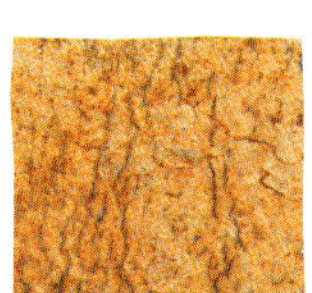

Recipe 2: Making imitation gold

Mix red bole (a red clay used as a pigment) with gum arabic and paint this over a surface as a deep ground. Once the red ground has dried, apply silver/tin leaf to the surface using glue made from isinglass, then leave to dry. Mix powdered turmeric root with either egg white or gum arabic to form a yellow glaze, then apply with a brush over the silver/tin leaf to create a golden color.

—Nabil's modern recipe

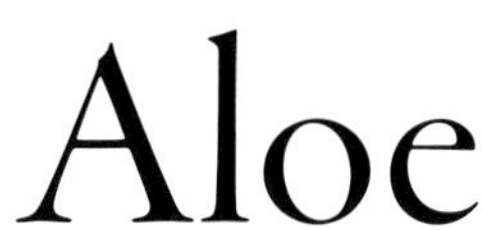

Aloe

Latin name: *Aloe buhrii, Aloe vera*

Other names: *Aloe buhrii* (Buhr's stemless aloe), *Aloe vera* (Barbados aloe)

Family: *Asphodelaceae*

Genus: *Aloe*

Native: *Aloe buhrii* (South Africa); *Aloe vera* (Northern Oman)

Type: Evergreen succulents

Aloe buhrii can be a challenging plant to cultivate. Originating from the hot, dry climate of South Africa's cape, it was named after Elias Albert Buhr, a farmer from Northern Cape Province, who first collected and cultivated it. This nontoxic succulent grows in a rosette formation, consisting of about sixteen leaves, and can withstand harsh growing conditions. It serves as a habitat for birds and a hiding spot for snails. When exposed to direct sunlight, the leaves turn red and the plant has displays of yellow to orange flowers from spring to fall. A large *Aloe buhrii* plant can be seen in the glasshouse at the Cambridge University Botanic Garden, where it has been growing since the 1980s.

There are hundreds of *Aloe* species, some of which are poisonous—for example, *Aloe ruspoliana*, native to East Africa, is used to kill hyenas; *Aloe buettneri* is used to make poison arrows in Mali, West Africa; and *Aloe ballyi*, again from East Africa, is known as the rat aloe. However, a handful of the plants have beneficial properties in folk medicine, while *Aloe megalacantha* and *Aloe confusa* are used as a yellow dye and ink that oxidize to purple.

Aloe vera is perhaps the most familiar species. The ancient Egyptians called it the "Plant of Immortality" in the 16th-century-BCE *Ebers Papyrus*, held in Leipzig University Library, and used it in combination with other plants. The earliest written mention of *Aloe* appears on Sumerian clay tablets from around 2200 BCE and it was also documented by the ancient Greeks and Romans, with Dioscorides stating that the plant can be used for binding wounds and swallowed with pine resin in water or hot honey to loosen the bowels. *Aloe vera* has beneficial medicinal and cosmetic properties, and the leaves can be processed into a soft, hemp-like fiber.

Aloe vera is depicted in the 6th-century-CE manuscript, the *Juliana Anicia Codex* (Codex Vindobonensis med. gr. 1), also called the *Vienna Dioscorides*, which contains part of Dioscorides' *De Materia Medica*, held in Vienna's National Library. This shows the leaves, roots, stalk, and flowers, with similar attributes illustrated in a 9th-century-CE Greek manuscript (BNF Gr.2179) in the Bibliothèque Nationale de France, Paris, and a simplified version included in a 13th-century Arabic manuscript (Cod. Arab. 2954) in the University Library of Bologna.

Barbados aloe (*Aloe vera*). Illustration by Nicolas-François Regnault from *La Botanique Mise à la Portée de Tout le Monde* (Botany Made Accessible to Everyone; 1774).

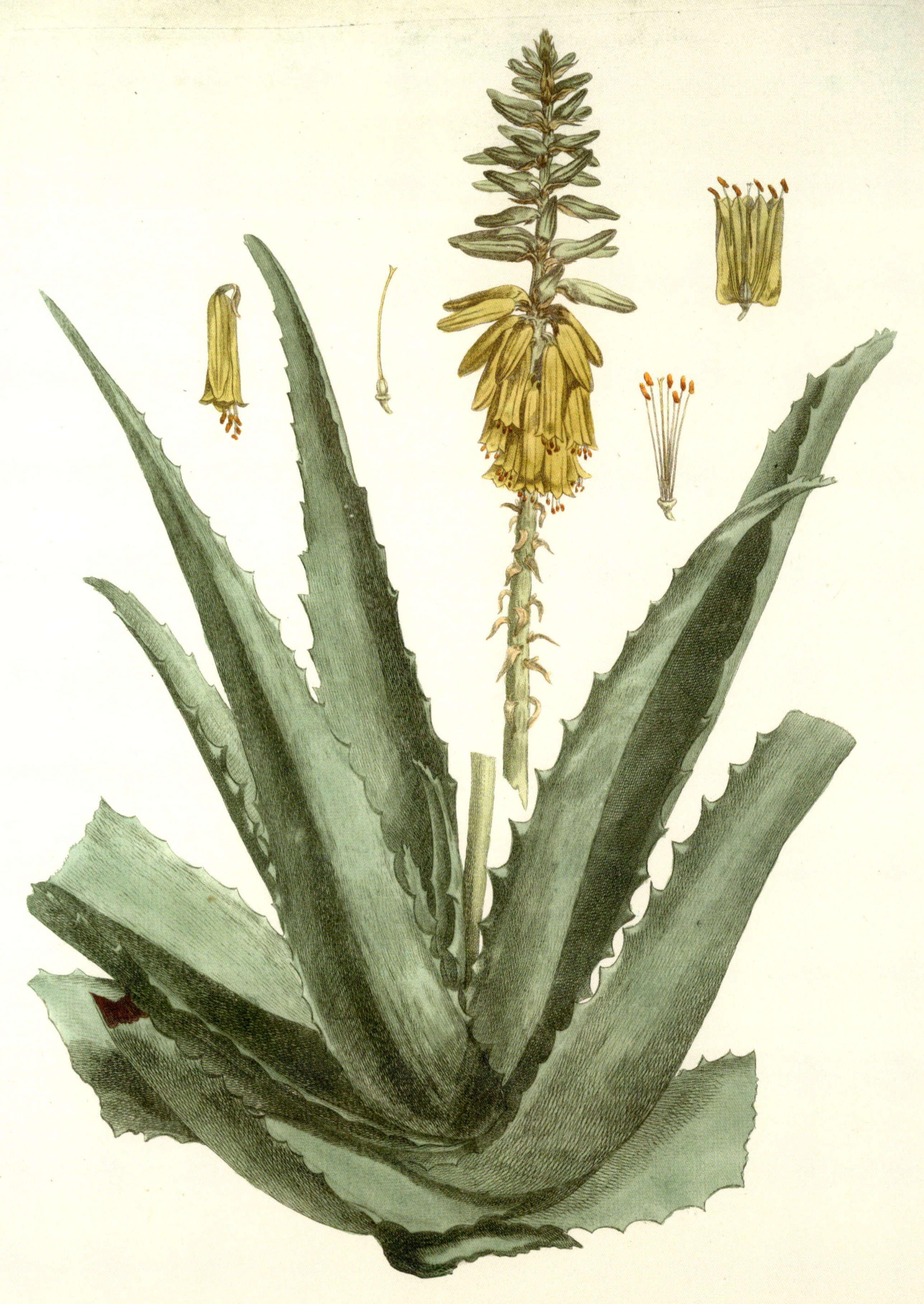

Recipes from Aloe

O Libro de Komo Se Fazen As Kores (Book on How to Make Colors), is a 15th-century Hebraized Portuguese manuscript (Parma MS 1959) held in Palatine Library, Parma, Italy. It contains a recipe for a varnish based on flaxseed (linseed) oil using *Aloe* gum, which can be fixed to tin or silver foil as a translucent colorant that resembles gold.

Leonardo da Vinci mentions that mixing verdigris with aloe will improve the color of green. Aloe was combined with other materials like flaxseed oil and saffron and used as a glaze to imitate gold. This can be seen in the 14th-century *Liber Diversarum Arcium* (Book of Various Arts), which includes several recipes that use dried aloe resin, called hepatic aloe (obtained from *Aloe vera*), produced through slow vaporation. In the *Liber Diversarum Arcium*, recipe 2.8.2 explains how to make a golden color: "take two parts of pine resin, two parts hepatic aloe—or horse-aloe—and a tenth of a part of dragons blood, and liquefy them together, and boil with flaxseed oil, mixed together, and cover well; and leave it alone until it cools." I reconstructed this recipe using resin from Cape aloe (*Aloe ferox*) and cold-pressed flaxseed oil and produced a reddish, yellow-brown glaze that worked well when applied over silver leaf.

During my work, it was pleasant to receive a message from Kathryn, one of the horticultural staff at Cambridge University Botanic Garden. She kindly asked if I would like some freshly cut *Aloe buhrii* leaves for my research, explaining that they were exuding a bright yellow sap. I collected the leaves, extracted some of the sap, and applied it as a glaze over silver leaf to produce an imitation gold (see *Recipe 2*). This resulted in a good translucent color that shimmered in the light. I took the remaining sap and placed it with some wool, chamois leather, and silk samples in glass jar and left them to soak for over a day or so, which resulted in a golden yellow shade.

The inner gel from *Aloe* leaves can be mixed into ink, acting as a thickener in a similar way to gum arabic. The yellow sap can also be left to evaporate to form a dry powder and used to tint a varnish that can be applied as a colored glaze.

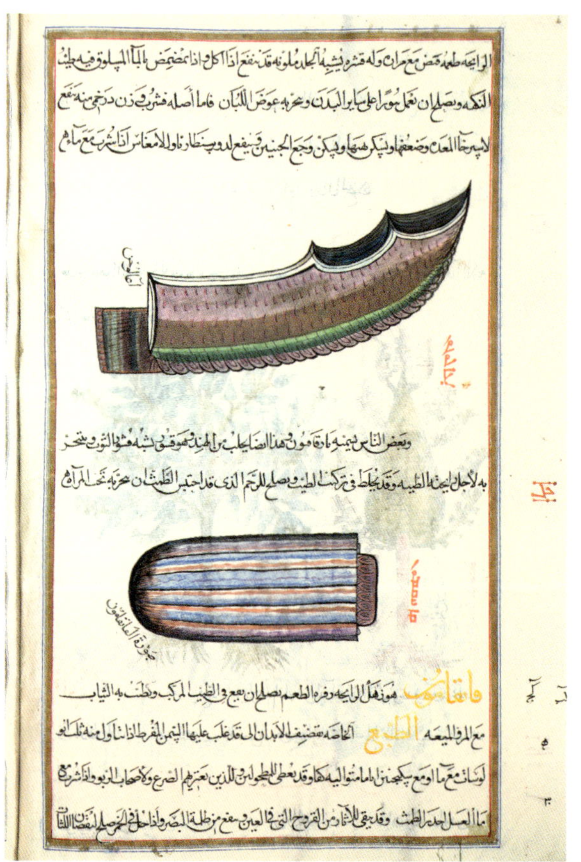

Barbados aloe (*Aloe vera*) (top). Illustration by Persian artist Mirzā Bāqir for an 1889–90 edition of a 9th-century Persian translation of Pedanius Dioscorides' *De Materia Medica* (1st century CE).

Recipe 1: To make giltinges upon leather

"And note, that the aloe is it that giveth the yellow colors to it and taketh it looks like golde, and the other things make it thycke."

—Girolamo Ruscelli, *The Secrets of the Reverend Maister Alexis of Piemont*, Chapter 72, translated by William Ward ca. 1562

Recipe 2: Making aloe red-gold

To make real silver or tin leaf appear like red-gold, cut a leaf from an *Aloe buhrii* plant and apply the fresh, bright yellow sap over the metal leaf and let it dry. Then coat the colored metal leaf with gum arabic to seal and protect the glaze from scratches.

—Nabil's modern recipe

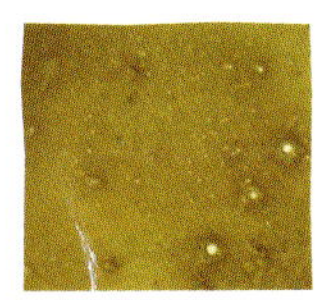

Recipe 3: Making seedpod yellow

A yellow can be made from the seedpods of *Aloe buhrii*. Crush the fresh seedpods and heat them in water with alum, which will produce a yellow dye, then reduce the dye by half. Add gum arabic to the dye to make a yellow ink wash. You can also mix the dye with chalk to produce a yellow pigment, then add gum arabic to make a yellow paint.

—Nabil's modern recipe

Colors from Nabil's Practice

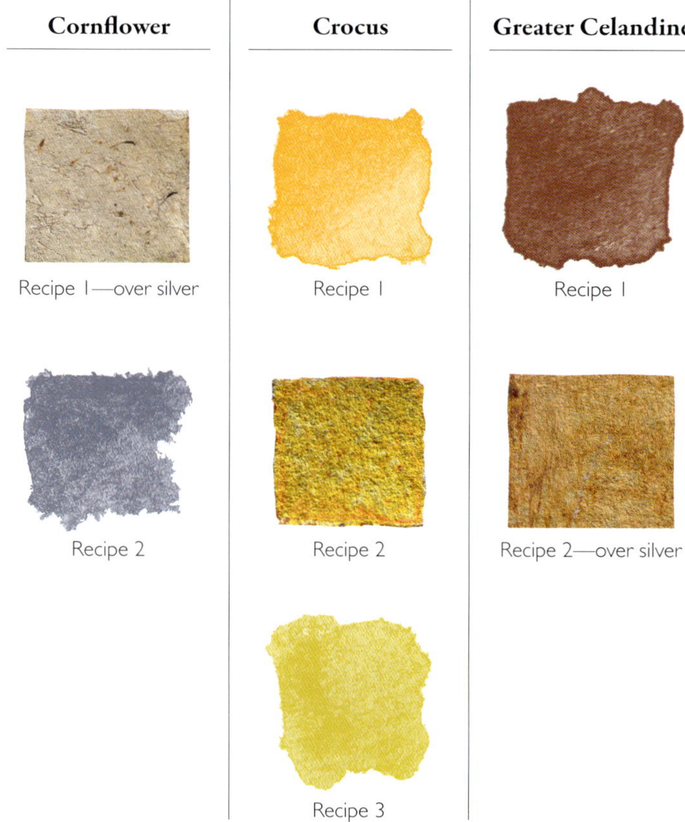

Making imitation gold from a variety of different plants is an historic practice documented in technical manuscripts. Saffron and turmeric are mentioned in medieval recipes, yet cornflower gives one of the most convincing imitations. The basic theory of making imitation gold is that if a plant can produce a concentrated yellow or orange dye, it can be mixed with gum arabic or fish glue and applied to silver or tin leaf.

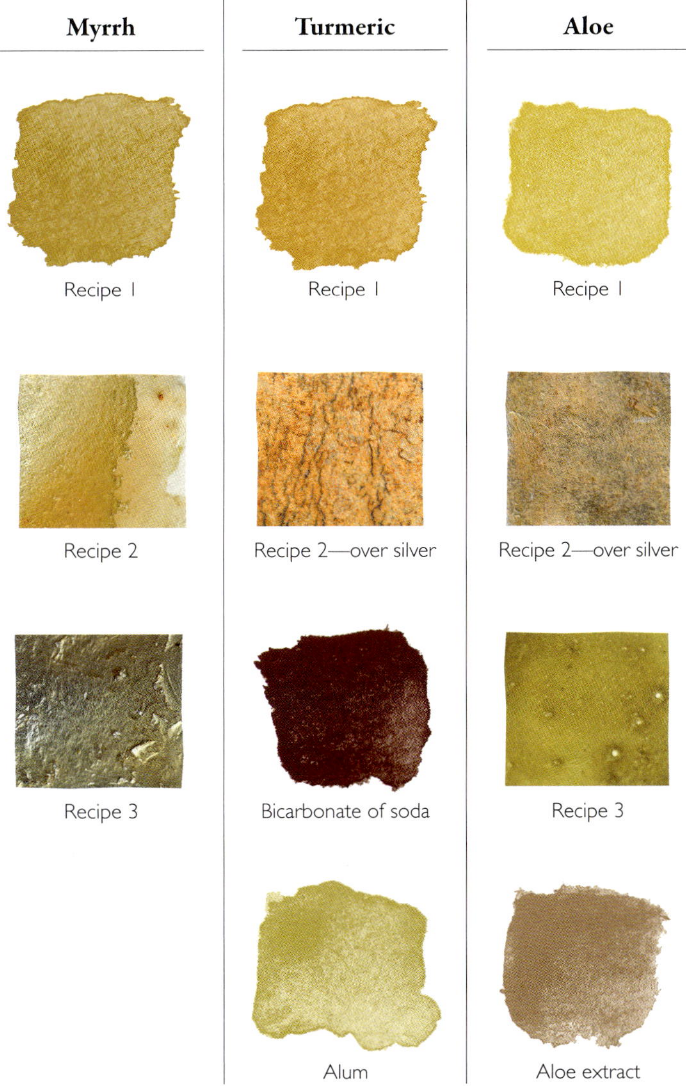

Chapter 2
Blue & Purple

Bog Bilberry

Latin name: *Vaccinium uliginosum*
Other names: Bog blueberry, bog whortleberry
Family: *Ericaceae*
Genus: *Vaccinium*
Native: Europe, North America
Type: Deciduous shrub

The European bilberry or whortleberry (*Vaccinium myrtillus*) is native to northern Europe and closely related to the North American wild blueberry (*Vaccinium angustifolium*). Bilberries have been used as a dye since ancient times, being grown, for example, in Roman Gaul to dye the garments of slaves purple. In the 1st century CE, Pliny the Elder noted in his *Naturalis Historia* that *Vaccinium* plants were used to drug captives.

There are around 450 species in the *Vaccinium* genus, including blueberries, cranberries (*Vaccinium microcarpum*), and lingonberries (*Vaccinium vitis-idaea*). Wild blueberries (*Vaccinium angustifolium* Aiton), which thrive in swampy conditions, have been used for thousands of years by the Indigenous peoples of North America. The Abnaki, for example, used blueberries as well as pokeweed (*Phytolacca americana*) as a color stain. The tribes also used them to make pemmican, a traditional food of dried meat with fat and berries. The plants were highly regarded by these tribes, who recognized the berries as a valuable source of food and for their health properties. In the early 1600s, French diplomat and explorer Samuel de Champlain visited the Indigenous peoples inhabiting the East Coast of America. He documented that they made a pudding with dried blueberries by grinding them into a fine powder and combining them with fresh water, cornmeal, and wild honey.

The blueberry was first commercialized in 1916 by botanist Elizabeth White from New Jersey and chief botanist of the US Department of Agriculture Frederick Coville, but work on this had already begun a decade before. Today America's main cultivated blueberry crops are the highbush (*Vaccinium corymbosum*), rabbiteye (*Vaccinium virgatum*), lowbush (*Vaccinium angustifolium* Aiton), mountain (*Vaccinium membranaceum*), and low blueberry (*Vaccinium palladium*). Other species are grown around the world, but blueberries primarily come from America.

Like most plants, the weather can have a huge influence on blueberry harvests, as humid conditions and reduced rainfall can affect the crop. The cultivated plants, according to Coville, are best grown in acidic soils consisting of a mixture of washed sand and rotted upland peat procured in *Kalmia* or laurel thickets, with good drainage from broken earthenware pottery and aeration keeping the soil moist.

Bog bilberry
(Vaccinium uliginosum). Engraving by Jacob Sturm from *Deutschlands Flora* (Germany's Flora; 1855).

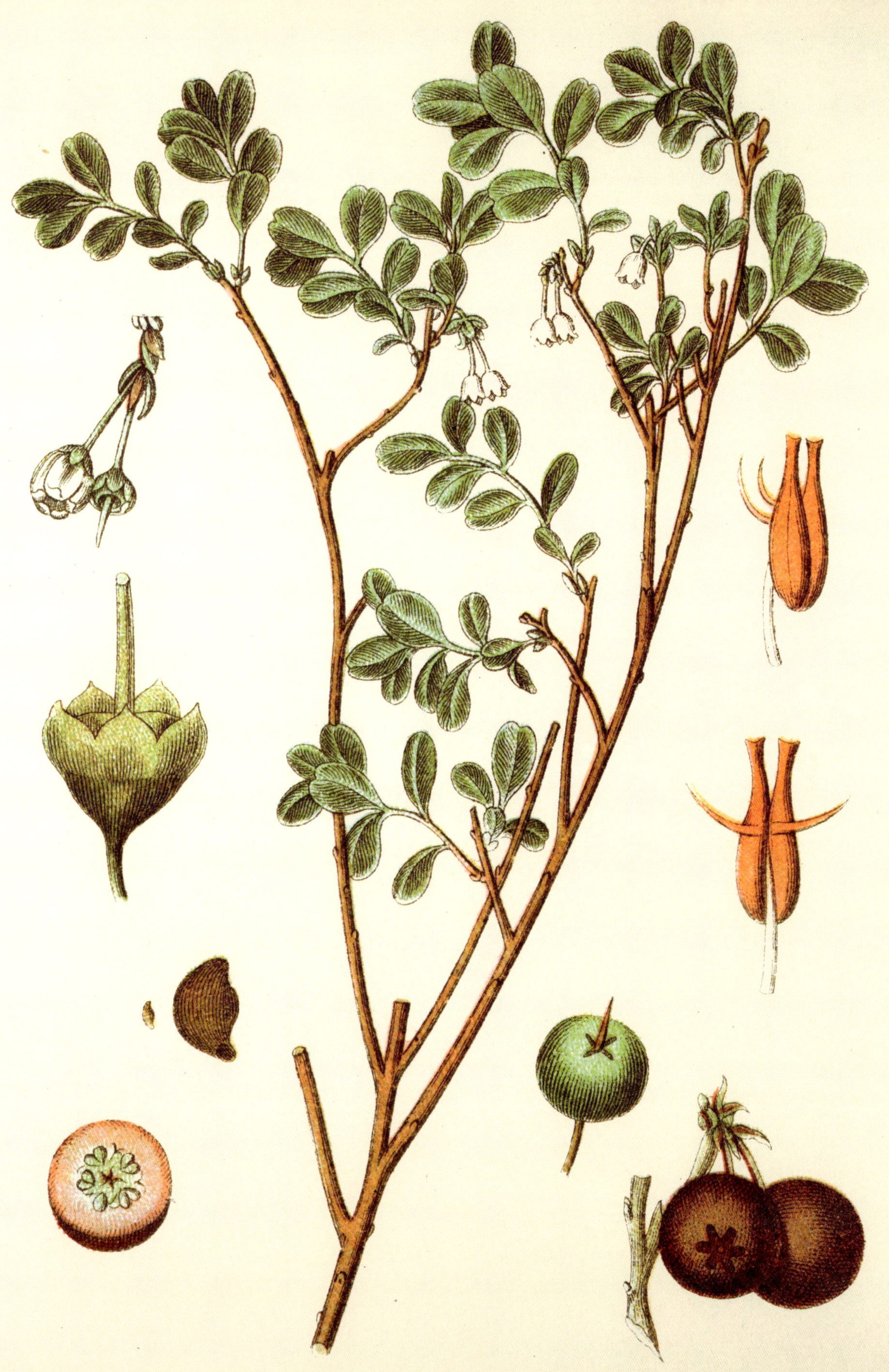

Recipes from Bog Bilberry

Occasional recipes in historical manuscripts relate to blueberries, with the name changing as it was recorded or copied. For example, the Roman architect and engineer Vitruvius, in the 1st century BCE, included a chapter on "Substitute Pigments" in his work *De Architectura Libri Decem* (Ten Books on Architecture). He explains preparing "blue-berries" by mixing the prepared pigment—made by heating berries in water—then straining the dye through a linen cloth before adding it to chalk to dry. The pigment was mixed with "milk" to create "an elegant purple" for wall paintings (see *Recipe 1*), which can also be used in contemporary artistic works. Vitruvius alludes to other plants for making different colors, but in general any bluish-purple to black berries can create a colorant—for instance, the European blueberry (*Vaccinium myrtillus*) will produce browns from the young shoots and purples and blues from the berries.

A 15th-century Middle English manuscript (MS Rawlinson C.506), held by the Bodleian Library, University of Oxford, includes a recipe for making "fine turnsole" with a range of berries. Turnsole is dyer's croton (*Chrozophora tinctoria*), but might also be a "color word" for plants like bilberries, blackberries, mulberries, crowberries, and even blueberries (see *Recipe 2*). The crowberry (*Empetrum nigrum*), as well as the bog bilberry (*Vaccinium uliginosum*), are in the same Ericaceae family as blueberries, with crowberries heated in alum—or aluminum-rich Scandinavian club moss (*Lycopodium* species) gathered in spring—to create a blue colorant. There are many methods for creating a berry-blue dye. But when you see the blue created by the berries of different species, it becomes less important which plant was used in the past. Drawing on historical recipes is a starting point for exploring the natural world and experimenting with a set of processing values.

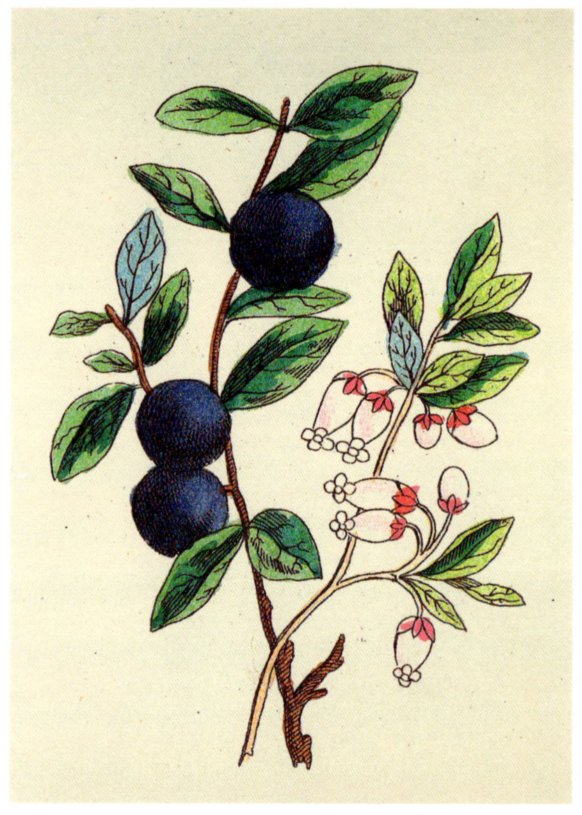

Bog bilberry (*Vaccinium uliginosum*). Handcolored copperplate engraving of a botanical illustration by J. Shaly from Gottlieb Wilhelm's *Unterhaltungen aus der Naturgeschichte* (Encyclopedia of Natural History; 1816).

Opposite: **Bog bilberry** (*Vaccinium uliginosum*). Illustration from Jan Kops and F. W. van Eeden's *Flora Batava* (Flora of the Netherlands; 1872).

Recipe 1: Making elegant purple

"Heat the European Blueberry [or Bilberry, *Vaccinium myrtillus*] in water, strain through a fine linen, and add the juice to chalk to create a pigment. Mix milk into the purple pigment and you should have an elegant purple color."

—Vitruvius, *De Architectura Libri Decem*, 1st century BCE

Recipe 2: Making fine turnsole (blue)

"Take bilberries, or blackberries, or mulberries. And take linen cloth and put your berries in there, and put it in horse dung for fourteen days. And then wash your cloth as nice and clean as you can. And then crush the berries in a mortar, and put that same cloth that you have washed in to there, until it is thoroughly soaked, and then let it dry. And again soak it and let it dry; and do so three times. And then keep it in bladders, closed from the air, and use it etcetera."

—Recipe 79, MS Rawlinson C.506, early 15th century, Bodleian Library, University of Oxford

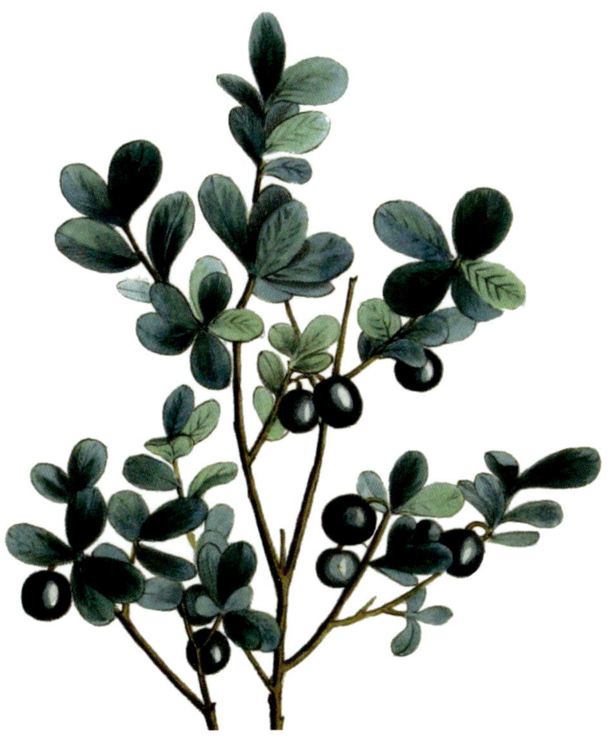

Danewort

Latin name: *Sambucus ebulus*	
Other names: Dwarf elder, deathwort, daneblood	
Family: *Adoxaceae*	
Genus: *Sambucus*	
Native: Europe, northwest Africa, West Asia, America	
Type: Perennial	

Danewort is a very common perennial in the wild and has juicy, dark purple fruits that can be used as a colorant. In her 2007 book *Natural Dyes*, Dominique Cardon cites archaeological evidence suggesting that dye from danewort berries may have been used by Neolithic villagers across Europe for dyeing linen from around 2700 BCE. Fragments of linen have been found, although it is difficult to determine whether they were once dyed or the color has simply vanished after thousands of years.

The berries of danewort (*Sambucus ebulus*) closely resemble those of the common elderberry (*Sambucus nigra*; see page 168). Both plants are in the same genus and have berries that are a dark purple-black when ripe. In his *Naturalis Historia*, Pliny the Elder explains: "Elder-trees have small, black berries with a sticky juice, chiefly used for a 'hair dye'; these [can] also be boiled in water and eaten." Danewort can similarly be boiled in water and eaten or used as a hair dye; Dioscorides points out in his 1st-century-CE *De Materia Medica* that "chamaiacte"—or danewort berries—have the same properties as elderberries.

Although there has been confusion over the centuries as to whether elderberries or danewort berries were used for dyeing purposes, English author Leonard Mascall indicated in his 1583 translations of an earlier Dutch manual that danewort was indeed used to color leathers—he describes the plant as the "low elder" since it only reaches a height of 3–6 feet (1–2 meters). As well as the Dutch dye recipe, the manual also provides methods for removing stains and grease from fabrics and paper and repelling moths with the flowers or leaves of wormwood (*Artemisia absinthium*), or lavender, dried orange peel, and Elecampane root (*Inula helenium*), which was cultivated extensively in monastery gardens and can itself be used to make a blue ink with iron sulfate. There are also recipes for making dyes with other plants using different techniques—for example, to "make a white leather blue" using danewort berries or "low elder." A later recipe in John Mozley's 1776 book *The School of Wisdom and Arts* explains how leather can be dyed blue by heating the berries of dwarf elder (another name for danewort) in water with an alum mordant, revealing very similar processes to recipes in earlier books and manuscripts.

Danewort (*Sambucus ebulus*). Handcolored copperplate engraving by James Sowerby from William Woodville and Sir William Jackson Hooker's *Medical Botany* (1832).

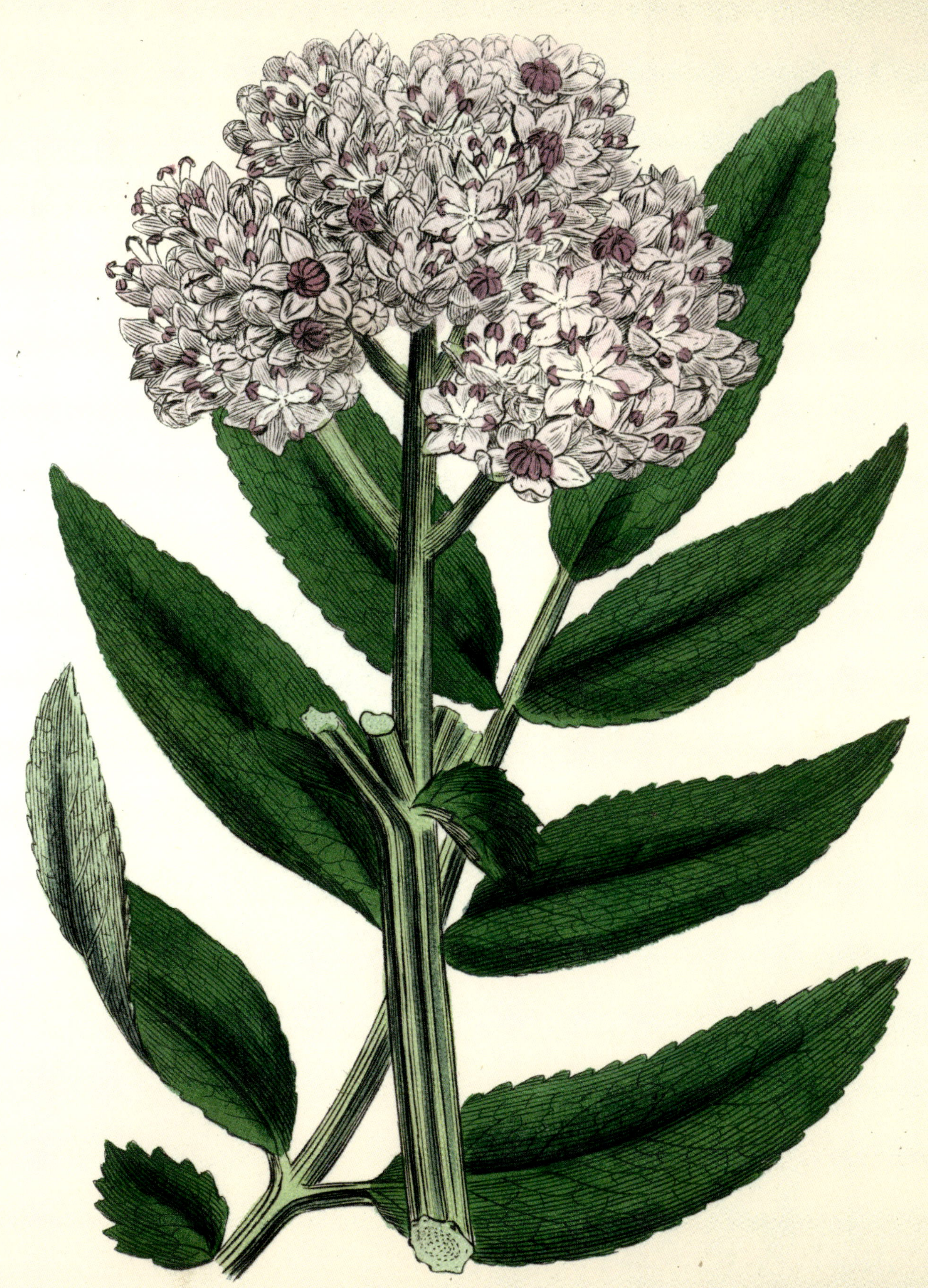

Recipes from Danewort

Danewort is among the richest sources of anthocyanins. These water-soluble compound pigments are found in purple, red, and blue fruits and can be used to make dyes in these colors.

Dark purple danewort berries can be picked in the fall when ripe, but not later in the year where there are early frosts. The berries have an unpleasant odor, but gloves will help to protect the hands. The berries are air-dried or placed in an oven at a low temperature to dry. They can then be crushed with a mortar and pestle to make a pigment for paint. Alternatively, the dried berries can be boiled with mordanted wool to create a lightfast golden color. The juice can be extracted from the berries by pressing them through a piece of cheesecloth (muslin) and stored in the fridge or freezer for later use. Recipe 1.21.1 in the 14th-century *Liber Diversarum Arcium* (Book of Various Arts) gives similar methods: "The juice of Dwarf Elder is carefully collected and dried in the sun, and pastilles are made from that which remains; and [tempered] with a little vinegar and wine, and are used like that." These instructions were copied from Recipe 97 for "making indigo [colored] pigment" in the earlier 12th-century *Mappae Clavicula* manuscript.

Another medieval example using dwarf elder appears in *De Coloribus et Mixtionibus* (Of Colors and Mixtures) an Anglo-Norman manuscript from 12th-century England, held in the British Library, London (MS Cotton Titus D.XXIV). It includes a reddish-purple leather dye recipe that uses the root of the madder plant *Rubia tinctorum* (see page 100), well established as a red dye, and dwarf elder berries. Adding some cream of tartar (potassium bitartrate/potassium hydrogen tartrate) brightens the colors by helping the fibers to absorb the alum, reducing the harsh effects on the leather (see *Recipe 1*).

Just as the berries can provide a colorant, so too can the fresh leaves in summer. To do this, the leaves are heated in alum to produce a yellow dye. Adding alum and iron together produces a green dye—iron is key for "saddening" or darkening the color. Making an alkaline solution with either potash (pH 12) or baking soda (pH 9) produces deeper colors, while using a vat bath with copper sulfate (pH 3–4) can produce a more stable dye.

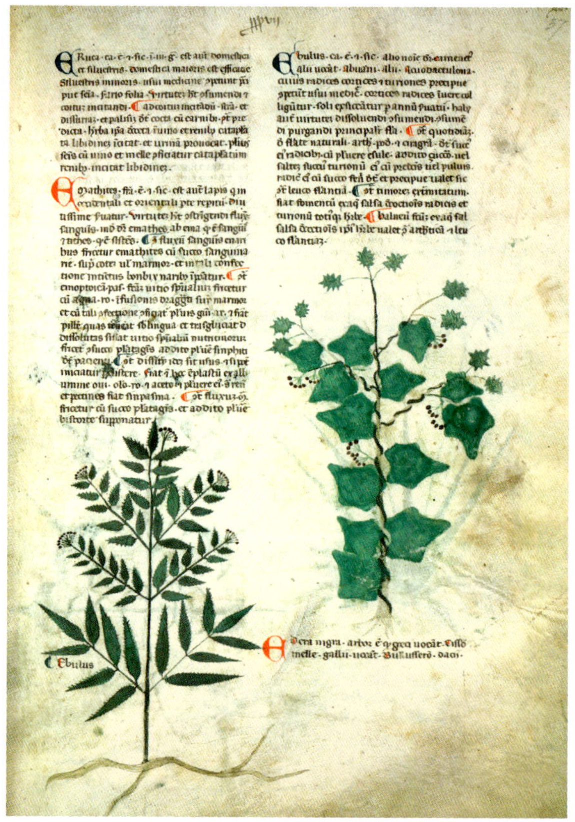

Danewort (*Sambucus ebulus*) (left). Illustrated page from *Bartholomaei Mini de Senis Tractatus de Herbis* (Bartholomaeus Mini de Senis's Treatise on Herbs; ca. 1300).

Recipe 1: Dyeing saddle leather

"Take cream of tartar and madder, sieve, and boil well in stale beer. Take the juice of danewort and three times the quantity of water and boil well. Then add lye in the same measure as the juice. And when you have removed it from the heat, dip the [leather] hide in to it when the temperature is such that you can put [your] hand in it."

—*De Coloribus et Mixtionibus*, 12th century (Hunt, 1995, p.209)

Recipe 2: Making danewort blue

Collect 200 grams of ripe purple berries in the fall, then crush them and leave to soak in 7 fluid ounces (200 milliliters) of water mixed with 3½ fluid ounces (100 milliliters) of clear vinegar. After several days, gently heat the purple liquid for about 60 minutes at a light simmer, then leave to cool and stand for half a day. Strain the liquid through a fine-mesh strainer and store in an airtight jar. You can put the fresh berries or the processed dye in the freezer for later use. No mordant is needed to produce a blue dye on leather and purple on cotton and silk, but you can also prepare fabric with the standard alum mordant or add it when heating the berries.

—Nabil's modern recipe

Flowering Currant

Latin name: *Ribes sanguineum*

Other name: American currant, winter currant

Family: *Grossulariaceae*

Genus: *Ribes*

Native: North America, Canada

Type: Deciduous shrub

There are over 200 wild and cultivated species of *Ribes* in the Northern Hemisphere and South America, including the gooseberry (*Ribes uva-crispa*), alpine and buffalo currant (*Ribes alpinum* and *Ribes odoratum*), and the European black and golden currant (*Ribes nigrum* and *Ribes aureum*). The flowering currant (*Ribes sanguineum*) is a small ornamental shrub, with hanging pink to rose-red flowers that bloom from mid- to late spring and attract pollinators including bees, butterflies, and hummingbirds. It bears bluish-purple berries covered in coarse hairs. These are high in pectin, a fiber that thickens when heated and is used in the food industry and as a natural glue substitute.

The species was discovered in around 1793 by Archibald Menzies, a Scottish botanist and Royal Navy surgeon. Menzies also made significant contributions to the study of lichens during his voyages, and his collections, along with those of William Lauder Lindsay, are housed at the Royal Botanic Garden in Edinburgh. These large collections are important for lichen dye color research, showing examples of the species with dyed fabrics and detailed notes, with Lindsay's comprehensive study recorded in his 1856 publication *A Popular History of British Lichens*.

In 1826, a handful of flowering currant seeds were collected around Fort Vancouver along the rocky shores of the Columbia River on the west coast of America by another Scottish botanist, David Douglas. The seeds were sent to the Royal Horticultural Society in London, where they were successfully cultivated. Due to Douglas's fieldwork, the flowering currant quickly became a popular garden shrub in the Victorian era, alongside several hundred other species that he collected. The RHS commissioned a watercolor painting of *Ribes sanguineum* in 1829, along with other botanical illustrations, by the English natural history and flower painter Augusta Innes Withers (née Baker), a member of the Society of Lady Artists.

No known recipes suggest that flowering currant berries were used as a colorant, but I processed them successfully to create a mid-blue dye. The berries and flowers contain anthocyanins, water-soluble pigments found in plants such as elderberries, blackcurrants, blueberries (see page 54), iris, hibiscus (see page 96), and geraniums, which are a few of the many examples that produce red, pink, and blue colorants.

Two varieties of **flowering currant**: (left) *Ribes sanguineum* and (right) *Ribes sanguineum* var. *album*. Illustration by Abraham Jacobus Wendel from Heinrich Witte's *Flora* (1868).

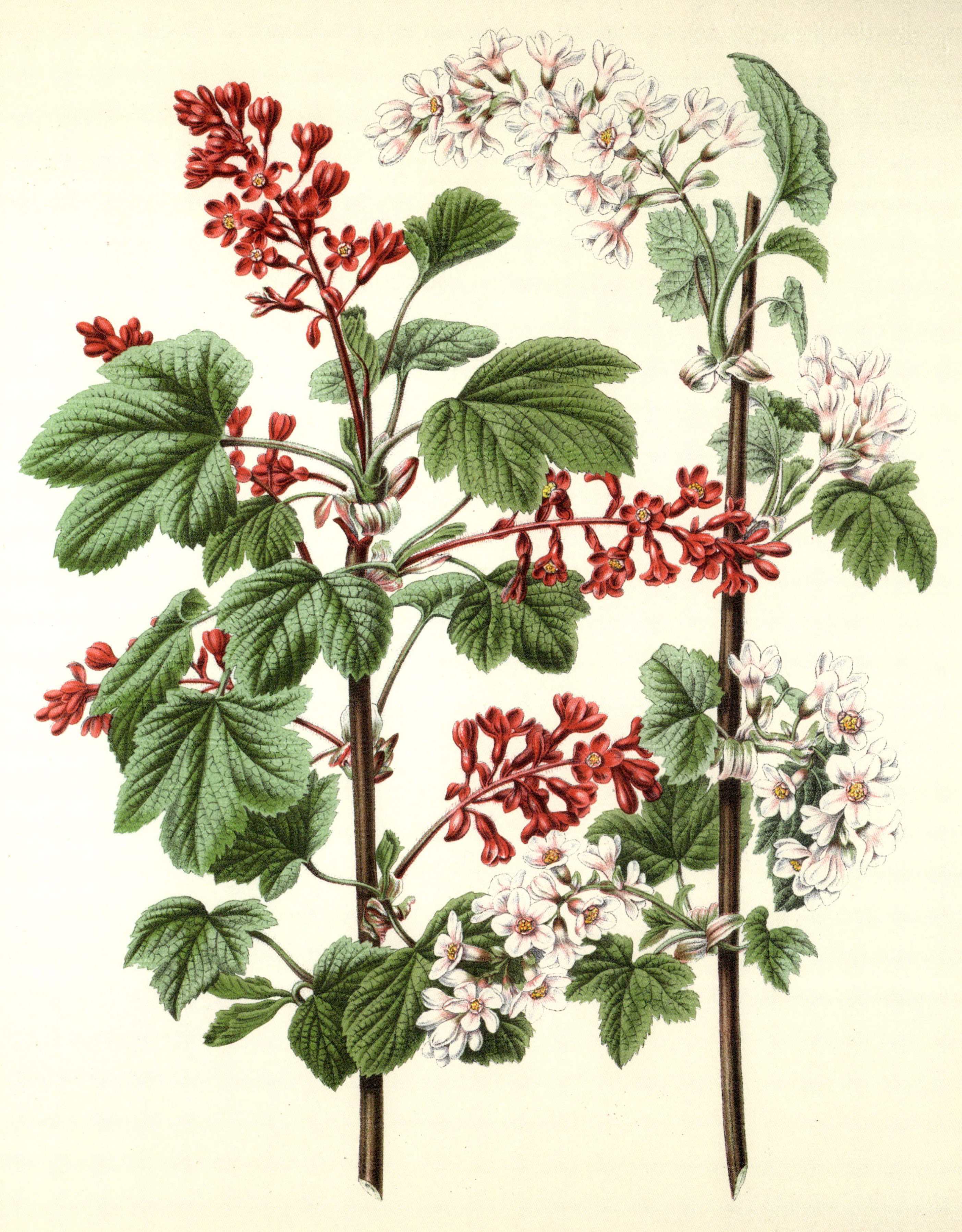

Recipes from Flowering Currant

Few medieval recipes in Western technical manuscripts document the use of flowering currant berries in organic alchemy. If the plant had been available to craftsmen, there may have been an inclination to use the berries to make a blue color. However, in my own reconstructions, I found that a range of colors can be extracted from the berries. In fact, careful experimentation reveals that different types of berries yield unique shades.

In my experiments, I focused primarily on the potential of flowering currant berries for making vibrant paints. I harvested the berries in late summer from my garden and used them to make several successful colorants. I stored half the berries in the freezer for later experimentation.

Heating the berries in water is standard practice, but several different mordants can be used: tin and alum yielded good solid colors. Both these mordants enhance the depth of the color, chemically reacting with the dye compounds and chalk substrates at regulated temperatures between 70 and 95°F (21–32°C). Using more berries enriches the color, while reducing the dye to a concentrated mix produces opaque colors if mixed with chalk (see *Recipes 1* and *2*). Further experiments were carried out, including blending clean recycled eggshells into the dye, which resulted in soft bluish-green. Using a chalk substrate gave the paint a pastel appearance, while mixing the dye with gesso (a white primer made from burned gypsum/plaster of Paris) and a tin mordant intensified the color, making a brighter and slightly deeper purple.

One recipe uses an alternative mordant as a substrate and filler. A total of 6 grams of concentrated berry dye is mixed with 6 grams of aluminum hydroxide powder, a natural mineral called gibbsite (also known as hydrargillite) with acidic properties. The combination creates a stable deep purple paint, which sparkles when light hits the dry surface due to the presence of aluminum. This method does not rely on the usual mordants (tin or alum) in the dye-making stages and can also be applied to other dyes to form slightly glittery paints.

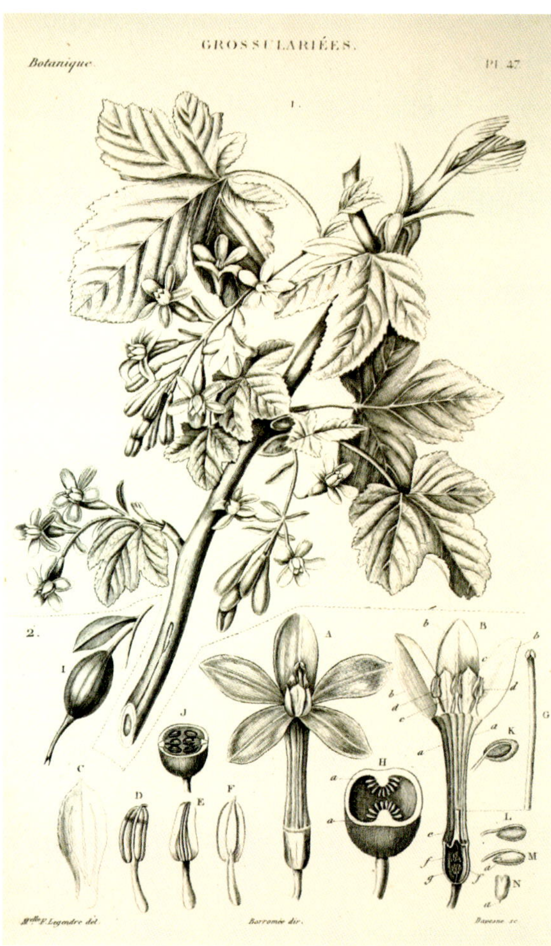

Flowering currant (*Ribes sanguineum*). Black-and-white illustration from Edouard Spach's *Histoire Naturelle des Végétaux* (Natural History of Plants; 1834–47).

Recipe 1: Making flowering currant blue

Collect 40 grams of fresh, ripe flowering currant berries in mid-summer, then crush them in 3½ fluid ounces (100 milliliters) of water. Heat in a saucepan to 70–95°F (21–32°C) with 3 grams of tin mordant and reduce the berry dye to a third—sieve the dye through a fine-mesh strainer to separate the cooked berries.

Add the concentrated dye to 5 grams of chalk or gesso and mix in well, then leave to air-dry for a day to form a blue pigment. Temper (mix) the pigment with gum arabic or egg white to form a bluish-green paint.

—Nabil's modern recipe

Recipe 2: Making flowering currant purple

Follow the basic method in Recipe 1, but use 3 grams of alum mordant to create a purple dye. After reducing and sieving the berry dye, mix it with chalk or other substrates like gesso or eggshells and leave to air-dry to make a purple pigment. Mix the pigment with gum arabic or egg white for a solid purple-colored paint.

—Nabil's modern recipe

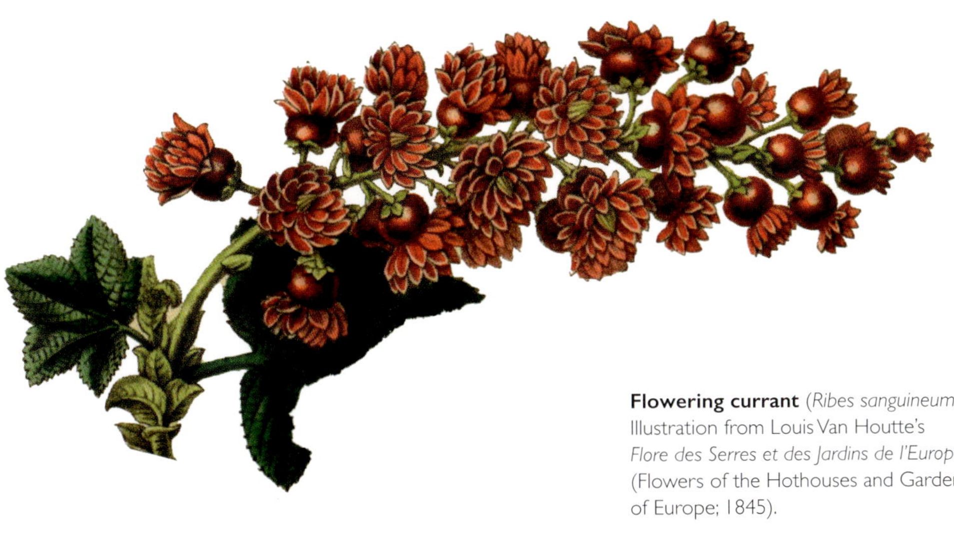

Flowering currant (*Ribes sanguineum*). Illustration from Louis Van Houtte's *Flore des Serres et des Jardins de l'Europe* (Flowers of the Hothouses and Gardens of Europe; 1845).

Common Ivy

Latin name: *Hedera helix*

Other names: European ivy, bindwood

Family: *Araliaceae*

Genus: *Hedera*

Native: Europe

Type: Evergreen climber

Ivy grows abundantly everywhere. According to the Greek philopher Socrates in the 5th century BCE, the Thracian culture relating to Dionysus, the god of wine and vegetation, used leaves inscribed with incantations as a remedy for headaches. Ivy leaves may have been used as magical charms before they were replaced with metal leaf shapes.

Ivy leaves feature extensively on ancient vases. One Etruscan example from around 540–530 BCE shows two heraldic roosters flanking an ivy leaf on an amphora. Centuries later, a Greek example shows a detailed garland of terracotta ivy leaves accompanied by clusters of berries. This was related to Dionysian thinking: wine-drinking and cult practices, explored in my publication *Goat's Blood, Tablets and Sacred Ivy* (see *Bibliography*, page 215). This focuses on ivy recipes in the *Liber Diversarum Arcium* (Book of Various Arts), which was used by medieval craftsmen, especially in metallurgy, who relied on forgers' alchemy to turn copper tablets a golden color with ivy berry juice and blood from an unfortunate goat.

The 6th-century herbal manuscript known as the *Juliana Anicia Codex* (Codex Vindobonensis med. gr. 1), or *Vienna Dioscurides*, held in the Austrian National Library, Vienna, provides good examples of early botanical drawings, including realistic representations of ivy in fruit and of the plant's root system. Later manuscripts contained similar drawing styles, but these became more stylized, especially in the European illuminated manuscripts of the 15th–16th centuries, where illustrated ivy leaves were usually gold-leafed to create decoration.

Ivy also decorated the illuminated pages of medieval bestiaries (books of real and imagined animals). One depicts a medieval knight in shining armor hanging from an ivy branch while defending himself against an oversized snail. Another shows the tongue of a fox being pecked out by a dark raven surrounded by ivy leaves.

Medieval stained-glass windows also featured ivy. The German monk Theophilus Presbyter provides an insight into the colors established in Europe by the 12th century in his treatise "The Art of the Worker in Glass" in the second book of *De Diversis Artibus* (On Diverse Arts). Stylized ivy can be seen in the stained-glass windows of the early 13th-century Regensburg Cathedral in Germany.

Common ivy (*Hedera helix*). Illustration by Pierre-Joseph Redouté from Henri-Louis Duhamel du Monceau's *Traité des Arbres et Arbustes que l'on Cultive en France en Plaine Terre* (Treatise on Trees and Shrubs Grown in France in Open Ground; 1801–19).

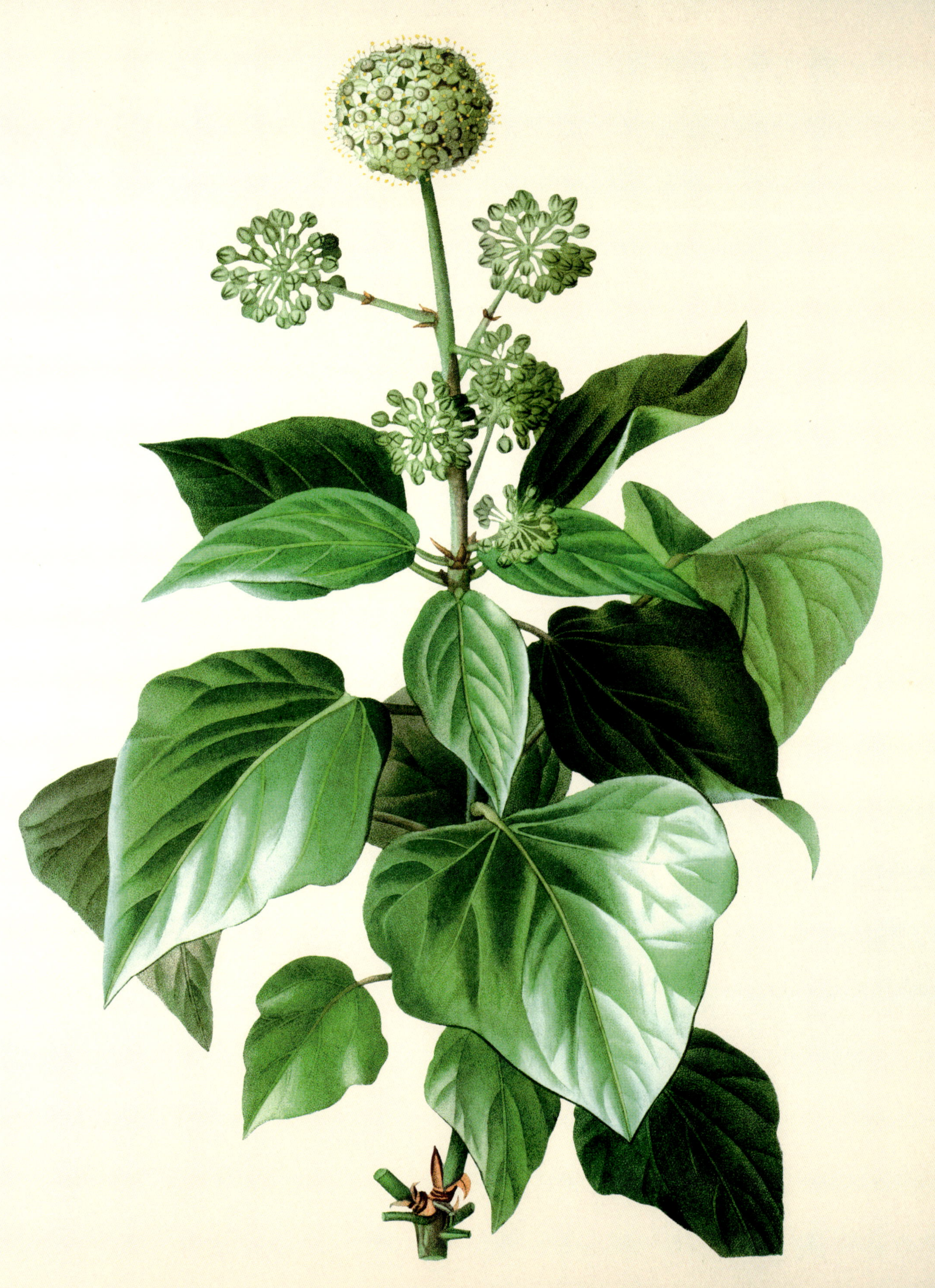

Recipes from Common Ivy

It might be surprising to learn that blue can be made from ivy berries, since historical sources do not mention this or include recipes suggesting it was common practice. My modern recipe for making "woad blue" with ivy berries shows that it is, in fact, possible (see *Recipe 1*). Other colors can be made with ripe ivy berries, and I experimented with a range of dye mordants. I used 40 grams of ripe ivy berries, then heated them in 80 grams of white wine, 7.5 grams of gum arabic resin, and 10 grams of alum. The wine was reduced to a total of 9 grams, then strained through a fine-mesh strainer, creating a purple-colored dye.

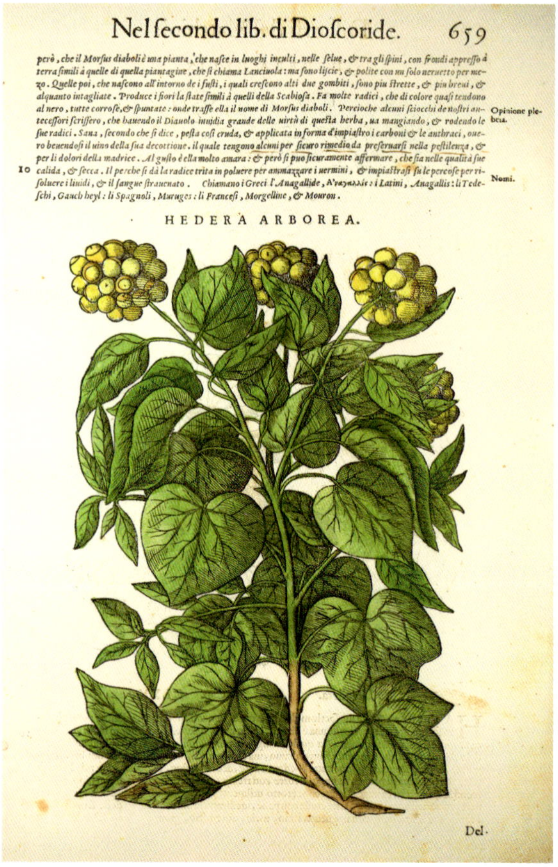

I repeated the process in spring with new berries, heated in 50 grams of white wine and reduced to a total of 10 grams. The mixture was combined with 2.5 grams of Italian gesso, which produced a green color. I repeated the process again using fresh berry dye mixed with chalk—or kaolin (China clay, a soft, white, natural clay) with alum and gum arabic—which also made a green color.

We can also draw on ideas from the past to make other colors. For example, Recipe 43 in the 15th-century manuscript MS Rawlinson C.506, held by the Bodleian Library, University of Oxford, describes a technique for extracting "red gum" from common ivy stems to make a red color (lake or lac) but does not stipulate using ripe ivy berries:

"To make lake, go in early or mid-spring where ivy grows upon oak, and take an axe and pierce the stalk [sic] *of the ivy as many times as you like; and then a month afterward, you shall see come out of the same piercings a good bright gum; then take that gum and cook it well in a new earthenware pan with the urine of a man, for a good while; and skim it well and then you will see a good sanguine color; and then take it off from the fire, and let it cool, and when it is cold, pour out the thin [liquid] above, and from the thick [material] in the bottom make small pellets, and dry them in the sun: and that shall be good lake."* (Clarke, 2016, p.170)

Common ivy (*Hedera helix*). Handcolored woodblock print by Wolfgang Meyerpick from Pietro Andrea Mattioli's *Discorsi di P.A. Mattioli ne i sei Libri della Materia Medicinale di Pedacio Dioscoride Anazarbeo* (Commentary on the *Materia Medica* of Dioscorides; 1568).

Recipe 1: Imitating "woad blue"

Collect 100 grams of ripe ivy berries in early spring before the pigeons eat them all and heat them in 5 fluid ounces (150 milliliters) of white wine or clear vinegar. Add 25 grams of tin mordant (stannous chloride) to the dye and reduce by half. Sieve the dye through a fine-mesh strainer, then add 25 grams of chalk, mix well, and leave to air-dry. Once dried, the color will resemble a light woad blue pigment, which can be tempered (mixed) with gum arabic or egg white to create a gouache paint (opaque watercolor).

—Nabil's modern recipe

Recipe 2: Making ivy purple

Collect ripe ivy berries in spring, heat in clear wine with alum, and reduce by half. Filter the ivy berry dye and reheat for five minutes, then add powdered gum arabic to make a usable purple colorant.

—Nabil's modern recipe

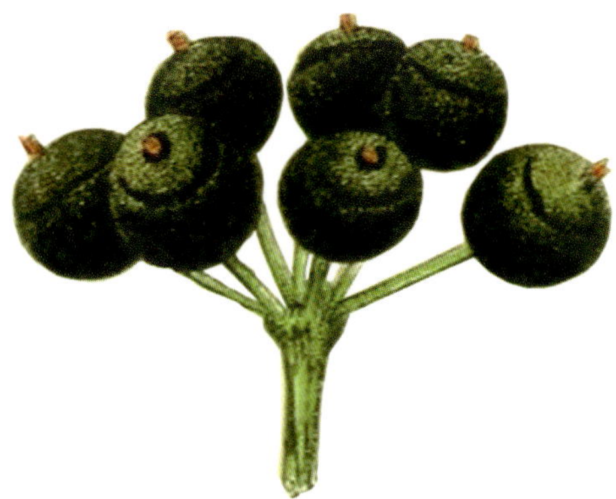

Detail of **common ivy** (*Hedera helix*) berries. Illustration from A. Mentz and C. H. Ostenfeld's *Billeder af Nordens Flora*, vol I (Pictures of the Flora of the North; 1917–23).

Portuguese Cherry Laurel

Latin name: *Prunus lusitanica*

Other name: Portugal laurel

Family: *Rosaceae*

Genus: *Prunus*

Native: North Africa, Europe

Type: Evergreen shrub or small tree

The *Prunus* genus was introduced to Britain after the Roman general and statesman Lucius Lucullus cultivated the sweet cherry (*Prunus avium).* In around 1259, William the Gardener, who worked in the rose garden at Westminster Abbey, became the first supplier of fruit trees and shrubs to London, stocking *Prunus* species and other plants.

The Portuguese cherry laurel (*Prunus lusitanica*) arrived in Britain in around 1648 and was imported from Portugal in 1719 by an English nurseryman by the name of Thomas Fairchild. Fairchild was a freeman of the Worshipful Company of Gardeners and the Worshipful Company of Clothworkers. He grew evergreens, fruit trees, flowering shrubs, flowers, and exotic plants and described the Morello cherry (*Prunus cerasus*) in his book *The City Gardener* (1722) as follows: "The Morello Cherry will live and thrive very well in London; and not only blossom, but bring Fruit to Perfection, in the most airy Parts of the City" (Fairchild, 1722, p.20).

The fruits of Portuguese cherry laurel ripen in late summer and fall and are deep purple; they can be used to create a dye. The gum that bleeds from the tree's trunk or branches can be used to make a varnish, as described in Recipe 246 for an oil-based "gold-colored transparent varnish" in the 12th-century *Mappae Clavicula*. Cherry gum was also used to repair the pages of manuscripts by joining two surfaces together, as outlined in Recipe 1.34.1 of the *Liber Diversarum Arcium* (Book of Various Arts), while Recipe 2.9.3A recommends it as an alternative medium to oil for panel painting to speed up the work.

According to Nicholas Culpeper in his *Complete Herbal*, first published as *The English Physitian* (1652), cherry gum should be dissolved in wine for medicinal purposes—this advice is similar to that in the "Fruit of the Cherry" recipe written centuries earlier by Greek physician Dioscorides, with Culpeper adding that berries from the winter cherry (*Prunus × subhirtella*) can be dried and "powdered [and] taken in drink." This technique can also be used to make paint by mixing the cherry pigment with a water-based binder (gum arabic, gum tragacanth, or animal glue). Mixing this with chalk, crushed cuttlefish, or white marble dust lightens the color and creates softer shades in a gouache paint.

Portuguese cherry laurel (*Prunus lusitanica*). Handcolored copperplate engraving by D.W. Campbell from Thomas Green's *The Universal Herbal, or Botanical, Medical and Agricultural Dictionary* (ca. 1824).

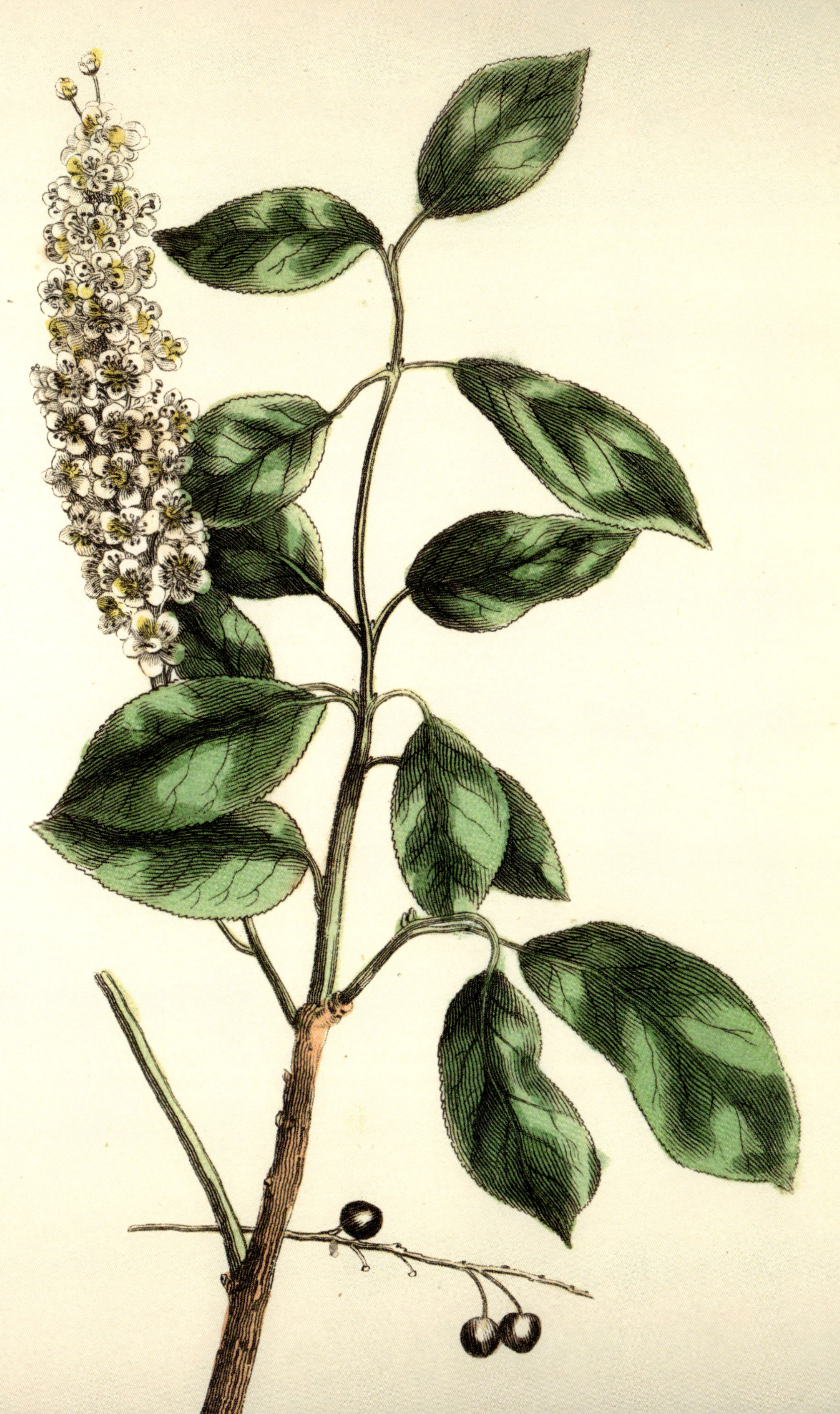

Recipes from Portuguese Cherry Laurel

The deep purple berries of Portuguese cherry laurel are plentiful in summer and ripen in the fall. They can be collected and then stored in the freezer for later processing and use. You can also use the fresh berries straightaway to create a solid mid- to dark blue and a solid purple (see *Recipe 2*). The fresh dye can be added to chalk or another substrate to make a pigment for paint. Drying the berries is also an option; the whole berries can be kept in a dark cupboard or ground into a concentrated pigment to be used for the same purpose.

To get an instant purple to blue color, place fresh berries inside a cheesecloth (muslin) bag with a very fine mesh, then squeeze and crush the berries until you have a pure juice that has been filtered through the cloth. This can be used immediately, either as a dye or as a brush wash on paper. The juice will stain the hands when extracted, so wear gloves for protection until the berries have been processed.

Once the juice has been extracted, the waste berry skins can be dried and reused. Using a pestle and mortar, grind the dried skins into a fine powder. Sieve the powder through a very fine mesh to form a finer pigment, then add it to an oil-based binder, such as flaxseed (linseed) oil or poppy oil, or a water-based binder like gum arabic or egg white to form a paint. You can add a little filler, such as chalk, gesso, or crushed eggshells to give the paint more substance when creating pastel shades.

Some historical recipes make mention of Portuguese cherry laurel berries, but these differ from my experiments. They include a recipe for making a wool dye, which is documented in the *Papyrus Graecus Holmiensis* (ca. 300 CE), often known as the *Stockholm Papyrus*, held in the National Library of Sweden in Stockholm (see *Recipe 1*).

Other trees in the *Prunus* genus can also be used. For example, the bark of sweet cherry (*Prunus avium*) produces mustard-yellow dyes with alum as a mordant and pink and mauve dyes if no mordants are used. A yellow dye can be made from the leaves, using an alum mordant, while an iron mordant results in olive- to lime-green dyes. Finally, heating the new shoots in water produces a bronze-brown dye or one with brownish orange-red tones if an alkaline modifier is used, such as a potash mordant or baking soda.

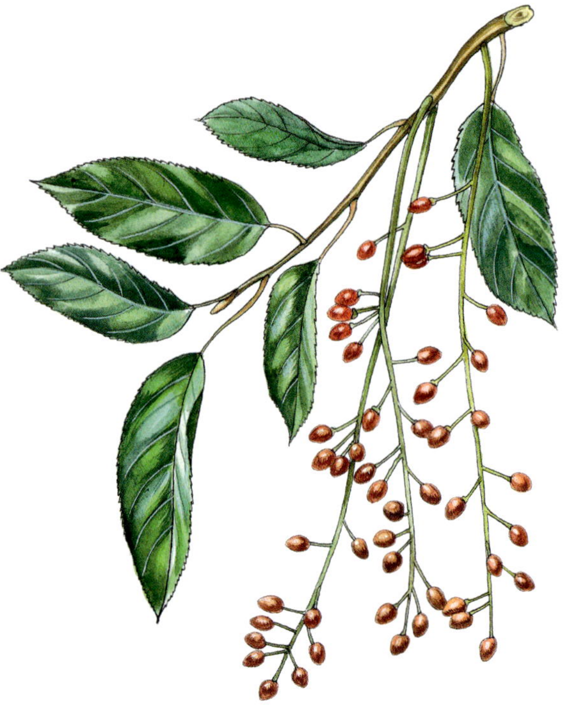

Portuguese cherry laurel (*Prunus lusitanica*). The ripe berries can be stored in the freezer or dried and processed later into a color.

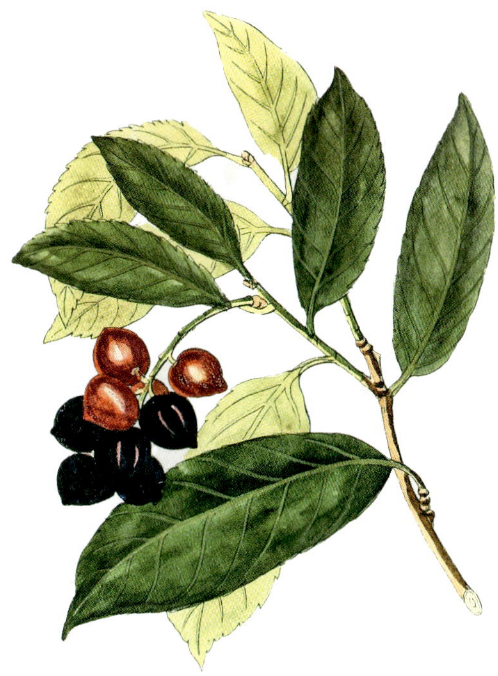

Fruit of the **Portuguese cherry laurel** (*Prunus lusitanica*).

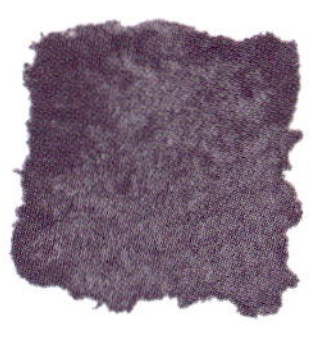

Recipe 1: Dyeing wool purple

"Take and boil grain weevils, dross of iron, and 'laurel berries.' Put in 2 minas [570 grams] of wool, which you have previously mordanted, and now have boiled. Take it out and let it cool off. Brighten the color with limewater."

—Recipe 98, *Papyrus Graecus Holmiensis*, also known as the *Stockholm Papyrus*, ca. 300 CE (Caley and Jensen, 2008, pp.71–72)

Recipe 2: Making solid blue or purple

Collect a handful of fresh ripe berries in the fall and crush and simmer them for an hour in clear white vinegar. Reduce the dye by half, then sieve this through a fine-mesh strainer. No mordant is needed when dyeing leather, silk, or cotton, and this produces a purple that oxidizes in the air, changing the color to a solid blue. If you add a tin mordant to the dye, it brightens the color, making a pleasant pinkish-purple.

—Nabil's modern recipe

Sweet Box

Latin name: *Sarcococca confusa*

Other name: Christmas box

Family: *Buxaceae*

Genus: *Sarcococca*

Native: China

Type: Evergreen shrub

The tiny, pure white blooms of this winter-flowering evergreen shrub scent the air with their vanilla-honey fragrance, a pleasing perfume that lingers for weeks. Sweet box is a resilient shrub that tolerates polluted areas, and will grow in dry soil and shade. As the name suggests, sweet box (*Sarcococca confusa*) is easily mistaken for other species. *Sarcococca hookeriana* var. *hookeriana*—a more upright variety, for example, which has narrower flowers and is hardier than *Sarcococca ruscifolia*—was introduced to English gardens in the mid-19th century by Sir Joseph Dalton Hooker, a botanist and explorer from Royal Botanic Gardens, Kew, and a close friend of Charles Darwin.

Sarcococca comes from the Greek *sarkos*, meaning "flesh," and *kokkos*, which means "berry." The species *Sarcococca pruniformis* was named by John Lindley and first published in the *Botanical Register* in 1826. Lindley was a 19th-century English botanist and the garden assistant secretary at the Royal Horticultural Society. He was also a skilled botanical artist and illustrated rare and exotic plants, detailing many other species in his 1821 publication *Collectanea Botanica*. In 1859, he was elected an honorary member of the American Academy of Arts & Sciences.

Another academy member was the English gardener, botanist, and plant hunter Ernest H. Wilson, who trained at Birmingham Botanical Garden and was a student at Kew. In 1899, he set sail for China on the first of four plant-hunting expeditions. There he found a wild *Sarcococca confusa* plant (possibly a hybrid), as well as other species such as *Sarcococca ruscifolia* and *Sarcococca hookeriana*. He documented these discoveries in his botanical papers, held at Arnold Arboretum Horticultural Library, Harvard University. Wilson's explorations of China made it possible to compare the flora of eastern continental Asia with that of North America at the time.

Sweet box produces small, shiny, black, inedible berries, which are poisonous. The plant is susceptible to honey fungus (*Armillaria* species), which attacks and kills the roots of many woody plants. When the ripe berries are crushed, they produce a rich purple juice that can be used as a dye. These uses are typically contemporary and there are few references to any historical application as a colorant, but this does not detract from the valuable properties of the plant.

Sweet box (*Sarcococca coriacea*). Handcolored copperplate engraving by Joseph Swan from William Jackson Hooker's *Exotic Flora* (1823–27).

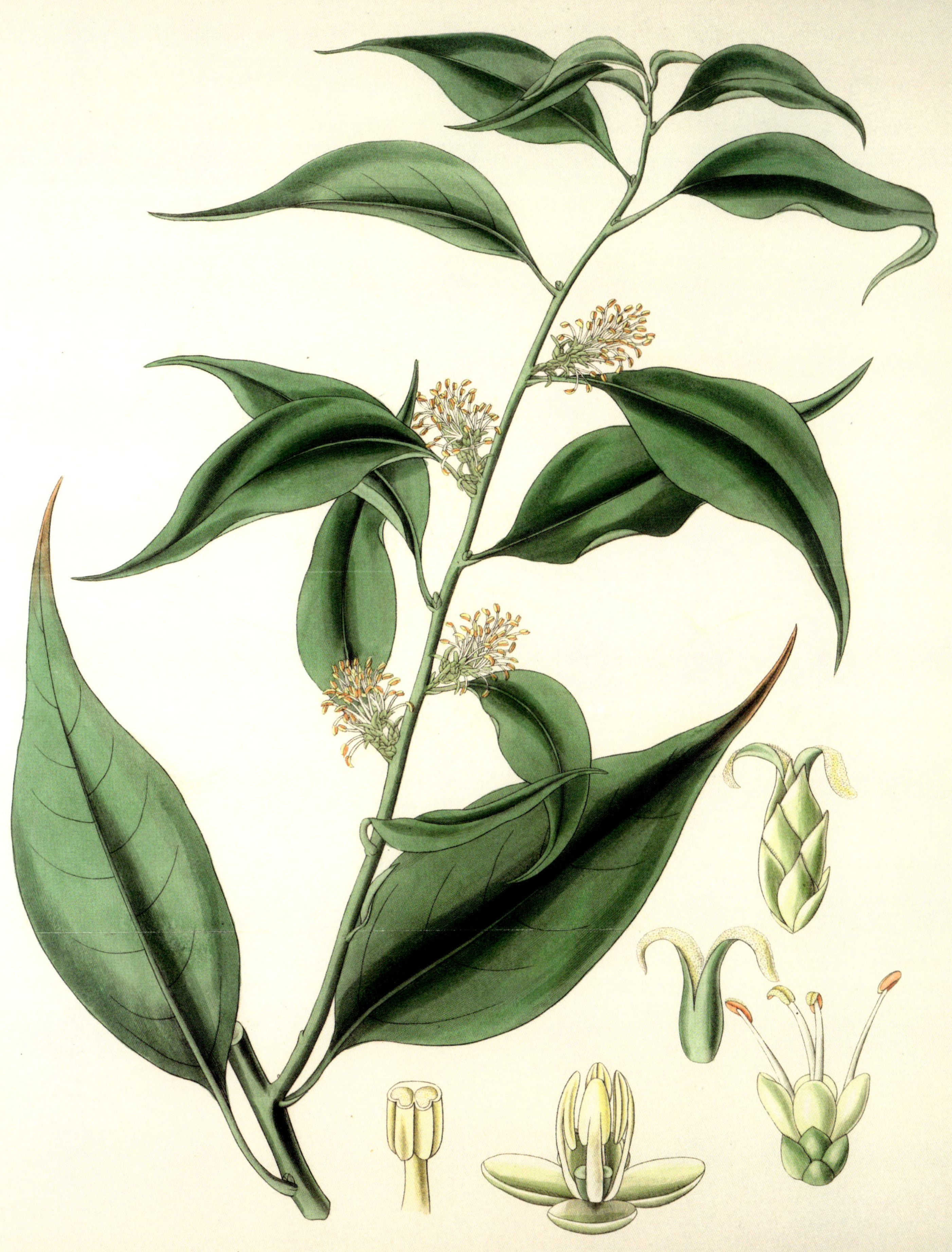

Recipes from Sweet Box

Sweet box grows abundantly in the grounds of Cambridge University Botanic Garden, its flowers scenting the air as I strolled around the garden. It is a pleasing plant to grow and the berries are a good source of blue and purple dye.

For my dye-making experiments, I collected some sweet box berries in mid-winter. Wearing gloves is encouraged when handling this plant to protect the hands from staining and any irritations caused by the berry juice. I wanted to develop a colorant through multiple testings, so once the berries had been collected, a third was stored in the freezer for later use. Another third was dried indoors, then placed in an airtight container or paper bag, ready to be crushed into a dry plant pigment, which I developed over a period of time for making paints (see *Recipe 2*). The remaining third was freshly processed to make a blue-purple dye (see *Recipe 1*).

The berries can also be collected in late winter or early spring. Although they may start to go moldy as the year progresses, they can still be used if you boil them in water first before making the dye. Adding grapefruit seed extract to the water can help combat the further growth of mold or bacteria. Other natural antibacterial ingredients can be added too: for example, rosemary (*Rosmarinus officinalis*) and oregano extract (*Origanum syriacum*), or essential oil made from the seed of white mustard (*Sinapis alba*). Alternatively, to reduce the chance of mold developing, making the dye with clear vinegar or an alcohol-based solution—especially vodka—is good practice. Once made, the vodka-based dye can be kept in the freezer to prolong its shelf life; the dye-ink can be removed from the freezer and used straightaway without defrosting. This method can be applied to most dye-ink processes using other plants or organic materials.

Many organic paints and ink colorants made from plants are best kept out of direct sunlight due to UV and other environmental conditions, which can affect the original shade and quality. Although the colors may change as the years pass, they will still be beautiful. Nevertheless, dyeing fabrics with sweet box or other plants can become an annual ritual, with the berries collected each year in winter or early spring, ready to make fresh dyes or to re-dye and darken those colors that have started to change in appearance.

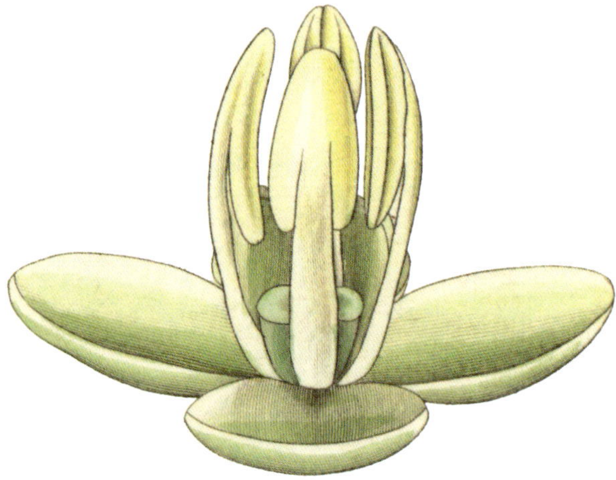

Detail of **Sweet box** flower (*Sarcococca coriacea*). Handcolored copperplate engraving by Joseph Swan from William Jackson Hooker's *Exotic Flora* (1823–27).

Recipe 1: Making berry blue and purple

Collect 50 grams of ripe, dark berries in winter: mid- or late winter is ideal. Crush the berries in 7 fluid ounces (200 milliliters) of water with a little clear vinegar, then add 5 grams of alum mordant. Gently heat the dye until the liquid has reduced by half, then sieve through a fine-mesh strainer. This dye is good for coloring chamois leather, turning it mid-blue when alum is used. If you replace the alum with a potash mordant, the leather becomes green, then yellow when dry. Tin mordant produces a reddish-purple to pink, which can be used to stain paper or dye fabric. Or place the dye in a small glass dish and leave it to evaporate to form a concentrated dye-pigment. Then scrape the pigment off and add to a binder to make a water-based stain.

—Nabil's modern recipe

Recipe 2: Making a purple paint

Mix sweet box dye from the first recipe with a filler such as chalk, gesso, marble dust offered by stone masons, washed and crushed eggshells, white clay, or crushed cuttlefish from the beach. Leave the mix to dry for a day or so until it becomes a pigment. Mix gum arabic or egg white with the pigment to create a paint, which will be lighter than the original dye color as a filler has been added.

—Nabil's modern recipe

Indigotin Blue

The main coloring agent derived from the Indian indigo plant (*Indigofera tinctoria*) is the compound indigotin, which is also produced by woad (*Isatis tinctoria*) and dyer's knotweed (*Persicaria tinctoria*). Indigotin forms when two indoxyl molecules combine and undergo oxidation in the presence of air, and was historically influenced by a bacteria called *Clostridium isatidis*, which is pivotal to making a soluble dyeable form at the reduction stage. These plants were used for thousands of years for dyeing textiles and leather, parchment, and paper; painting; in cosmetics and medicine; and as a colorant for medieval illuminated manuscripts.

Traces of indigotin have been found on ancient stone tools in Dzudzuana Cave in Georgia and have been analyzed by archaeologist Laura Longo together with scientist Elena Badetti, who has suggested the use of blue, whether for a dye or a medicine, dates back 32–34,000 years. Later civilizations used indigo-woad colorant for dyeing textiles and leather, parchment, and paper; painting; in cosmetics and medicine; and as a colorant for medieval illuminated manuscripts.

Scientific analysis of artifacts from Tutankhamun's funerary objects has revealed the presence of indigotin. For example, blue kerchief linen from Tutankhamun's embalming cache (ca.1327 BCE) can be seen at New York's Metropolitan Museum of Art. Earlier textile fragments found at the Peruvian prehistoric site of Huaca Prieta date to around 2400 BCE and feature a blue-striped woven cotton. This was likely dyed with the native *Indigofera suffruticosa*, also known as añil, which can be mixed with a natural clay called palygorskite to create the Maya blue paint used in Mayan and Aztec art.

The insoluble scum or froth that forms on top of an indigo dye vat was scraped off by craftsmen, as described by Pedanius Dioscorides in the 1st century CE in his *De Materia Medica*. It is known as "flory" (or flower) and would be skimmed from the dyer's woad pot for use as a blue pigment, as mentioned in the *Papyrus Graecus Holmiensis*, also known as the *Stockholm Papyrus*, from around 300 CE. It was standard practice to cook woad leaves in "de-frothed urine," a method documented in the 12th-century *Mappae Clavicula*, in order to make an "azure" (blue) color. The 11th-century Arabic *Book of the Staff of the Scribe* by al-Muʿizz ibn Bādīs describes the preparation of a beautiful blue using greater celandine (*Chelidonium majus*; see page 34) and indigo, which in practice would produce a deep blue with a hint of green.

One Latin recipe in a manuscript from around 1400 (MS Lat.235) in the Houghton Library at Harvard University, emphasizes the importance of preserving flory, stating that it is "valuable for painters and illuminators." In the 15th century, Italian painter Cennino Cennini describes in his *Il Libro dell'Arte* (The Craftsman's Handbook) the process of coloring paper with indigo bagdadel—that is, indigo from Baghdad, in Iraq—mixed with lead white pigment, bonded using portion glue (animal glue) "made from clippings from muzzles, bottom of the animal legs, sinews, and lots of hide clippings of kids."

By the late 19th century, natural indigo and woad dyes faced challenges from synthetic indigo developed by the German chemist Adolf Von Baeyer, which was later used in Japanese woodblock prints and primarily to color blue jeans. This innovation altered our perception of indigotin-based plant dyes, leading to their rapid decline, yet the plants are, indeed, experiencing a revival in the 21st century.

Indigo Blue Dye

This is my modern recipe for indigo blue dye, inspired by traditional techniques, the Maiwa School of Textiles in Vancouver, Canada, and Jill Goodwin, author of *A Dyer's Manual*.

1. Prepare indigo or woad leaves
Cut the leaves from a plant in its first year into small pieces and place in a dye vat. Pour boiling water over the leaves and let stand for about an hour, or until the liquid takes on a sherry color.

2. Raise the pH level
Add washing soda (sodium carbonate), slaked lime (calcium hydroxide), or stale urine to raise the water to pH 10 (alkaline)—if using indigo, raise to pH 12. Then aerate the mixture by pouring the liquid back and forth between two buckets for five minutes to promote oxidation.

3. Reduction stage
Heat the water to 167°F (75°C), then introduce a reducing agent such as fructose to reduce the oxygen levels. Fructose is a natural reducing sugar found in overripe fruits like pears, apples, and figs, or mashed bananas. This helps the blue pigment dissolve in the water. It lowers the oxidation state, changing the liquid from a yellowish-green to a tea color. The unreduced indigo bubbles floating centrally on the surface are known as the flory or flower.

Indigo blue dye.

4. Dyeing stage
Skim off the flower and give it to a passing artist, then carefully dip the cloth, or paper, in the dye vat before slowly removing with a pair of tongs. The cloth will oxidize in the air, changing from yellow to blue. Re-dip the cloth in the dye to create deeper shades. To brighten the color, add 10 percent clear vinegar to clean water and soak the dyed fabric again.

Pink-flowered indigo (*Indigofera tinctoria*). Handcolored copperplate engraving of a botanical illustration by J. Shaly from Gottlieb Tobias Wilheim's *Unterhaltungen aus der Naturgeschicte* (Encyclopedia of Natural History; 1817).

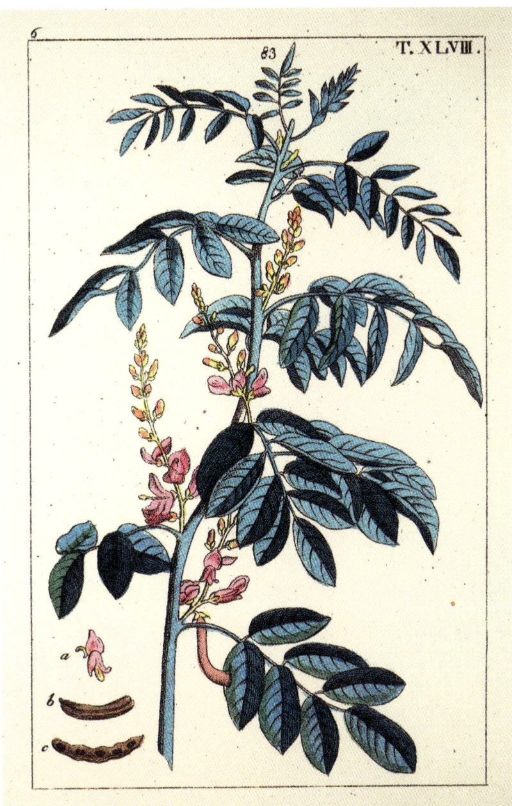

Colors from Nabil's Practice

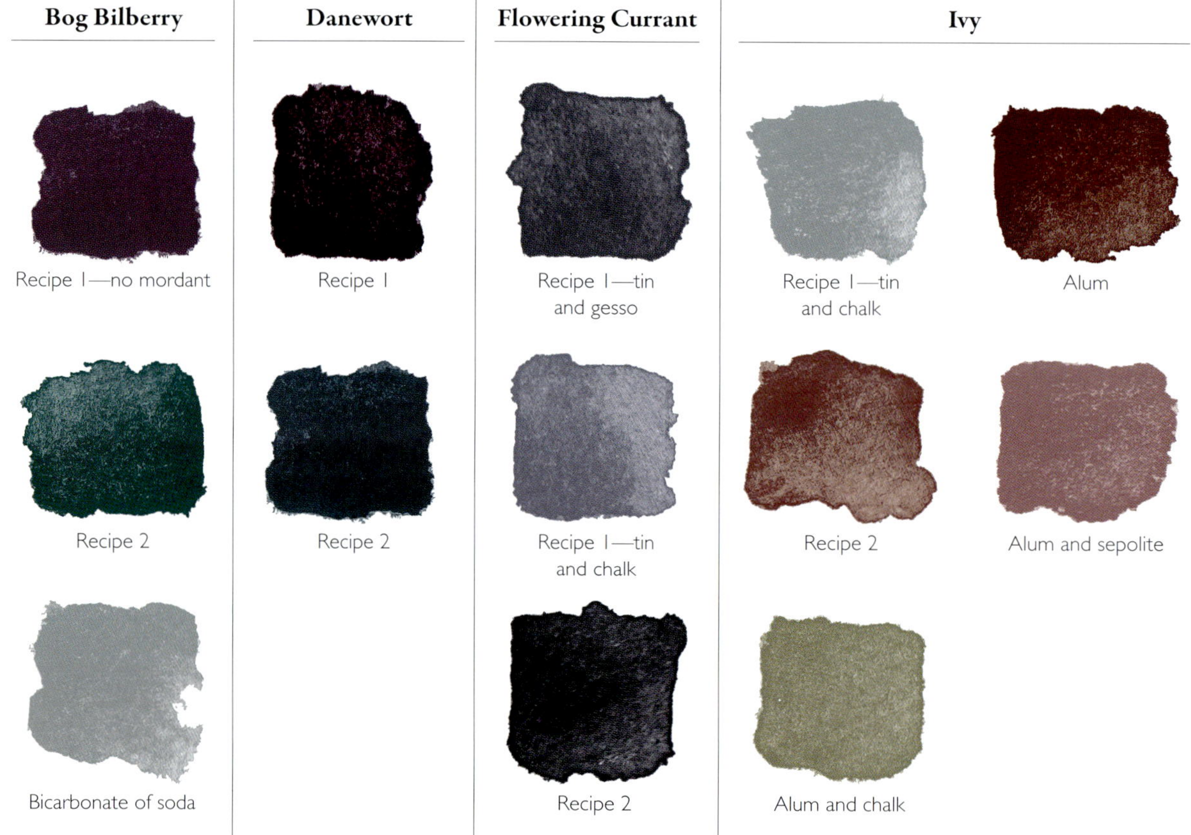

Blue colorants are fairly common in nature, with indigo-woad being one of the oldest organic blue dyes used across ancient civilizations. Bog bilberry can be purchased from supermarkets. Danewort and Portuguese laurel both produce solid colors through oxidizing in the air. Flowering currant and sweet box, like many of the berry colors, are easy to make and store in the freezer and can be harvested at different times of the year.

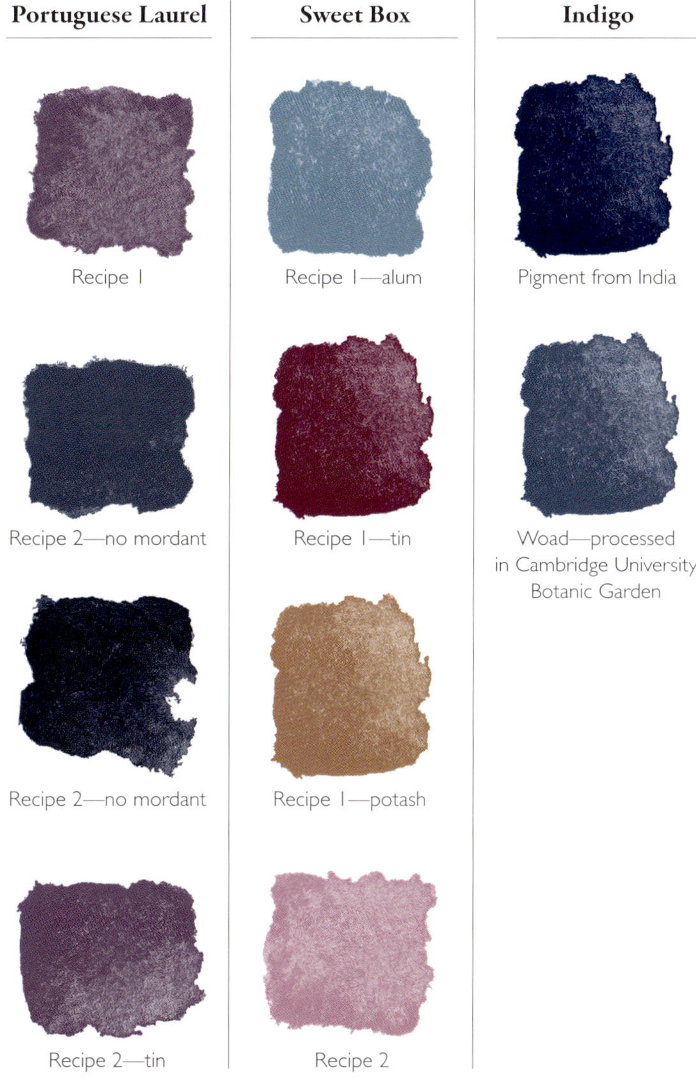

Chapter 3

Red & Pink

Brazilwood

Latin name: *Biancaea sappan*
Other names: Indian redwood, sappanwood
Family: *Fabaceae*
Genus: *Biancaea*
Native: South and Southeast Asia
Type: Evergreen tree

Portuguese explorers in South America, in the early 1500s, recognized the native Pernambuco tree (*Paubrasilia echinata*) as similar to one they already knew from the Old World, the brazilwood or sappanwood (*Biancaea sappan*)—formerly in the *Caesalpinia* genus—the wood of which was obtained in the Middle Ages through trade with Southeast Asia and exported from India to China as early as 900 BCE. From then on, *Paubrasilia echinata*, or pau-brasil, became the main source of brazilwood in Europe.

The wood is heavy, hard, and flexible; indeed, professional 18th-century luthiers discovered that Pernambuco wood was ideal for violin bows and other stringed musical instruments, due to its responsiveness to vibration. It is a strong wood and was used to make wooden bowls and spiritual spears by Indigenous South American peoples, who are today involved in conservation programs to regrow *Paubrasilia echinata*, with laws protecting the species from commerical activities worldwide.

Before the 1500s, the Pernambuco tree grew abundantly up and down the Atlantic Coast of South America, from Cabo de São Roque in the north to Guanabara Bay in the south. It was first exploited for its highly prized timber, with the Portuguese using Indigenous Indian and African slave labor for logging. By the late 1500s, the Dutch had also established a strong influence in South American logging. The logs were exported to the Netherlands and processed by prisoners at the Rasphuis prison in Amsterdam—they were forced to provide cheap labor and rigidly disciplined. The Rasphuis held a monopoly over brazilwood in the Netherlands, with the prisoners working intensively, two men to a log, to grate the logs into a fine powder with long rasps. The powder was then sold to the dyeing industry to produce expensive cloth.

Brazilwood from the Pernambuco tree has now been replaced as a dyewood by the less endangered sappanwood tree (*Biancaea sappan*), which is native to southern India, the Indonesian archipelago, China, Taiwan, Hawaii, and Thailand. The small, shrubby trees are managed in plantations and naturally produce a rich claret-red dye called brazilin, which is used for dyeing fabrics and making red paints and inks. The dye is extracted from the tree's heartwood and large branches, which cannot be harvested more than every six to eight years until the heartwood matures.

Sappanwood (*Caesalpinia sappan*). Illustration from *Plants of the Coast of Coromandel: Selected from Drawings and Descriptions Presented to the Hon. Court of Directors of the East India Company* (1795–1819).

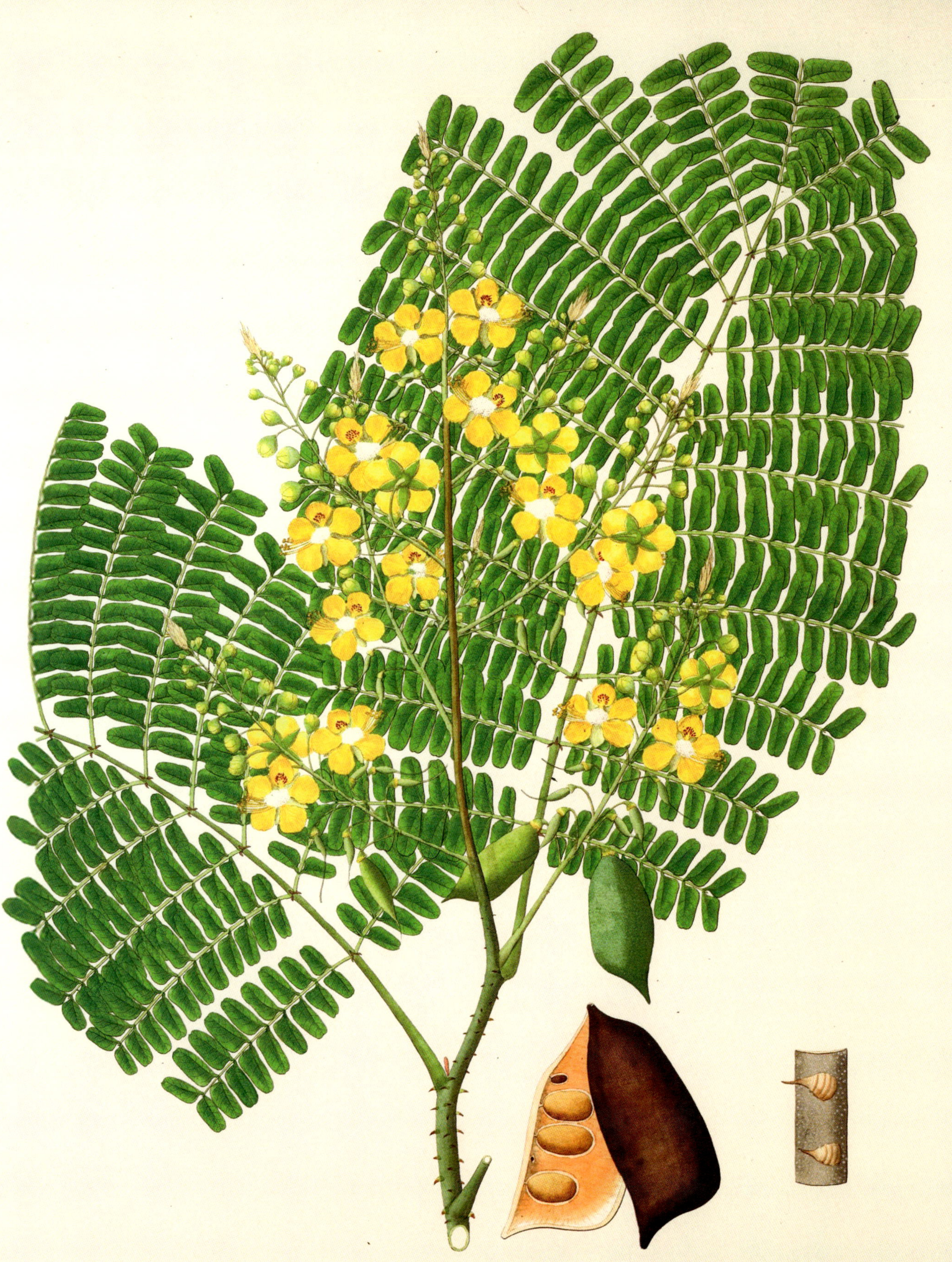

Recipes from Brazilwood

Brazilwood recipes are found in many medieval manuscripts, which describe processes for making a red dye. The dye was typically used on its own, but occasionally mixed with other pigments to create different shades or as a glaze (see *Recipe 2*).

Liber de Coloribus Illuminatorum Sive Pictorum (Book of Colors for Illuminators or Painters) is an Anglo-Norman manuscript, from around 1300, held in the British Library, London (Sloane MS 1754; ff.1v,102v–265v). It describes making a tawny color by mixing indigo (a natural blue dye from *Indigo tinctoria*), brazilwood, and Apulian white (a chalk-based substance) in equal proportions with egg white. More basic recipes make "couleur de rose" from brazilwood powder mixed with chalk and alum, heated in urine and then strained through a linen cloth. Finally, the dye is poured into a hole in a large lump of chalk, then left to dry before being stored in a leather bag. This recipe is found in the 14th-century *Trinity Encyclopedia* (MS O.9.39, Recipes 56 and 57), held at Trinity College, Cambridge.

The 14th-century *Liber Diversarum Arcium* (Book of Various Arts) contains recipes for a red colorant, including Recipe 1.8.3, "On the Nature and Tempering and Recognition of Brazil or Sanguine Color," on which I based my own experiment (see *Recipe 2*). The recipe describes extracting the dye from powdered brazilwood heated in urine and water, then sieved and washed in water to remove unpleasant smells. Similar recipes are found in the *Trinity Encyclopedia*.

Recipe 1.28.74 in the *Liber Diversarum Arcium* explains how to apply a red brazilwood "glaze" over other colors: "If then you want to color your work with gum water or glaire colored with brazil, the work when dry is coated with a brush, and [you] draw the brush once across the place [the surface]." This makes an effective glaze when applied over yellow ocher, orpiment, saffron, or weld (from the *Reseda luteola* plant; see page 144), giving a natural orange when dried.

Italian painter Cennino Cennini's 15th-century *Il Libro de'll Arte* (The Craftsman's Handbook) describes using brazilwood on parchment: "There is also a color made of brazil boiled with lye and rock alum; and then, when it is cold, it is ground with quicklime, and makes a very lovely pink."

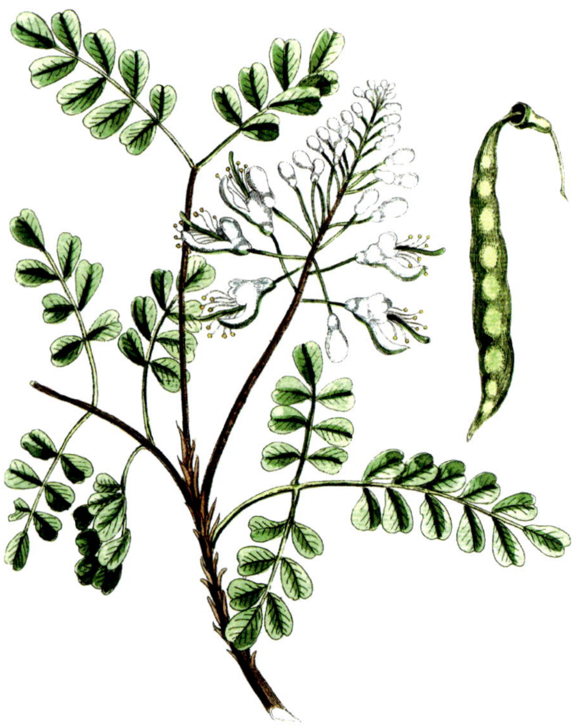

Brazilwood (*Caesalpinia brasiliensis*). Illustration from Andreas Friedrich Happe's *Botanica Pharmaceutica* (1790–1810).

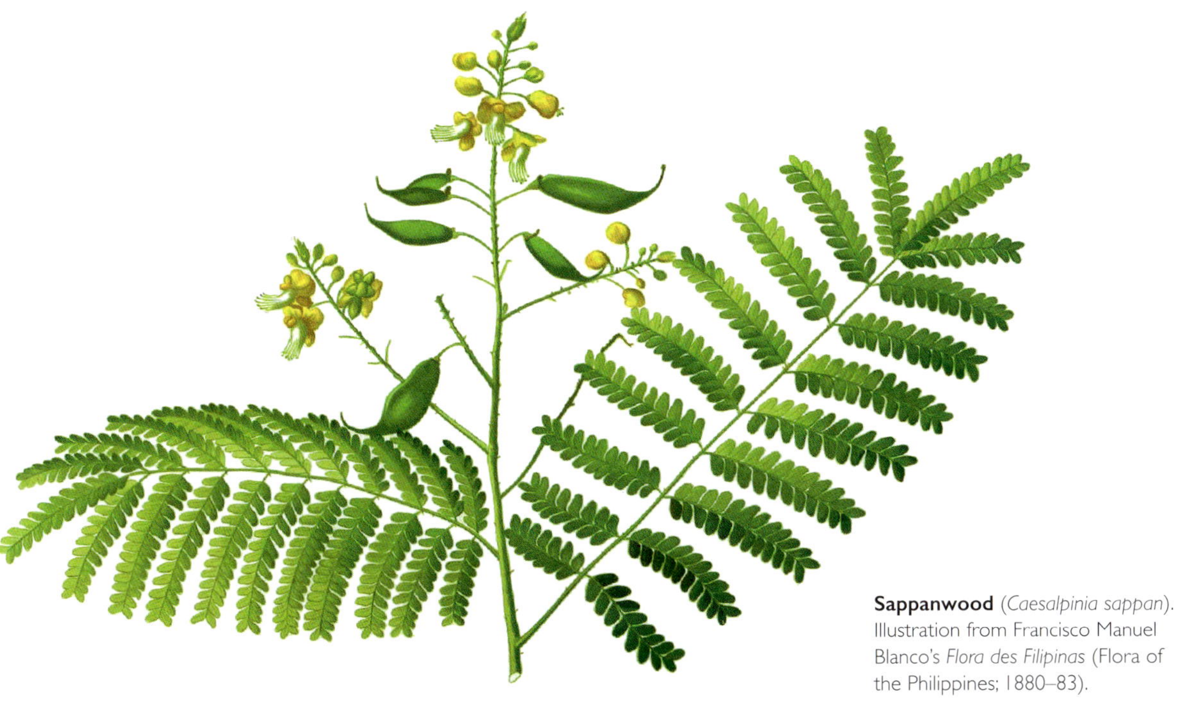

Sappanwood (*Caesalpinia sappan*). Illustration from Francisco Manuel Blanco's *Flora des Filipinas* (Flora of the Philippines; 1880–83).

Recipe 1: Dyeing leather red

"Put sapanwood in a vessel and let it soak in water. A hair brush is put in the sapanwood or a piece of felt is wound on the head of a rod and immersed in the dye, then rubbed on the leather. This is done twice or thrice. Then the leather is pressed and the dyeing is repeated."

—Al-Muʿizz ibn Bādīs, Royal Patron of the Arts, 11th century (Levey, 1962)

Recipe 2: Making a red glaze

Heat 10 grams of brazilwood (or sappanwood) in 4 fluid ounces (120 milliliters) of water with 16 grams of powdered gum arabic for 15–30 minutes and add 1 gram of alum to brighten the red dye. Sieve the dye through a fine-mesh strainer and add a small amount of egg white, then mix thoroughly. This red transparent glaze works well when applied over yellow or blue paint.

—Nabil's modern recipe (inspired by Recipe 1.8.3, *Liber Diversarum Arcium*, 14th century)

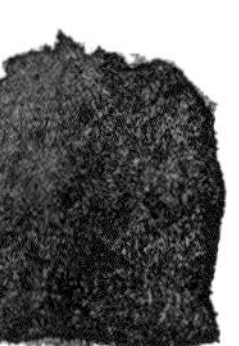

Corn Poppy

Latin name: *Papaver rhoeas*

Other names: Common poppy, corn rose, field poppy, Flanders poppy, red poppy, Odai

Family: *Papavaraceae*

Genus: *Papaver*

Native: North Africa, Eurasia

Type: Annual

The corn poppy's distinctive blood-red flower is famous as a symbol of remembrance of fallen soldiers of the First World War. It thrives in disturbed soil and is seen in swathes of red on farmland where herbicides are not in use. This resilient plant also grows in wastelands and on roadsides, blooming from its scattered seeds into a sea of swaying scarlet.

In the Thebes area of Deir el-Medina, Egypt, there are vibrant and exquisite paintings of red poppies in the tomb of Sennedjem, dating back to the 19th Dynasty (ca. 1295–1186 BCE). These depict scenes of the deceased plowing fields of golden wheat in the Underworld—a reflection of the ancient belief in the afterlife as described in the iconic *Book of the Dead*. Symbolic of life and death, poppies were cherished by the ancient Egyptians and are portrayed growing in their gardens, appearing alongside plants such as the mandrake (*Mandragora officinanum*) in intricate stone reliefs. Both the poppy and the blue cornflower (*Centaurea cyanus*; see page 26) were used in funerary wreaths and in the embalming process.

The influential ancient Egyptian herbal known as the *Ebers Papyrus*, from around 1550 BCE, references the corn poppy, known as shepenn or shepenon, a name later associated with the opium poppy (*Papaver somniferum*). The use of the corn poppy as a mild medicinal remedy persisted into the 1st century CE, when Dioscorides described its use for inducing sleep when heated in wine infusions, as a sedative, and for treating bronchitis, coughs, and hoarseness. The poppy also leaves its mark in the pages of Pliny's *Naturalis Historia* of 77 CE, where he describes the benefits of heating honey with wild corn poppy for treating throat infections.

Corn poppies are depicted in wall garden paintings in the House of the Orchard in Pompeii and amid serpents in the Hospitium of Pulcinella. Two silver vessels made by a 1st-century Greek silversmith, uncovered in the Villa della Pisanella, in Rome, portray scenes of hunting snakes, butterflies, and lizards, with cranes among wheat and corn poppies. In ancient Greece, the poppy was sacred to Aphrodite, the goddess of love, beauty, and vegetation.

Corn poppy (*Papaver rhoeas*). Illustration from John Stephenson and James Churchill's *Medical Botany* (1836).

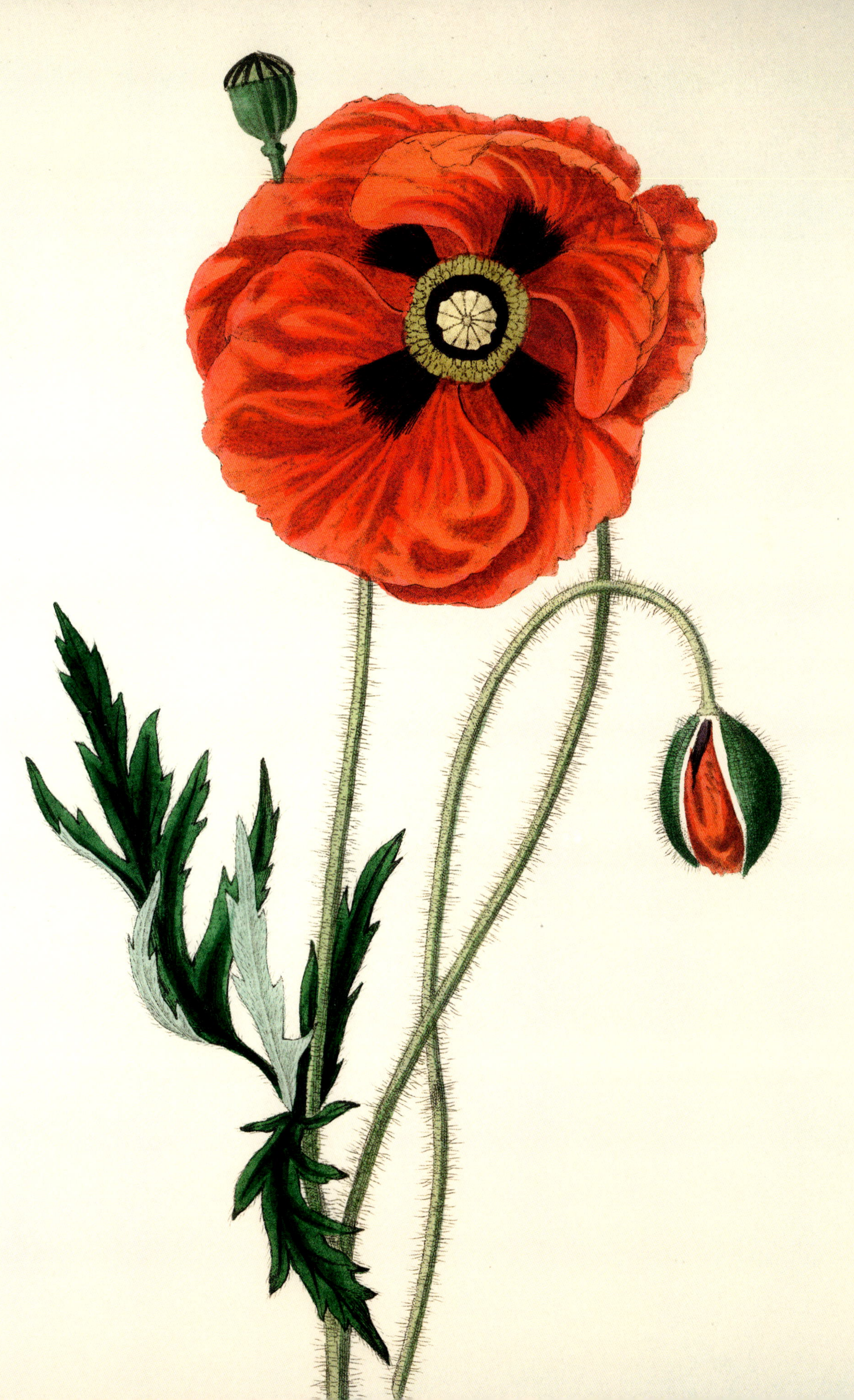

Recipes from Corn Poppy

The Greco-Roman Egyptian *Leiden Papyrus V*, dating from the 3rd to 5th century CE, provides a recipe for "Typhonian red ink" from the petals of the corn poppy (*Papaver rhoeas*; see *Recipe 1*). Adding poppies was more ritualistic in intent, as red ocher was the dominant colorant. During annual festivals, ancient Roman statues of Jupiter were painted with goat's blood to which red ocher, vermilion, or minium (red lead) had been added, possibly to prolong the color of the stain.

The corn poppy was later documented in the 12th-century Greek-Latin *Mappae Clavicula*, which describes grinding the juice of the flowers with cinnabar (mercury sulfide that turns red when ground, known as vermilion in its artificial form). Today, cinnabar is artificially synthesized by combining mercury and sulfur, the two principal substances used in alchemy, but it can be found naturally near areas of volcanic activity, being a common ore that has oxidized with mercury. It has been mined for thousands of years for its fiery red color and used in ritual practice.

The most substantial and comprehensive medieval painters' recipe book is the 14th-century Venetian-Latin *Liber Diversarum Arcium* (Book of Various Arts). The manuscript contains a straightforward recipe for making red ink from poppy petals, which was probably added as a note indicating that poppies can make a red ink or colorant (see *Recipe 2*).

The same manuscript contains an extraordinary recipe for creating an "azure" color from the poppy flower by carefully collecting and moving the delicate stamens with a clean feather, so as not to damage them. The stamens are mixed with egg white and isinglass—a glue from the swim bladder of the sturgeon fish—along with a little warm water, making the mixture alkaline and helping to maintain the hue. A small amount of cinnabar can be added to create a blue-purple color.

I often collect red poppy petals to make red ink and leave the seedpods on the stem, so the seeds can germinate into new plants. I add the petals to white wine in a glass jar and store it in the freezer for later use. This method preserves the red color, and the mixture can be taken out months, sometimes years, later and made into a red ink by heating the wine with gum arabic, isinglass, and alum mordant to enrich the color.

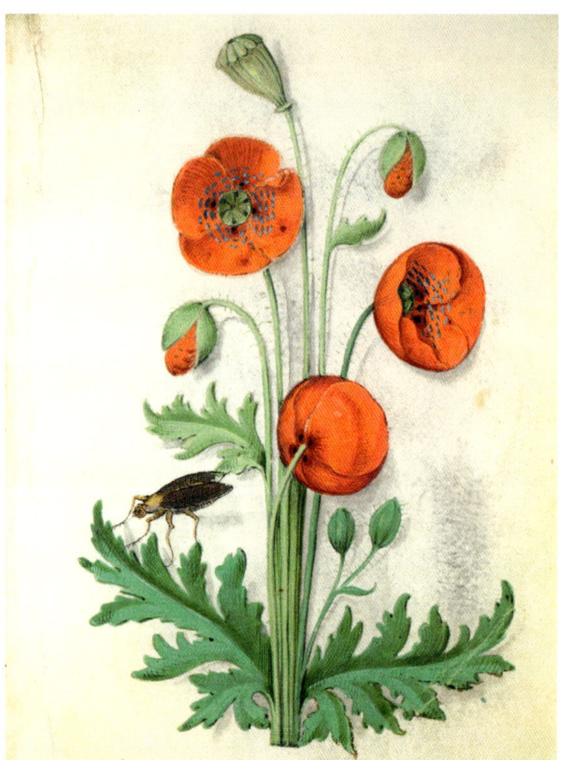

Corn poppy (*Papaver rhoeas*). Illustration by Master of Claude de France from the *Book of Flower Studies*, made in Tours, France (ca. 1510–15).

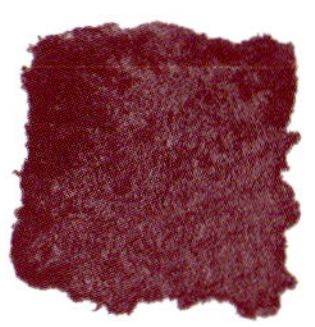

Recipe 1: Making Typhonian red ink

"A fiery red poppy, juice from an artichoke, seed of the Egyptian acacia, Typhons ocher, asbestos, quicklime, with a single stem of wormwood, gum [arabic], and rainwater."

—Recipe PGM X11, *Leiden Papyrus V*, 3rd–5th century CE (Betz, 1986, p.156)

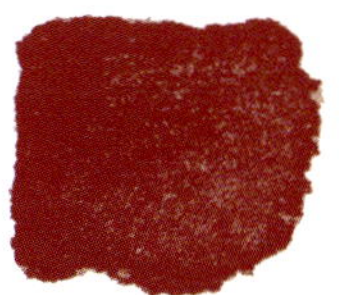

Recipe 2: Making red ink

"Another method [for making ink] in May is to grind flowers of the red poppy and heat some of the expressed juice and strain it, store for a month, then use."

—Recipe 1.6.3, *Liber Diversarum Arcium*, 14th century (Clarke, 2011, p.106)

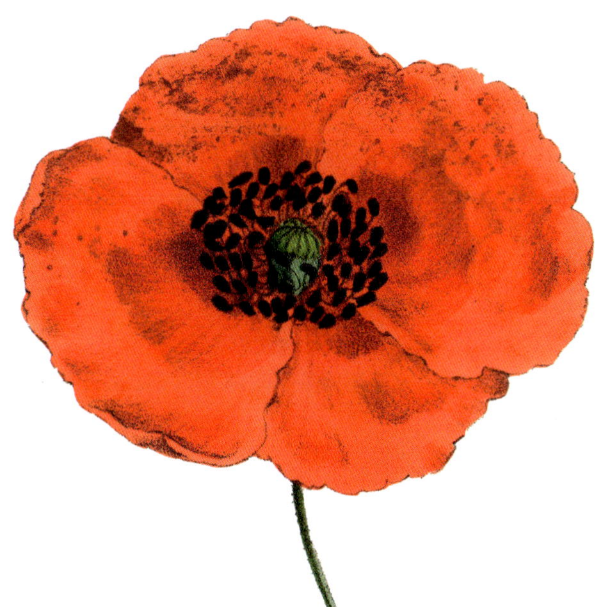

Corn poppy (*Papaver rhoeas*). Handcolored woodblock print from Konan Tanigami's *Seiyou Sokazufu* (Pictorial Album of Western Plants and Flowers; 1917).

Dragon's Blood Tree

Latin name: *Dracaena cinnabari*
Other names: Socotra dragon tree
Family: *Asparagaceae*
Genus: *Dracaena*
Native: Socotra (Indian Ocean)
Type: Evergreen tree

The dragon's blood tree (*Dracaena cinnabari*) is found growing extensively in woodland high in the Hajhir (or Haggeher) Mountains on Socotra, a Yemeni island in the Indian Ocean. Unfortunately, it is becoming vulnerable as a result of overgrazing by goats and a rapidly changing world climate, with a drying of the environment having a negative impact.

The tree's red resin was exported from the island for over two thousand years. The *Periplus Maris Erythraei* (Periplus of the Erythraean Sea), a 1st-century-CE Greek-Egyptian handbook listing ports and coastal landmarks for merchants trading between Roman Egypt, Africa, Arabia, and East Asia, refers to the export of the tree's dark red resin for use as a red pigment and a drug. Writing at the same time, Dioscorides also noted in his *De Materia Medica* that cinnabar was used by painters for costly wall decorations and medicinally for eye treatments.

From classical times dragon's blood was called Indian cinnabar, which was well known by the time Pliny the Elder alluded to it in his *Naturalis Historia*. He describes how the Indian elephant fought ferociously with the dragon (probably referring to the crocodile that drifts under the surface of rivers ready to snatch from the riverbank). When the dragon entwined itself around the elephant, they bit and clawed each other until blood spilled from both animals. The weight of the elephant crushed the dragon and their combined blood became dragon's blood.

This story recalls the wounds made in the bark of the evergreen climbing palm *Calamus draco* (formerly *Daemonorops draco*)—from which most commercial dragon's blood is produced today—and similar species in the *Calamus* genus when incisions are made to extract the red resin, which "bleeds" out as a result. The resin, which is indeed the color of blood, can be used as a red pigment, mainly for glazes and in paint, or to varnish wood and stain violins. The fruits of *Calamus draco* also produce on their surface a deep red resin known as dragon's blood. *Calamus draco* is native to Sumatra but is also found in Malaysia, Thailand, and Borneo, from where it is imported for use as a medicine, incense, and dye—the resin balls are transported wrapped in cylindrical rolls made from palm leaves. The dragon's blood tree (*Dracaena cinnabari*) has characteristics and uses similar to those of *Calamus draco*.

Dragon's blood tree *(Dracaena cinnabari)* (Left) and areca palm (*Areca catechu*) (right). Handcolored copperplate engraving after C. Franck from Friedrich Johann Bertuch's *Bilderbuch für Kinder* (Picture Book for Children; 1795).

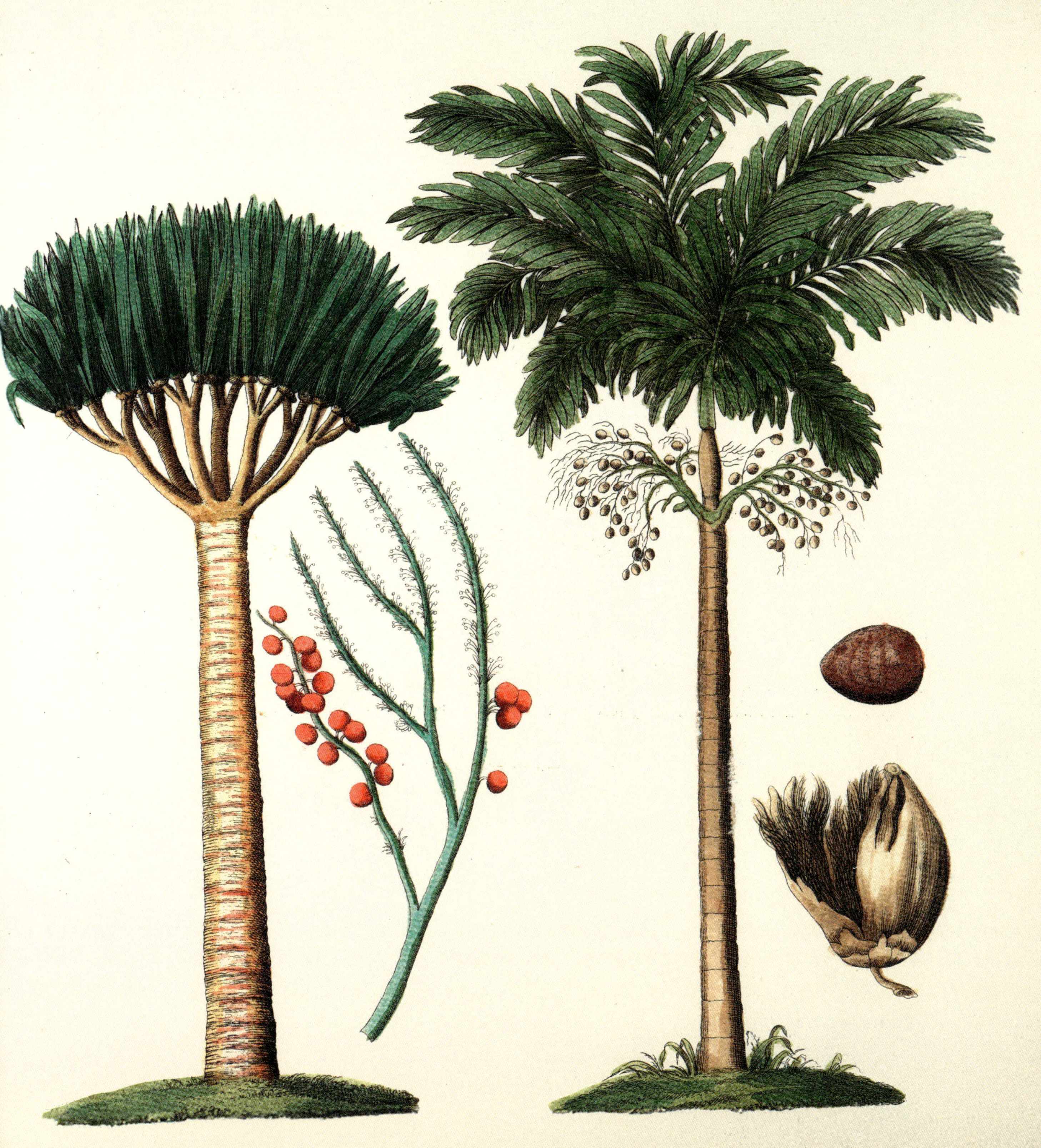

Recipes from Dragon's Blood Tree

The pigment made from the red resin of the dragon's blood tree was used to improve the appearance of gold and to act as a red glaze over metals. The material was favored by artists more for illustrations in illuminated manuscripts than for panel painting due to the pigment's sensitivity to light. In fact, in the 1390s Cennino Cennini warned in his treatise *Il Libro dell'Arte* (The Craftsman's Handbook) to avoid using this pigment for paint. However, Recipe 41 in the 12th-century *Mappae Clavicula* describes mixing Indian dragon's blood (dye) with powdered gold to make a writing ink, and also mentions mixing dragon's blood with red mercuric sulfide vermilion to darken the pigment, while using the bright golden yellow arsenic sulfide mineral known as orpiment to lighten the red, making it more orange.

The 14th-century *Liber Diversarum Arcium* (Book of Various Arts) gives recipes for using dragon's blood, mainly as a glaze when mixed with other resins. Recipe 2.8.2 describes how to produce a gold color that can be used over metal leaf as a varnish using pine resin mixed with aloe and dragon's blood in flaxseed (linseed) oil. Although the pigment is red, once applied the translucent quality of the glaze over the metal leaf produces a color that resembles a ruddy imitation gold.

Jehan Le Bégue's technical manuscript *De Coloribus et Artibus Predictis* (On the Aforesaid Colors and Arts) includes a recipe for using the juice of dragon's blood and "sandis"—the madder plant, *Rubia tinctorum* (see page 100). It suggests mixing madder dye with red chalk or using it as a red dye or glaze.

Later, the 16th- or 17th-century Venetian-Paduan manuscript *Ricotte per Far Ogni Sort di Colori* (Recipes for All Kinds of Colors) explains how to "make a color of dragon's blood" by grinding the resin with sal ammoniac (gum resin extracted from plants in the *Ferula* genus) and then using it with minium, a red lead pigment (see *Recipe 1*).

Red pigment from dragon's blood can also be mixed with lead white or other white substrates, such as gesso or chalk, to make a solid rose-pink color when combined with gum arabic. Using the pigment on its own, as described in the *Liber Diversarum Arcium* (see *Recipe 2*), produces a glaze. Dragon's blood can produce a deep warm red when the resin is ground into a powder, then mixed with egg glaire and gum arabic and a small amount of white gesso (see *Recipe 3*).

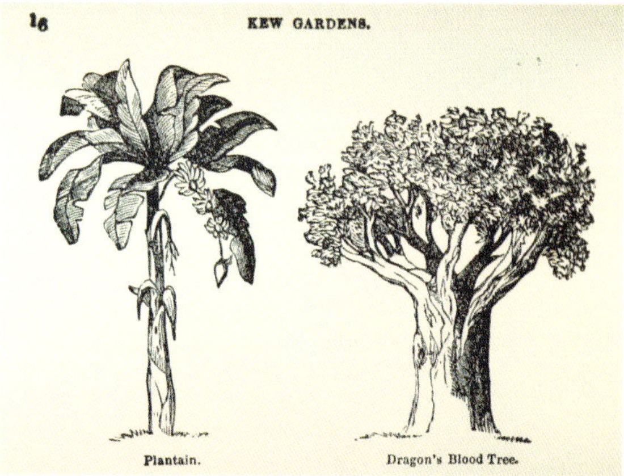

16 KEW GARDENS.

Plantain. Dragon's Blood Tree.

Dragon's blood tree *(Dracaena cinnabari)* (right). Illustration from *Kew Gardens With the Pleasure Grounds and Park: A Handbook Guide for Visitors* (1860).

Recipe 1: Tempering "dragonsblood"

"We shall now proceed to the nature and tempering of dragonsblood. Some say that dragonsblood is juice of a herb: that is silly; it is, on the contrary, the 'gum' of a tree which originates in Persia and in India; they call it dragonsblood, because it imitates blood, and truly that which is reddish, and with an interior like minium, is chosen."

—Recipe 1.12.1, *Liber Diversarum Arcium*, 14th century (Clarke, 2011, pp.110, 140)

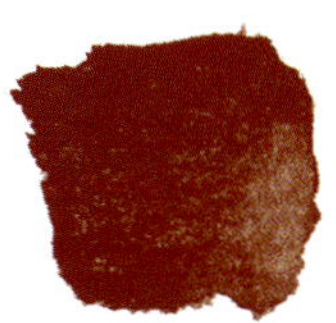

Recipe 2: Making "carmine red"

"On 'carmine' [cermino]. Take dragonsblood and liquefy it, and boil with flaxseed [linseed] oil, and with that same oil in a pan temper cerminium [a red vermeil or glaze]."

—Recipe 2.8.4, *Liber Diversarum Arcium*, 14th century (Clarke, 2011, pp.110, 140)

Recipe 3: Making dragon's blood red

Grind dragon's blood resin using a pestle and mortar until it is very fine. Then mix the red pigment with gum arabic or egg glaire or both together, and add a little gesso white to make a pleasant warm red gouache paint.

Adding sodium bicarbonate and a little tin mordant will produce a pink color.

—Nabil's modern recipe

Hibiscus

Latin name: *Hibiscus* species	
Other names: Hibiscus	
Family: *Malvaceae*	
Genus: *Hibiscus*	
Native: Central Africa	
Type: Annual or perennial	

There are several hundred species in the *Hibiscus* genus, which is a member of the mallow family, *Malvaceae*, and related to the marsh mallow (*Althaea officinalis*). Roselle (*Hibiscus sabdariffa*) is well known as a tea with a deep reddish color, and Dioscorides noted in his *De Materia Medica* in the 1st century CE that marsh mallow flowers were used medicinally in wine to treat various illnesses. *Hibiscus* flowers were used in exorcisms and for treating epidemics and disease in Malaysia, with the country adopting the red bloom as its national flower in 1960 to symbolize the courage and vitality of its people. Other *Hibiscus* species are also linked to medicine and ritual practice in East Asia and the Pacific Islands, where *Hibiscus sabdariffa* has been domesticated for thousands of years. By the 11th century, Chinese artisans were using *Hibiscus* bark, along with other plant-based materials for paper-making.

Many *Hibiscus* species, such as *Hibiscus titiaceus* and *Hibiscus cannabinus*, were cultivated in India and the West Indies in the 1800s for making rope and sack cloths and as a substitute for hemp. *Hibiscus esculentus* (now known as *Abelmoschus esculentus*) was used for paper-making and to produce gunny sacks.

Numerous *Hibiscus* species were listed in the 1840 *Indian Hand-Book of Gardening* by the horticulturalist and headmaster George Thomas Frederic Speede. He mentions that the bark of *Thespesia populnea* (formerly *Hibiscus populneus*), commonly known as the Indian tulip, was used to wash infected wounds. The fruits discharge a yellow juice that can be utilized as a dye resembling "gamboge"—a yellow resinous gum produced by *Garcinia* trees and used by early Flemish oil painters in their work and for spirit varnishes and transparent gold lacquers. When gamboge is mixed with indigo, it produces "Hooker's Green," a paint developed by the 19th-century English botanist and illustrator William J. Hooker.

The Chinese hibiscus (*Hibiscus rosa-sinensis*) was cultivated in 19th-century India and also listed by Speede. Its fragile petals were once used as a dye, producing a lilac color. The petals could also be used as a shoe polish to color leather—hence the plant's other name: "shoe flower." They were also used to darken the eyebrows and hair in China and other parts of the Far East.

Hibiscus (*Hibiscus syriacus* syn. *Althaea frutex*). Illustration by Pierre-Joseph Redouté from *Choix des Plus Belles Fleurs et des Plus Beaux Fruits* (Selection of the Most Beautiful Flowers and Fruits; 1827).

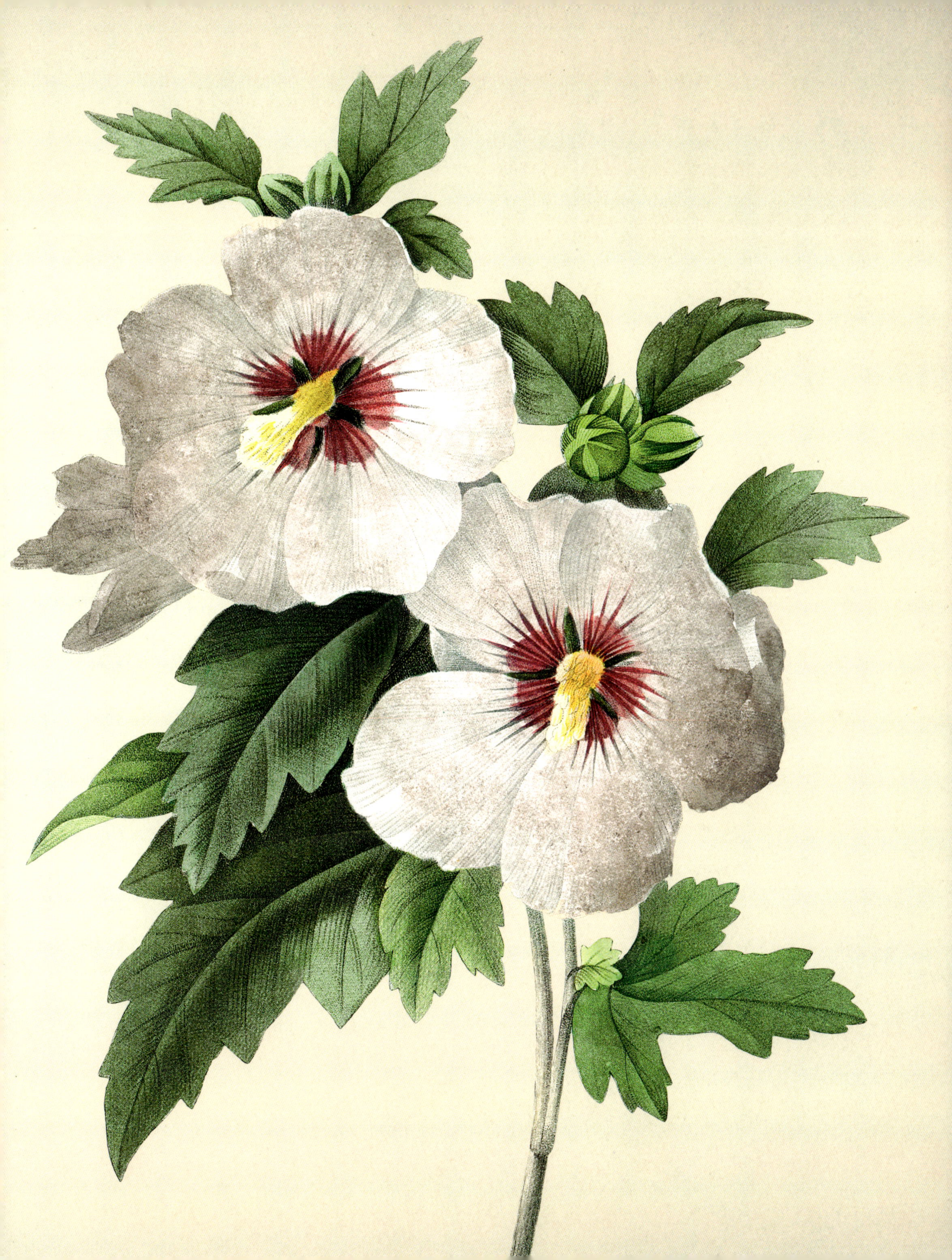

Recipes from Hibiscus

Few historical technical recipes document the use of *Hibiscus* as a colorant. Yet many plants absent from historical documentation can produce a dye or paint, depending on the processes used. If such plants were not available or traded with Europe, it's unlikely they will have been mentioned. But it's not impossible, as George Speede points out in the 1800s that *Hibiscus* species in India were useful as fiber and food plants, and as a local dye.

For my experiments in dyeing, I collected purple flowers from rose of Sharon (*Hibiscus syriacus*) in late summer, with permission from the Cambridge University Botanic Garden. I heated the fresh flowers in water with an alum mordant and then reduced the water to a quarter. (You can also mordant the fabric with alum water first and leave overnight, then place in a dye bath made with just the flowers.) The dye produced an olive-green color, and a pale yellow when potash was used as the mordant.

I experimented with *Hibiscus syriacus*, but red or deep purple flowers from other species can also be used. For example, the flowers of *Hibiscus sabdariffa* and *Hibiscus rosa-sinensis* can be heated in water to create a good magenta color and used as a transparent glaze on paper. Multiple layers can be applied to deepen the color, or the glaze can be used over primary colors like yellow or blue to create multiple shades. Lime or lemon juice can be added to the dye-ink to brighten the color, while adding eggshells produces a purple or lilac pigment. This is a fairly easy recipe to follow and produces a satisfactory dye-ink. It is proving promising for dyeing fabric and as a paint.

You can make other adjustments to create different effects. For example, reducing *Hibiscus* dyes by evaporating the water on the hotplate produces a more concentrated dye-ink. Adding powdered gum arabic or gum tragacanth (a natural polysaccharide from the dried sap of *Astragalus* species) to the dye can ease the flow of the dye-ink when applied to paper, enriching the color. Adding isinglass helps to prolong the ink's color but is not necessary if you're happy with the color for the moment. Dark red or purple *Hibiscus* flower petals produce stronger shades, deepening the color.

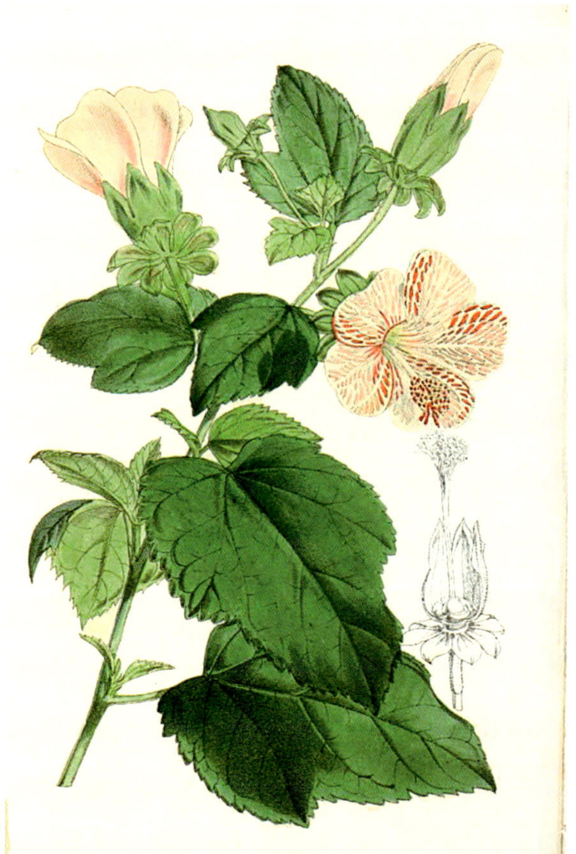

Hibiscus lavateroides. Illustration by Walter Hood Fitch from *Curtis's Botanical Magazine* (1868).

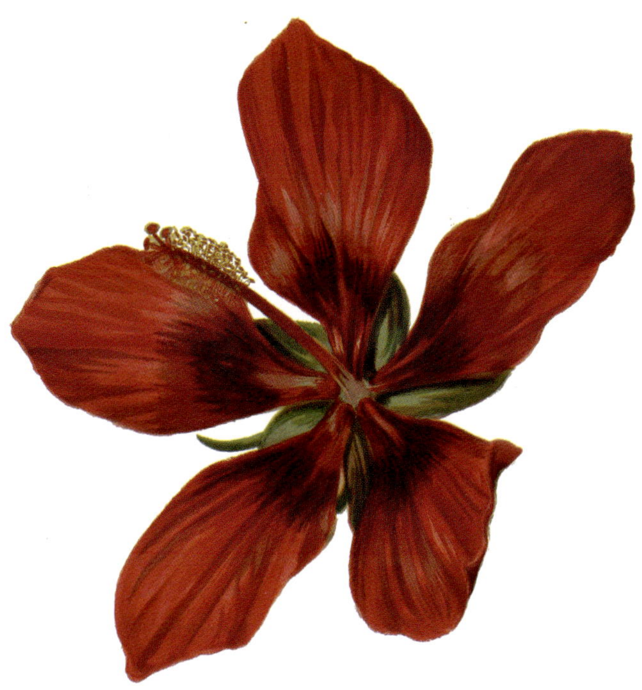

American scarlet rose mallow (*Hibiscus coccineus*). Chromolithograph by Alois Lunzer from Thomas Meehan's *The Native Flowers and Ferns of the United States in their Botanical, Horticultural and Popular Aspects* (vol.2, 1879).

Recipe 1: Making a glossy magenta glaze

Heat a handful of dried red *Hibiscus* flowers—*Hibiscus sabdariffa* or *Hibiscus rosa-sinensis*—in water until the red tannins are released. Reduce the water by half, add powdered gum arabic, and mix well. Add a small amount of lime or lemon juice to brighten the hue.

This color works well as a glossy magenta glaze that can be applied over yellow or blue paint to form oranges and purple shades—the more glaze layers you apply, the deeper the color becomes.

You can add chalk, kaolin (China clay), crushed cuttlefish, or powdered eggshells to the magenta glaze, which will change the color characteristics from pinks to lilac purples.

—Nabil's modern recipe

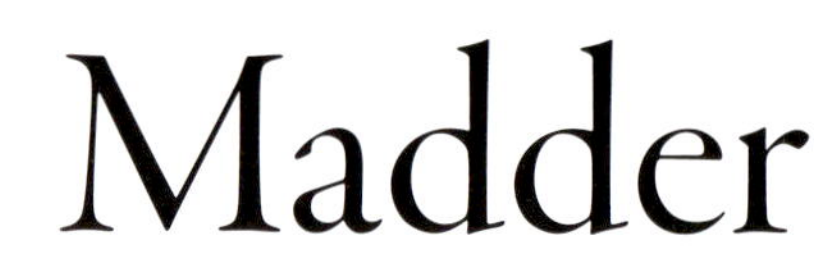

Madder

Latin name: *Rubia tinctorum*

Other names: Rose madder, dyer's madder

Family: *Rubiaceae*

Genus: *Rubia*

Native: Middle East, Asia, Europe

Type: Perennial

Prickly-stemmed madder (*Rubia tinctorum*) has been cultivated since Egypt's 18th Dynasty (1550–1295 BCE) and was used as a dye in Persia, Arabia, and Mesopotamia. The dyed red cloth was recorded in the early 2nd millennium BCE in cuneiform tablets from the reign of Amar-Sin of the Neo-Sumerian Empire relating to trade (held by the National Museum of Denmark, Copenhagen).

After scientific analysis, traces of bright pink madder paint have been identified on Egyptian papyri and on Greek terracotta figures from the 3rd century BCE. Madder paint made from red dye mixed with gypsum was also found on Roman bowls, and the palette of an ancient Roman fresco painter held by London's British Museum. The museum also houses one of the earliest written artifacts referencing dyeing with madder: a Neo-Babylonian clay tablet from the 17th century BCE that contains two distinct dye recipes. These document three species of madder: *hat-huritu* (for purple), *huratu* (normal madder), and *inza-huritu* (a madder imported from various other countries).

New York's Metropolitan Museum of Art also houses numerous examples of objects dyed with madder. These include an Egyptian red leather fragment from the 11th Dynasty (2124–1981 BCE) and an early 1st-century-CE Roman marble statue of an old woman. Textile fragments from a beautiful Byzantine scroll (4th–6th century CE) depict a stylized vine dyed with indigo mixed with madder, forming a deep color, and a child's dress (7th–9th century CE) displays a solid madder red color.

Centuries later, Theophilus wrote in his 12th-century craft manual *De Diversis Artibus* (On Diverse Arts) that madder could be a useful red dye for staining ivory, stag's horn, and fish bones. The Metropolitan holds an excellent example of a 12th-century German ivory game piece showing Samson slaying the Philistines.

Vitruvius noted in *De Architectura Libri Decem* (Ten Books on Architecture) ways to make a purple pigment using dye from the root of the madder plant mixed with chalk. The practice of mixing plant dye with chalk was used throughout the Middle Ages and is still current today among artists and colormen, with forty original notes on madder production in the notebooks of Windsor & Newton (an English company specializing in fine art products) dating back to the 19th century. French painter Claude Monet used madder lake on his paint palette.

The alizarin extracted from the roots of **madder** (*Rubia tinctorum*) produce the famous rose madder and alizarin crimson-lake pigments. Illustration from *Köhler's Medizinal-Pflanzen* (Köhler's Medicinal Plants; vol.3, 1898).

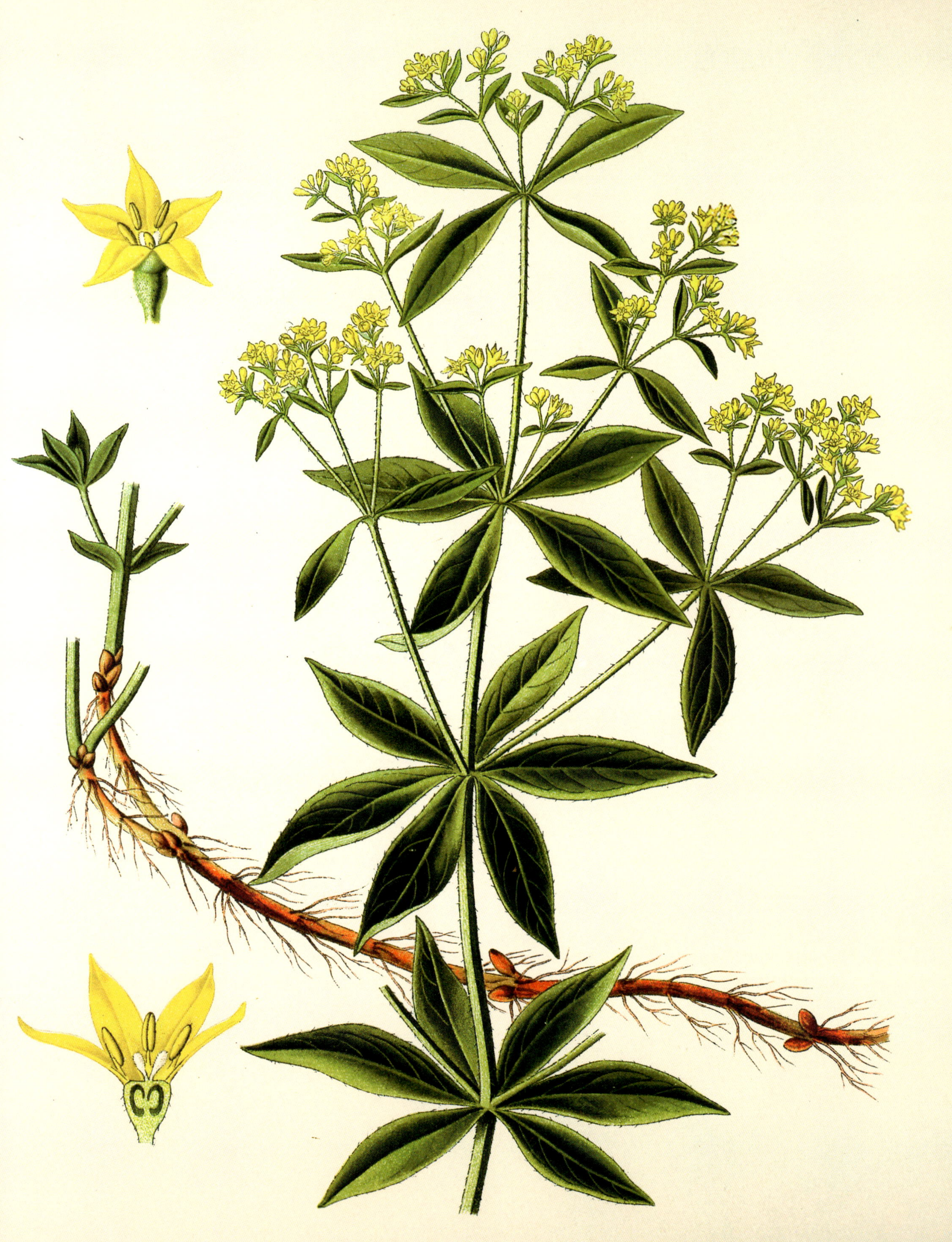

Recipes from Madder

Madder can be used for a wide range of colors, from yellow through orange to red, to purple or brown, with red most often used to color textiles, such as historic British military uniforms and hunting jackets, and to make organic paint.

The 12th-century *Mappae Clavicula* includes paint recipes for making multiple colors. Recipe 188 describes making a red color from the poisonous cinnabar (vermilion) pigment—red mercuric sulfide—mixed with murex (mucus from *Murex* snails). Recipe 189 gives instructions for extracting dye from madder roots cooked in urine and alum.

The 14th-century *Liber Diversarum Arcium* (Book of Various Arts) contains recipes for dyeing wool red. Recipe 4.17.1 outlines heating madder root in water at a low temperature with oak galls and sieved ashes for a red dye. Today it is standard to heat a dye vat no higher than 60°F (16°C), as impurities from the madder give a dull brownish color at higher temperatures (see *Recipe 2*).

In the 16th century, Gioanventura Rosetti's *Plictho de Larte de Tentori* (Instructions in the Art of Dyers) provides many madder recipes. There are instructions on coloring horse manes with "morello," a purplish blue-black color, by dyeing first with woad, then with madder. Another recipe for the "dyeing of red berets" describes mordanting the beret with alum, then heating madder roots for an hour before adding lye (a strong alkaline substance) and mixing well. The beret is thrown into the dye, heated for half an hour, and rinsed. It is then placed in a pan of clean water with more lye and reheated, before being dried in the sun.

In the late 1800s, William Morris employed techniques similar to those in Rosetti's *Plictho* to create purple dye. He experimented with double-dyeing synthetic blue silk in cochineal and madder. He was among the few individuals in Europe attempting to revive natural madder dye in an era dominated by synthetic alizarin.

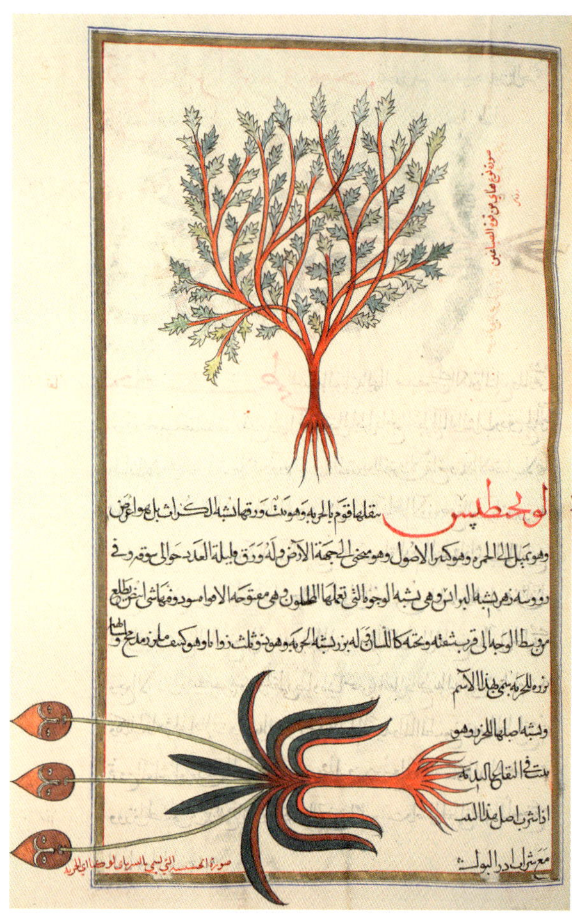

Levant madder (*Rubia peregrina*) (top). Illustration by Persian artist Mirzā Bāqir for an 1889–90 edition of a 9th-century Persian translation of *Pedanius Dioscorides' De Materia Medica* (1st century CE).

Although many Western technical manuscripts mention madder for dyeing cloth, guidance on using it for staining paper can be found in the 19th-century Arabic treatises *Majmu 'al-Sanaye* (A Collection of Crafts), along with many other techniques.

Recipe 1: Making madder red

"Take madder broth, and add 3 ounces of gallnut [oak galls] and grind it properly; take 2 pounds of madder broth and put it in a glass pot with the ground gallnut, and leave to soak for 2 days: after this, strain it and add 1 ounce of calcothar [artificial red ocher from the calcination of iron sulfate] and 2 solidi [9 grams] of cinnabar [and] grind them both, and put them with the above-mentioned things [ingredients from Recipe 188], and cook down until it is reduced to a third."

—*Mappae Clavicula*, 12th century (Smith and Hawthorne, 1974, p.54)

Recipe 2: Making red and pink

Pour boiling water over dried madder roots in a pan and leave for half a day, then throw the dye away. Reheat the roots with alum in rainwater and simmer at 60°F (16°C) for an hour, then sieve the dye and the color will be a glorious red.

To make pink paint, add 6 grams of madder dye to 3 grams of chalk, mix well, and leave to air-dry. Once dried, scrape off the pigment and crush with a pestle and mortar until you have a fine pigment. Add gum arabic or egg white to the pigment to form a pink paint.

—Nabil's modern recipe

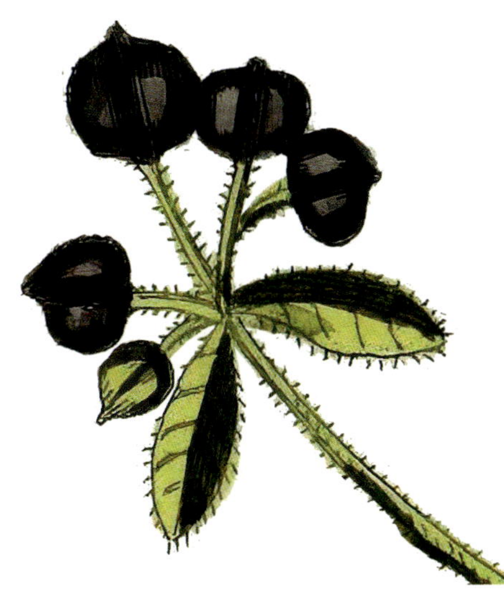

Madder (*Rubia tinctorum*). Illustration from *Plantarum Indigenarum et Exoticarum Icones ad Vivum Coloratae* (Icons of Indigenous and Exotic Plants Painted from Life; 1788).

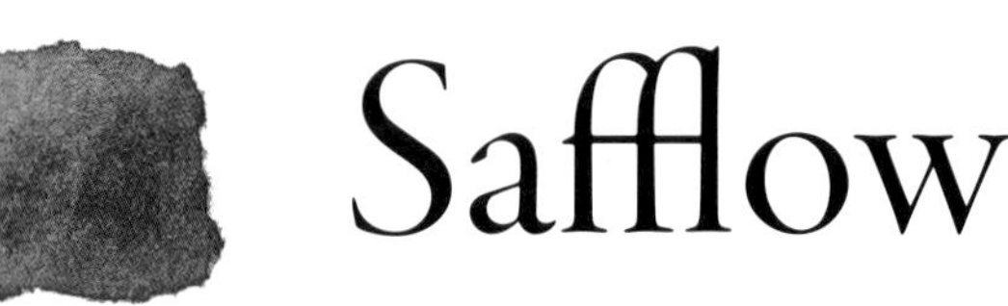

Safflower

Latin name: *Carthamus tinctorius*

Other names: Dyer's thistle, fake saffron

Family: *Asteraceae*

Genus: *Carthamus*

Native: Iran, Turkey

Type: Annual

Safflower (*Carthamus tinctorius*) grows in sandy places and was used in eastern Mediterranean regions and ancient Egypt from the 11th and 12th Dynasties (2124–1802 BCE) to make red and yellow dyes. Threads dyed with safflower have been found on mummy wrappings, and garlands containing safflower blossoms and offerings have been discovered in ancient Egyptian tombs, indicating that safflower was important for religious ceremonies. Flower heads and seeds dating from 2500 BCE have been identified in Tell Hammam et-Turkmen in northern Syria. These are believed to have been used as a dyestuff, making safflower one of the oldest known dye plants. It has been used not only for colorants but also for the oils extracted from the seeds and is suitable for paint varnishes and as a glaze since the color does not yellow.

Like turmeric, safflower was used to dye medieval Islamic paper orange and yellow, as described in the 16th-century manuscript *Golzār-e Ṣafā* (Garden of Joy) by Alī Ṣeyrafī, held in the Bibliothèque Nationale de France, in Paris (No.S.P.1656). The text describes the dyes and pigments used in calligraphy for poetry and explores how feelings can be expressed through the use of color symbolism. It also explains that coloring paper with dye is better for the eyesight due to the relationship between the color of the paper and that of the ink. The earliest examples of recipes devoted to Islamic arts date to around 1025 CE and are attributed to al-Muʿizz ibn Bādīs (see *Recipe 1*) and later, in about 1185, to Jamali-ye Yazdi. Although fragments of illuminated Persian manuscripts have been found in the Turfan region, East Turkistan (the present-day Xinjiang Uyghur region of China), from the 6th or 7th century CE, such manuscripts are more commonly dated centuries later.

Safflower is widely used by many cultures, with flower offerings and pieces of dyed silks documented in Buddhist monasteries in the late 8th and 9th centuries. It was used on Persian and Afghani woolen rugs. The plant was also used to produce a red pigment for Japanese block printing and for facial makeup in China, where the powdered dye was mixed with talc to create reds and pinks. Similar processes were followed for making organic paint by fixing plant dyes to a substrate such as chalk, as described by Vitruvius, or mixing them with gesso or eggshells.

Safflower (*Carthamus tinctorius*). Handcolored steel engraving by Oudet from Pierre Oscar Reveil, A. Dupuis, Fr. Gérard, and François Hérincq's *Le Règne Végétal: Plantes Agricoles et Forestières* (The Plant Kingdom: Agricultural and Forestry Plants; 1864–71).

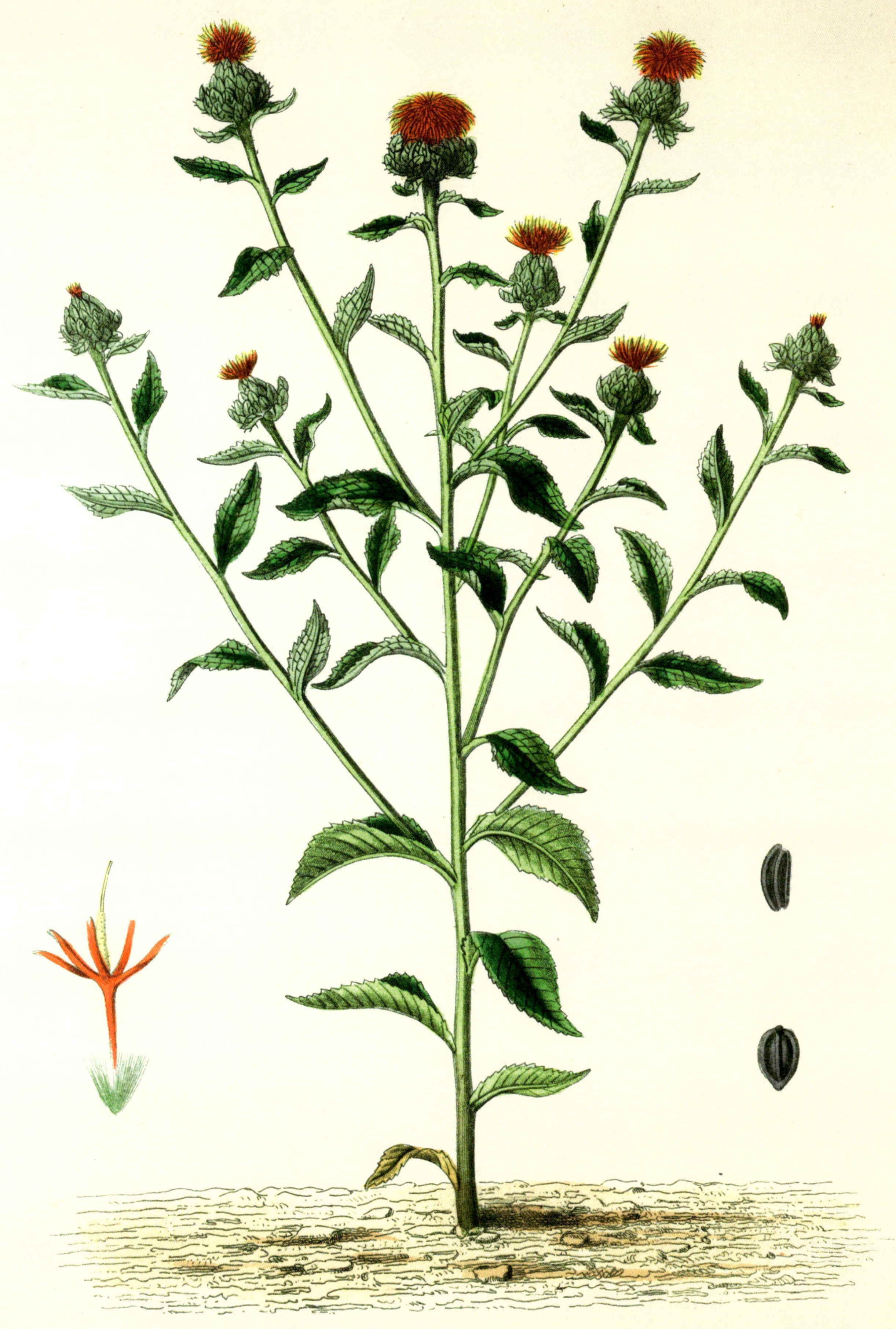

Recipes from Safflower

Safflower produces a yellow dye, or green dyes when mixed with dyer's indigo (*Indigofera tinctoria*). Many recipes indicate this, especially Recipe 12 for how "to dye yellow linen cloth" in the *Trinity Encyclopaedia* manuscript and recipes for dyeing paper in early Arabic manuscripts. Dried safflower petals can also produce pink and cherry colors and poppy red using complex dye processes, with carthamin, the natural red pigment in safflower, being the main dye compound.

Safflower is classified as a direct dye, meaning it does not require a mordant to fix the colors. Instead, you can fix the colorant directly to fabric fibers after washing away the yellow dye from the petals. However, the dye must be activated in an acidic dye bath first. To direct-dye with safflower, mix baking soda (bicarbonate of soda) or another alkaline solution such as potash with the petals, then place in a filter with warm water washing over them to produce a dye bath. Lemon juice or vinegar is added to the dye bath to make it acidic, which precipitates the yellow alkaline solution until it causes a color change, ranging from oranges to reds. Shades of pink and coral are created on silk, while continuous dyeing in a safflower dye bath can produce a deep red colorant, or the fabric can be left overnight to absorb the red dye.

An anonymous 16th-century-CE (10th-century AH) Arabic manuscript called the *Resāleh dar Bayān-e Ṭarīqeh-ye Sāk̲tan-e Morakkab va Kāghaz-e Alvān* (A Treatise about the Technique of Preparation of Colored Paper and Inks) describes a color called *lāk*, which in this recipe is a color word for "red." It explains crushing safflower blossoms until they become soft, pouring a little wine over them until the yellow dye is removed, and then leaving to dry in piece of cheesecloth (muslin). The dried blossoms are then heated in water with some salt rubbed in with the fingers and left while the dye drips out of the cloth. The recipe offers a choice of old vinegar, lemon or orange juice, sour pomegranate juice, or any available sour fruit juices. The main purpose of this safflower recipe is to dye paper, with the advice being to dye for no less than six hours to obtain a true color. Colors from safflower other than red have also been documented in Arabic recipes, including a peach, reddish-yellow, olive, and, of course, the classic yellow.

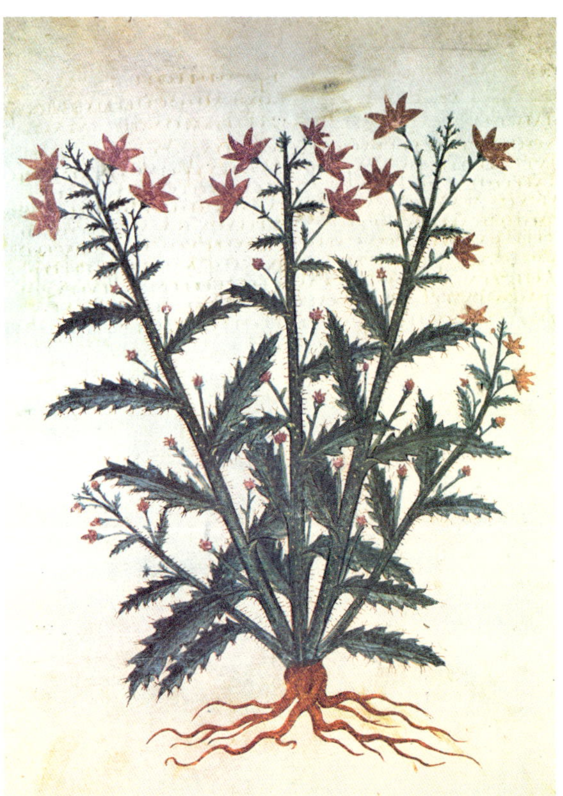

Safflower (*Carthamus tinctorius*). Illustration from the herbal manuscript known as the *Juliana Anicia Codex*, or *Vienna Dioscurides* (6th century).

Recipe 1: Making gleaming ruby red

"Three ratls [6½ gallons/24.5 liters] of *Carthamus* [safflower] are pulverized a day in the sun. It is then pulverized and sieved and smaller than a regular sieve. Then it is hung in a cloth above a wide filter on a support used by dyers. While it is hanging, about sixty ratls [27 gallons/102 liters] of water are poured over it. Let it drip into a basin so that not a drop of water remains in it. Then that water which it was washed [in] is poured over it. The *Carthamus* in its cloth is then taken. For it ten dirhams [30 grams] of black alum of the dyers are pounded. It is sprinkled over it more than once while it is being rubbed well with the hands. This is done each time until the palms are dyed with its redness. It is then hung again and twenty ratls [9 gallons/34 liters] of pure water poured on it. It is allowed to drip until not a bit of water remains. What has dripped is the essence of the needed *Carthamus*. With it are mixed a rate of 1 ratl [2¼ gallons/8.5 liters] wine vinegar and some water of gum. It is used the same day. Its color comes out well. Nothing else mixes with it. It can be applied to gold and silver and tin. It comes out well. When used on paper or parchment, it comes out a wonderful red color."

—Al-Muʿizz ibn Bādīs, Royal Patron of the Arts, 11th century, (Levey, 1962, pp.30–31)

Recipe 2: Making safflower red dye

Place some safflower petals in cold water for several days to extract and decant the yellow dye into a separate jar. Return the petals to clean water and wash the yellow away four or five times until the water becomes clear. Add potash to clean water to make it more alkaline and soak the safflower petals for several days. Then add alum (or clear vinegar) to the dye bath to intensify the color, making the water less alkaline and the dye a coral or pinkish color.

The discarded yellow dye can be mixed with chalk or Italian marble dust and left to dry to form a yellow pigment for paint, or it can be used as a golden-colored dye.

—Nabil's modern recipe

Colors from Nabil's Practice

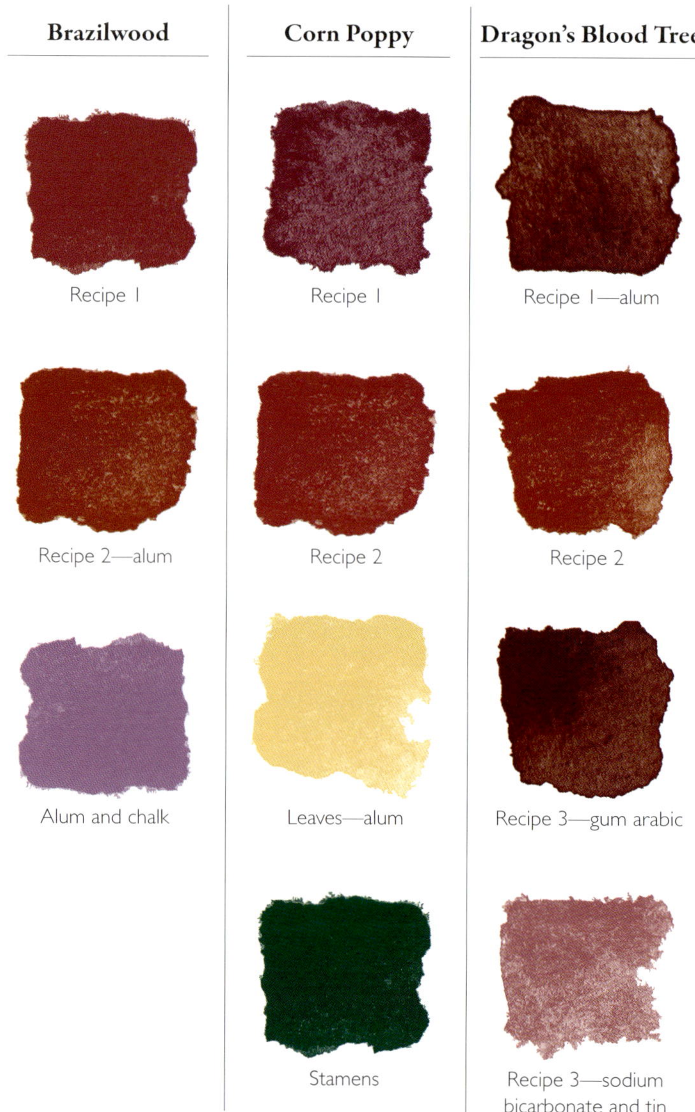

Brazilwood and dragon's blood ideally need to be purchased from a supplier, but hibiscus and safflower are seasonal and can be harvested and dried. Poppy dye ink is the most seductive and easy shade of red to make, resembling blood. It is a good choice for a momentary color and can be kept in the freezer for more than ten years. Madder root can be grown in large pots or in the ground; the plant needs to be more than three years old before the root can be dug up.

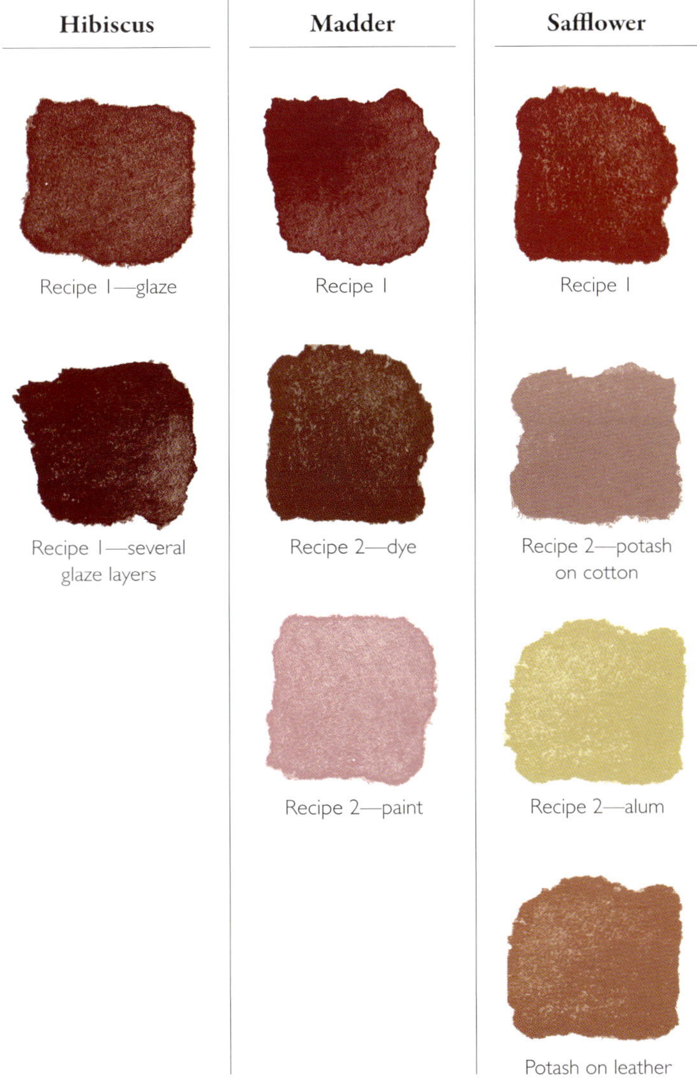

Chapter 4

Yellow & Orange

Apple

Latin name: *Malus domestica*

Other name: Orchard apple, domestic apple

Family: *Rosaceae*

Genus: *Malus*

Native: Central and South Asia, northwest China

Type: Deciduous tree

According to the Royal Botanic Gardens in Kew, the modern domesticated apple was cultivated from a wild apple known as *Malus sieversii* from Central Asia, which is now listed as vulnerable to extinction due to habitat loss. The beauty of apple trees in full bloom is a sure sign of spring, when the delicate flowers are pollinated by bees and other insects. Soon after, the white to pinkish-red petals fall to the ground like confetti, blown by the breeze, leaving behind swelling ovaries that develop into apples.

With more than 7,500 apple varieties grown worldwide, the choice is overwhelming. The Romans brought some varieties of cultivated apples to the shores of Britain from around 43 CE and created orchards, which declined when they left in about 410. St. Augustine of Canterbury reestablished apple orchards in monasteries from about 597, and these were developed over the centuries.

The Costard apple (*Malus domestica* 'Costard') became popular in England in the 13th century. This greenish-yellow cooking apple was favored for cider and remained the national favorite until the 17th century. The sour crab apple (*Malus sylvestris*), which is native to Britain and produces small fruits, could not compete with the Costard, although it was good for apple verjuice and sour cooking cider.

In England and Ireland, the crab apple was used in traditional apple-bobbing or snap-apple, which took place in fall at the Samhain festival. Unmarried maidens cut identifying marks into apples and placed them in a bucket of water. Local swains looking for a potential bride would then try to retrieve an apple without using their hands to secure a future wife. This led to the Halloween activity in which children try to grasp apples with their teeth, with hands behind their backs.

Cutting an apple crosswise reveals a five-pointed star, which can be associated with the Roman fertility goddess Venus—or the Greek goddess Aphrodite. In Greek myth, Paris, the prince of Troy, gave a golden apple to Aphrodite for her outstanding beauty instead of Hera, Zeus's wife, or Athena, the goddess of battle. Aphrodite then gave Helen, the wife of Menelaus of Sparta, to Paris in return for the golden apple. The apple was claimed by all three goddesses, sparking the start of the Trojan War, which supposedly took place in the 12th or 13th century BCE.

Apple (*Malus domestica*). Illustration from Dame Ann Hamilton's *100 Drawings of Plants* (1752–66).

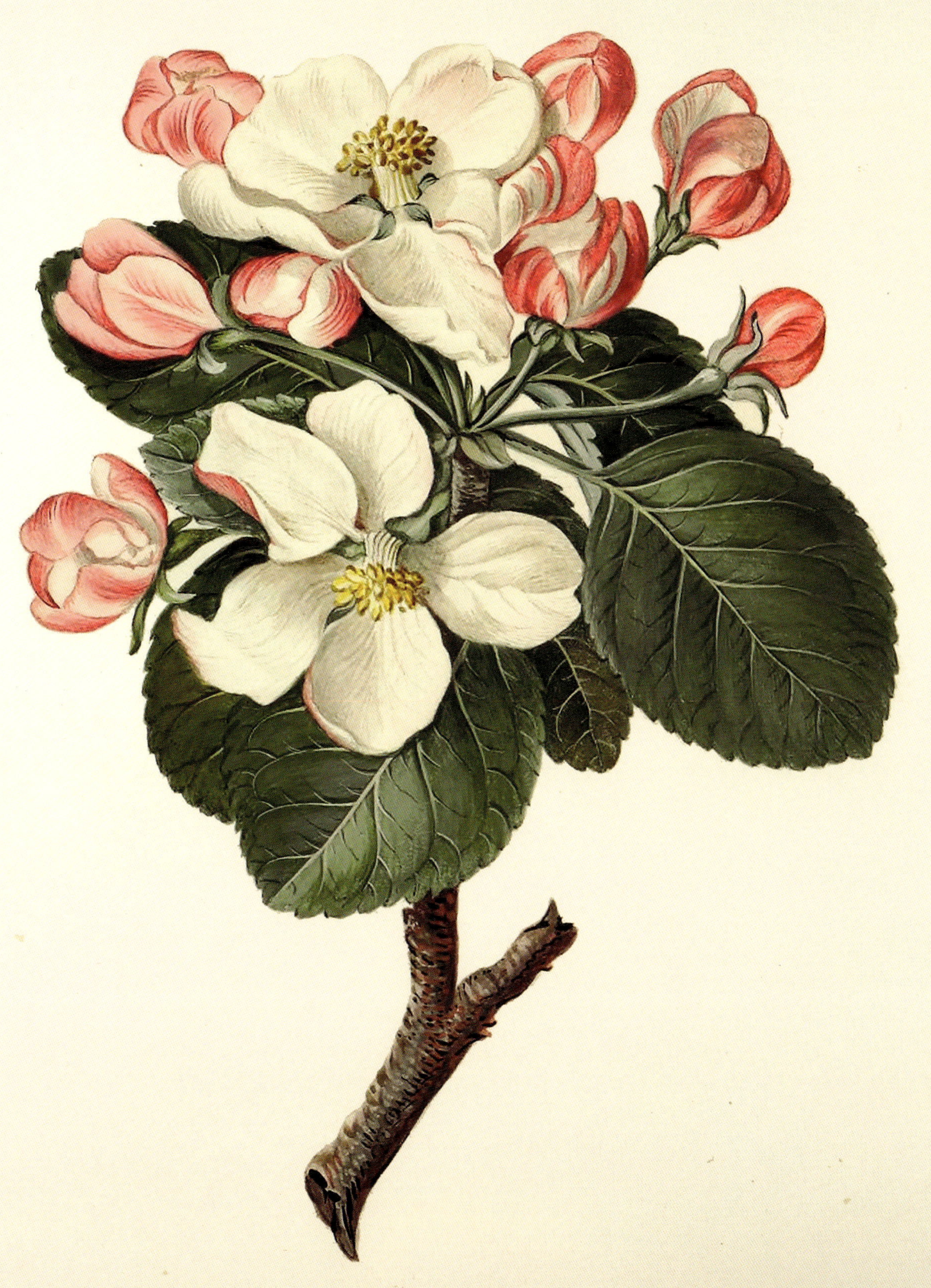

Recipes from Apple

Creating a yellow dye from apple tree bark is a centuries-old practice. You can make several colors from apple wood shavings or sawdust, by using different mordants, either mordanting the fabrics first or adding the mordant to the dye. Using alum produces a yellow color, and a brilliant yellow is created if you only use apple twigs. Tin mordant gives a bright orange-yellow, and an iron mordant will give soft to mid-greens and greenish browns from just the twigs. Mixing the dyes together can result in unique shades and mustard-golden colors if a large amount of bark is used.

Crab apple bark is mentioned in by Swedish botanist Engelbertus Jörlin in his 1759 list of dye plants *Plantae Tinctoriae* (Dye Plants), which was published in Carl Linnaeus's *Amoenitates Academicae* (Academic Pleasures) in around 1760. This indicates that the bark was used to make a yellow dye. Although noted as a plant of minor importance, it was clearly known as a dye source by the 18th century. Apple bark contains quercetin, a flavonoid compound that can be used as a yellow dye, as mentioned in the 19th-century Wiesel Collection in Dr. Georg Kremer's library, in Aichstetten, Germany. This is similar to quercitron, another yellow pigment, which is extracted from the inner bark of the black oak (*Quercus velutina*).

An earlier 15th-century manuscript (MS Sloane 3548), held by the British Library in London, contains a recipe for adding rotting apple juice and verdigris to the blue scum that floats on the surface of a woad vat to make deeper green inks. A later 16th-century manuscript (MS Sloane 4), also in the British Library, includes a similar recipe for turning rotting apple juice into a green ink by mixing it with verdigris and stamens from saffron crocus (*Crocus sativus*; see page 30).

Using the same method as Recipe 1, with which I made a yellow dye, the inner bark of the Costard apple tree can be used to produce a red dye. Similarly, the early 16th-century *Liber Illuministarum* (Book of the Illuminator), held at the Bavarian State Library in Munich, Germany (MS BSB Cgm. 821), suggests using crab apple leaves to dye leather red: "Take the red leaves from the crab apple tree around St. John's Day [June 24] and dry them; after that boil them in wine and dye the leather with it."

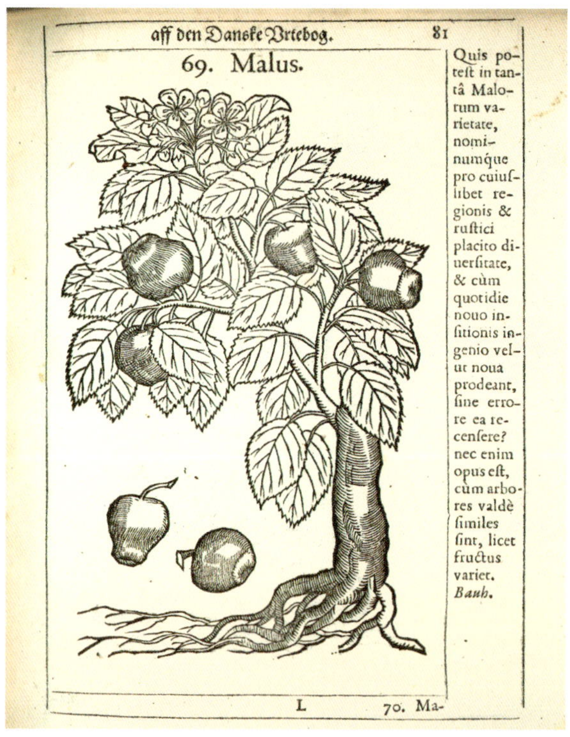

Apple (*Malus domestica*). Illustration by German-Danish botanist Simon Paulli from *Flora Danica* (Flora of Denmark; 1648).

Recipe 1: Making yellow dye

"Take the barke of an Apple tree, not the outward harde and rough barke, but the inward, cut it in small loppins [pieces], and poure some water upon them, and then put in your woode, bone, or horne, to it with Alome [alum], And let it seeth well together."

—Girolamo Ruscelli, *The Secrets of the Reverend Maister Alexis of Piemont*, Chapter 59, translated by William Ward ca. 1562

Recipe 2: Making apple yellow/Newton's Gold

Collect some apple branches after the fall and scrape the inner bark in little pieces into a saucepan. Soak the bark for 24 hours or so in water, then heat for an hour or more. Add leather that has been soaked overnight in alum to the dye bath and it will make a yellow.

—Nabil's modern recipe (using bark from Newton's apple tree, Cambridge, UK)

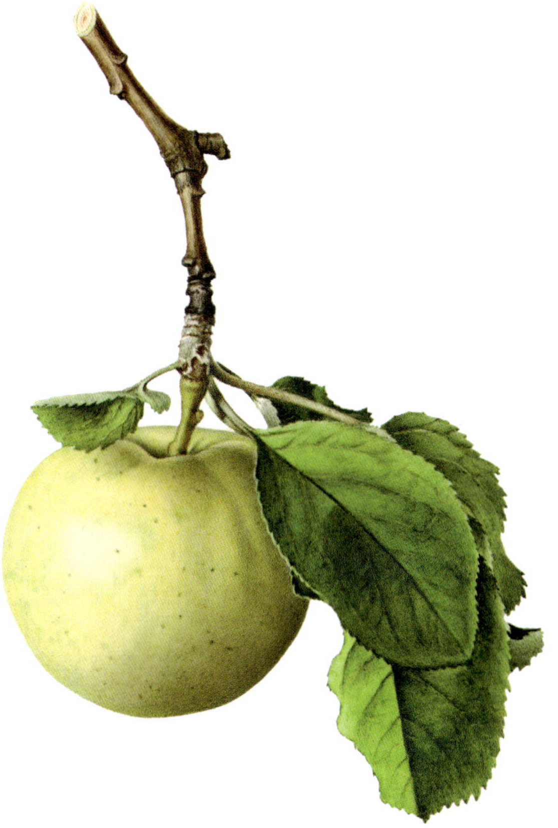

A ripe **apple** with twigs and leaves. The dry leaves of the apple (*Malus domestica*) can be used in midsummer to dye leather red. Vintage apple illustration in the Rare and Special Collections of the US National Agricultural Library (1923–1928).

Dyer's Broom

Latin name: *Genista tinctoria*

Other names: Dyer's broom, dyer's greenweed, woad-waxen

Family: Fabaceae

Genus: *Genista*

Native: Europe to western Siberia and Afghanistan

Type: Deciduous shrub

In the 1st century CE, Pliny the Elder included a chapter on plants for making color in his *Naturalis Historia*, noting that "genistae" among other plants can be used to dye cloth yellow. Centuries later, dyer's broom (*Genista tinctoria*), which grows on dry, sandy heaths and in woodlands, was one of the earliest plants used for dyeing textiles in Anglo-Saxon England, then later in the early Middle Ages before weld (*Reseda luteola*; see page 144) was imported. Plant matter from 9th-century Viking archaeological sites in England indicates that dyer's broom was used to dye silk and wool yellow. Green could be produced if the fabric was over-dyed with dyer's indigo (*Indigofera tinctoria*) or woad (*Isatis tinctoria*). Dyer's broom could also be mixed with other yellows such as weld.

Fulk V, Duke of Anjou (1092–1143), wore broom on his helmet in battle. The plant became symbolic of the Plantagenet Dynasty through his descendants, with his son Geoffrey V of Anjou wearing a sprig of "planta-genista" in his hat. The plant can also be seen in the 14th-century portrait of Richard of Bordeaux (who became Richard II in 1377), with the open broomcods (seedpods) featuring on his cape to signify the end of the king's reign when he was deposed in 1399. Yellow broom flowers were an important symbol in the French courts, and during the 13th century they were worn on livery alongside an enameled fleur-de-lis, which dates back to the Frankish king Clovis I (ca. 466–511). This practice was associated with the medieval Ordre de la Coste de Genest (Order of the Broomcod), which was founded by Louis IX of France in 1234 to commemorate his wedding to Margaret of Provence, with the broompods appearing as an emblem on livery collars.

The flowers and stems of dyer's broom are very similar in appearance to those of *Cytisus scoparius*, the Scotch broom (or common broom in England). Also in the Fabaceae family, this was introduced to New England in the early 1800s and has become established in the eastern side of the United States. Similarly, the Gloucestershire clothier and dyer William Partridge, who emigrated to the United States in the early part of the 1800s, took seeds of *Genista tinctoria* with him, planting them in New Jersey to grow for yellow dye. He gained a reputation as an experienced dyer for numerous textile mills between Delaware and Rhode Island.

Dyer's broom (*Genista tinctoria*). Illustration from Jan Kops and F.W. van Eeden's *Flora Batava* (Flora of the Netherlands; vol.14, 1872).

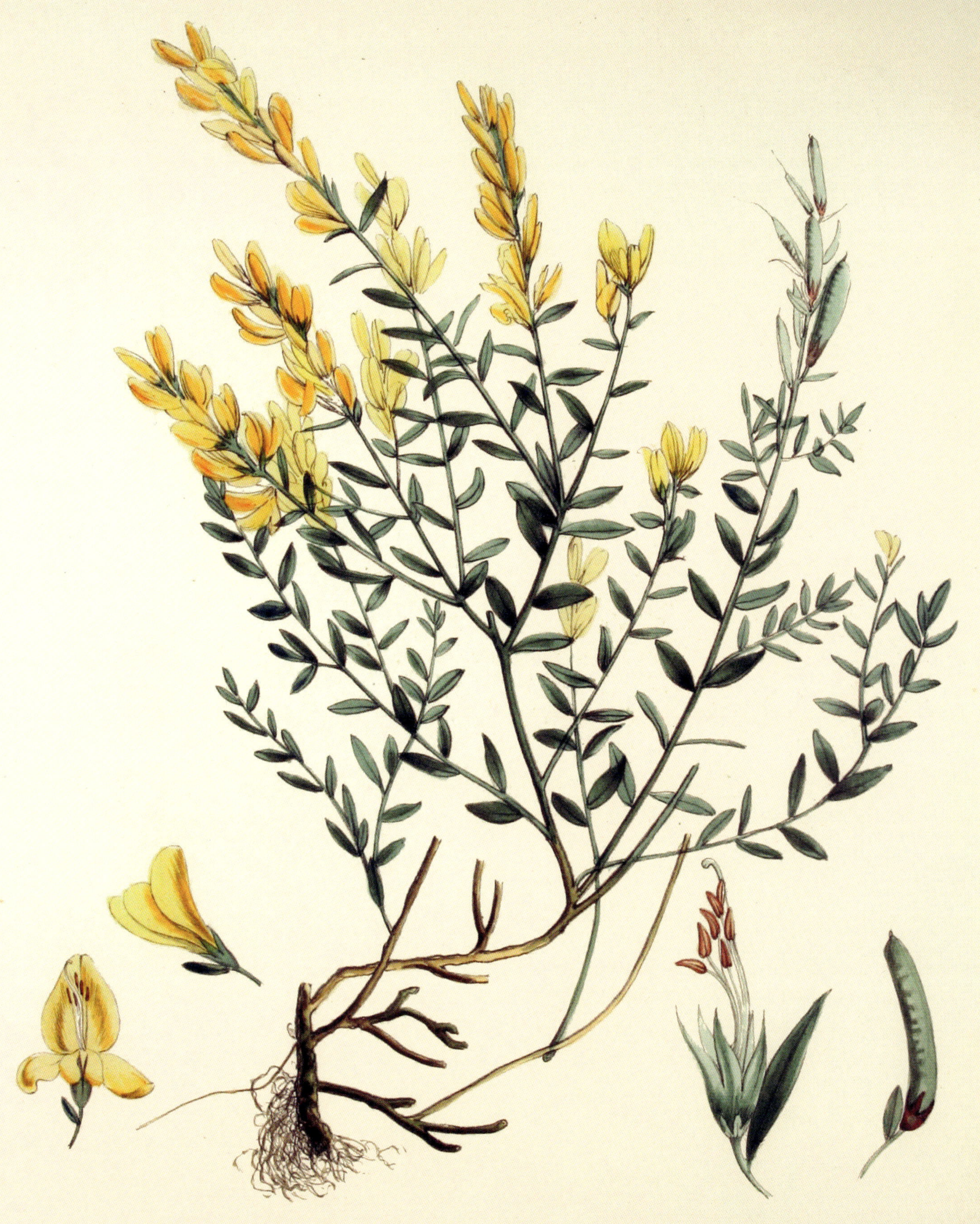

Recipes from Dyer's Broom

Dyer's broom was one of several yellow dyes used by the Anglo-Saxons, in combination with the weld plant (*Reseda luteola*). They were also used in double-dyeing with woad or dyer's indigo to produce a beautiful green, known in medieval times as "Kendal Green," reflected in the common names dyer's greenweed and woad-waxen. It was used in the 14th century by Flemish weaver John Kempe, who established a business in England around 1331.

Dyeing with dyer's broom is documented as "Wdewise" in *De Coloribus et Mixtionibus* (Of Colors and Mixtures), a 12th-century Anglo-Norman manuscript, housed in the British Library, London (MS Cotton Titus D.XXIV). The manuscript gives instructions for making yellow from broom: "Take wood[ash] lye and pick dyer's broom and boil it in the lye. Then remove the broom [plant] and dye the cloth and it will be yellow." The use of woad was well established by the 12th century, so it is likely that over-dyeing also took place as this practice dates back to ancient Egypt—in fact, it is still recommended in contemporary practice.

In the 16th century, recipes for dyeing silks with herba giuletta and herba corniola—both names for dyer's broom—can be found in Italian Gioanventura Rosetti's 1548 *Plictho de Larte de Tentori* (Instructions in the Art of Dyers). The dye was widely used in Europe, particularly England. The dyeing process is straightforward and can be accomplished quickly, using either fresh or dried flowers that have been stored away from light.

Spanish broom (*Spartium junceum*) is quite similar to dyer's broom, with both plants producing yellow flowers that can be used to create a dye. I collected fresh Spanish broom flowers from the Cambridge University Botanic Garden in mid-summer and experimented by cooking a handful in water. I tried several different mordants to alter the color and fix it to soft chamois leather. When alum was added, there was minimal color change. However, introducing a potash mordant raised the pH level to highly alkaline, resulting in a deeper shade of yellow. Using baking soda, which makes the water mildly alkaline, produced a bright golden yellow that is particularly pleasing to the eye.

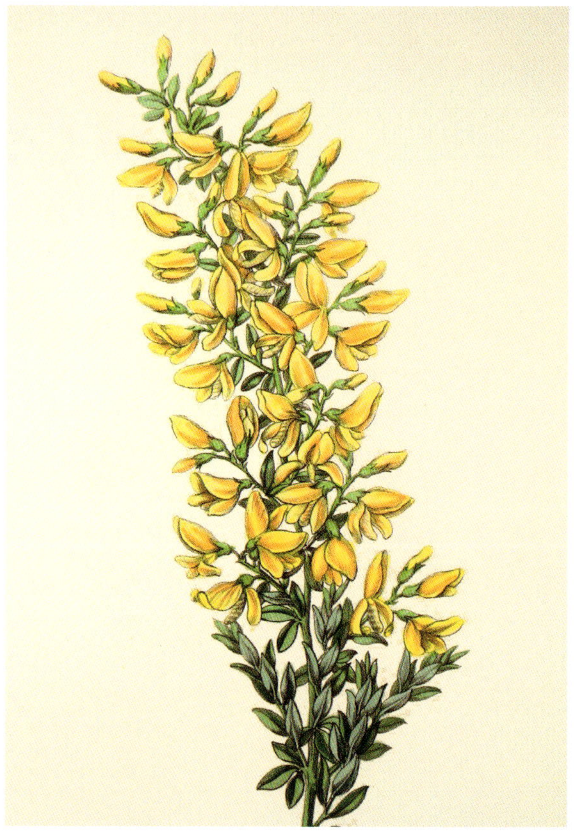

Dyer's broom (*Genista tinctoria*). Illustration from *Edwards' Botanical Register, or, Ornamental Flower-Garden and Shrubbery* (1829–47).

Recipe 1: To make yealwe turnesoll (yellow turnsole)

"Take þe flowres of brome when they be most fresch, and pyke leue be leue clene as þu mayst and | serve hem in all wyse as þe reed."

Translation: "Take the flowers of broom when they are most fresh, and leaf by leaf [that is, petal by petal] pick it as clean as you can, and treat it in all ways like the red [see Recipe C1 for making red turnsole from poppies]."

—Recipe C3 (a Middle English cookery dye recipe), Sloane MS 122, 16th century (Clarke, 2016, p.332)

Recipe 2: Making broom yellow

Add 100 small dyer's broom flowers with alum to 7 fluid ounces (200 milliliters) of water, then heat and reduce to ⅔ fluid ounce (20 milliliters) to produce a concentrated yellow dye. Sieve this through a fine-mesh strainer, adding 10 grams of gum arabic powder to thicken the dye and make a translucent ink. Add a few drops of grapefruit seed extract to the dye as a preservative to prevent mold growth if you are using water. Other preservatives can be used, including a little vinegar, camphor, or rosemary extract. Another option is 3% Preventol® (Sodium-2-phenylphenolate), but this must be used cautiously.

Mixing the dye with 3 grams of chalk and leaving to dry naturally produces an additional pigment, which can be tempered with liquid gum arabic to create a yellow paint.

—Nabil's modern recipe

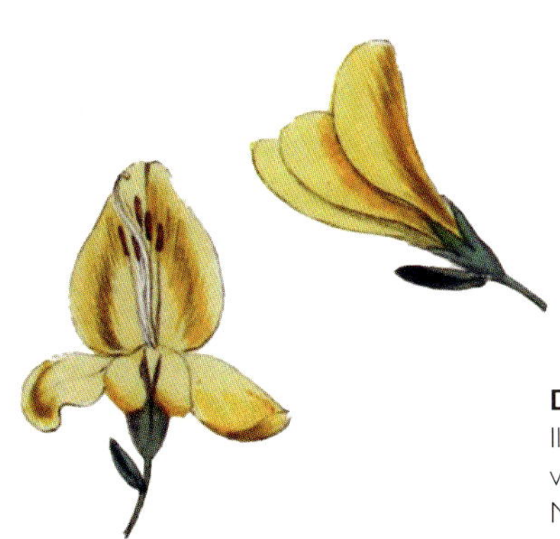

Dyer's broom (*Genista tinctoria*). Illustration from Jan Kops and F.W. van Eeden's *Flora Batava* (Flora of the Netherlands; vol.14, 1872).

Dyer's Tickseed

Latin name: *Coreopsis tinctoria*

Other names: Garden tickseed, golden tickseed, Pahuau

Family: *Asteraceae*

Genus: *Coreopsis*

Native: Canada to eastern Mexico

Type: Annual

Many *Coreopsis* species grow wild in Peru and North America, with dyer's tickseed (*Coreopsis tinctoria*) being particularly common. *Coreopsis* are tough plants that self-sow and are easy to grow from seed. The genus name is derived from the Greek words *koris*, meaning "bedbug," and *opsis*, which means "appearance."

Coreopsis is native to the Americas and was documented around 1615 in the Spanish manuscript *El Primer Nueva Corónica y Buen Gobierno* (The First New Chronicle and Good Government), written by Felipe Guamán Poma de Ayala (or Falcon Puma)—a Quechua nobleman, skillful artist-illustrator, and writer—and addressed to King Philip III of Spain. The manuscript noted the importance of Incan culture in the pre-Columbian period in what is now Peru and references Pahuau plants (*Coreopsis* species), attributing these and other herbs to the Incan sun cult—the golden yellow to dark red flowers of *Coreopsis* were thought to resemble the sun. The flowers were collected by the Palau Pallac, a group of young female flower gatherers who used them for dyeing fabric yellow. The flowers were also used by Indigenous North American peoples, including the Navajo, for ceremonial chant lotions, which were applied to the body during chanting, and the Zuni and Apache used them for dyeing. The roots were used by the Cheroke for a medicinal tea.

The apothecary and botanist James Sherard, who was taught at Chelsea Physic Garden, established an exotic garden in Eltham, near London. He introduced *Coreopsis lanceolata*, another species with its origins in North America, to London in around 1725. *Coreopsis* and other genera were later illustrated by the Belgium painter and botanist Pierre-Joseph Redouté, who worked as the official artist at the French courts of Marie Antoinette. The flowers are illustrated in Redouté's *Choix des Plus Belles Fleurs et des Plus Beaux Fruits* (Selection of the Most Beautiful Flowers and Fruits), published between 1827 and 1833, which notes that *Coreopsis tinctoria* was brought from the banks of the Arkansas River in the United States in around 1823. Redouté's botanical watercolor drawings are of the highest standard and include the flowers of *Coreopsis tinctoria*, which, with their golden yellow petals and dark scarlet centers, are perfect for bringing color to flowerbeds.

Dyer's tickseed (*Coreopsis tinctoria*). Handcolored engraving by William Clark from Richard Morris's *Flora Conspicua: A Selection of the Most Ornamental Flowering, Hardy, Exotic, and Indigenous Trees, Shrubs, and Herbaceous Plants, for Embellishing Flower-Gardens and Pleasure-Grounds* (1826).

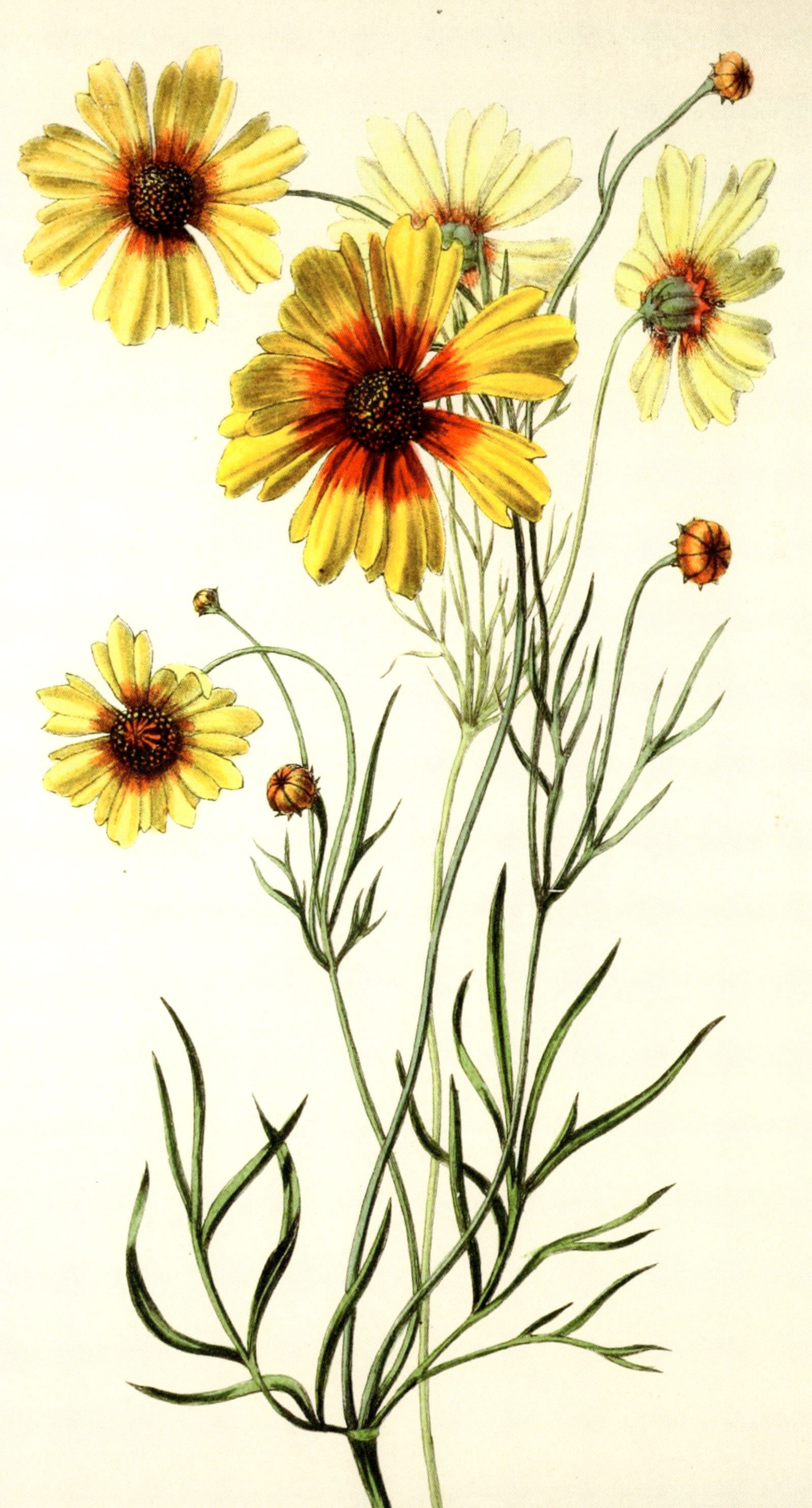

Recipes from Dyer's Tickseed

Dyer's tickseed thrives in a variety of garden settings, making it a popular choice among avid gardeners and organic artists. Characterized by vibrant yellow and red blooms, it has more than just aesthetic appeal, being renowned for dye production. The flowers can be used either fresh or dried, with both producing a colorant. One notable advantage of using dried flowers is their extended shelf life, as they can keep for years if properly stored in an airtight container, retaining their quality and potency until needed.

The dye extraction process is simple yet effective: gently heating the dried flowers in water releases rich orange and yellow colors, resulting in a beautiful, natural dye that can be used for various crafting and textile purposes. In early summer, I explored Cambridge University Botanic Garden where many flowers were in bloom. I collected the yellow and red flowers of dyer's tickseed to process into an orange dye (see *Recipe 1*). The vibrant hues of these flowers inspired me to experiment with dyeing, leading to the successful production of several striking color samples that showcase the plant's potential. Lots of flowers were left growing in the garden to ensure there would be new plants the following season through self-seeding and to maintain the living collections for others to enjoy.

Like many other yellow dyes, dyer's tickseed can be used as a base for making green colorants by re-dyeing yellow cloth in a blue dye vat made from woad (*Isatis tinctoria*), dyer's indigo (*Indigofera tinctoria*), or other blue-producing plants (see *Chapter 2: Blue & Purple*).

The yellow from dyer's tickseed becomes deeper and stronger with multiple dips in a dye bath, which is the general approach when building up a dense, solid color and makes a good yellow or orange pigment when mixed with an inert substrate such as marble dust, kaolin (China clay), or another white pigment like chalk that can hold the dye (see *Recipe 2*). Introducing an iron mordant gives shades of greenish-brown, while acid mordants like alum (pH 3–4) produce reds. Using different mordants to control the pH level of the dye (that is, making it more acidic or alkaline) can influence the color, with baking soda (pH 8–9), for example, making the dye slightly alkaline.

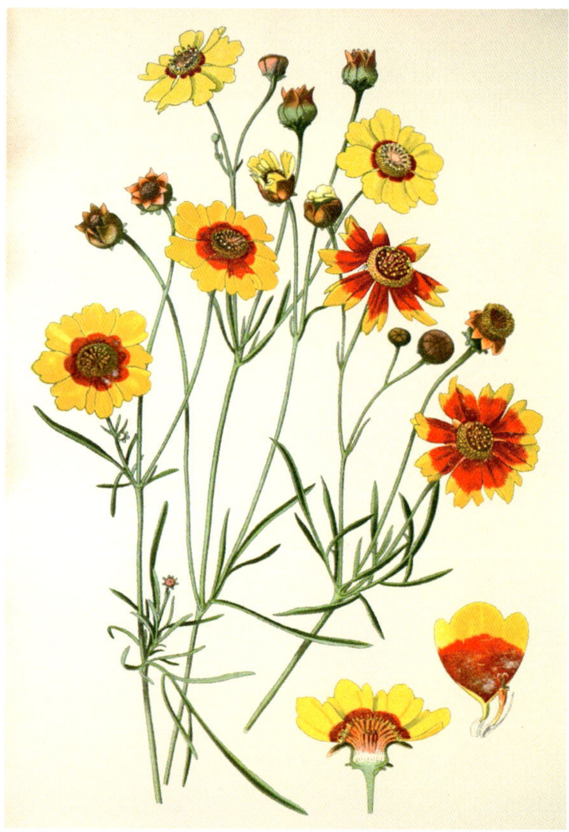

Dyer's tickseed (*Coreopsis tinctoria*). Vintage chromolithograph by Frederick Edward Hulme (1897).

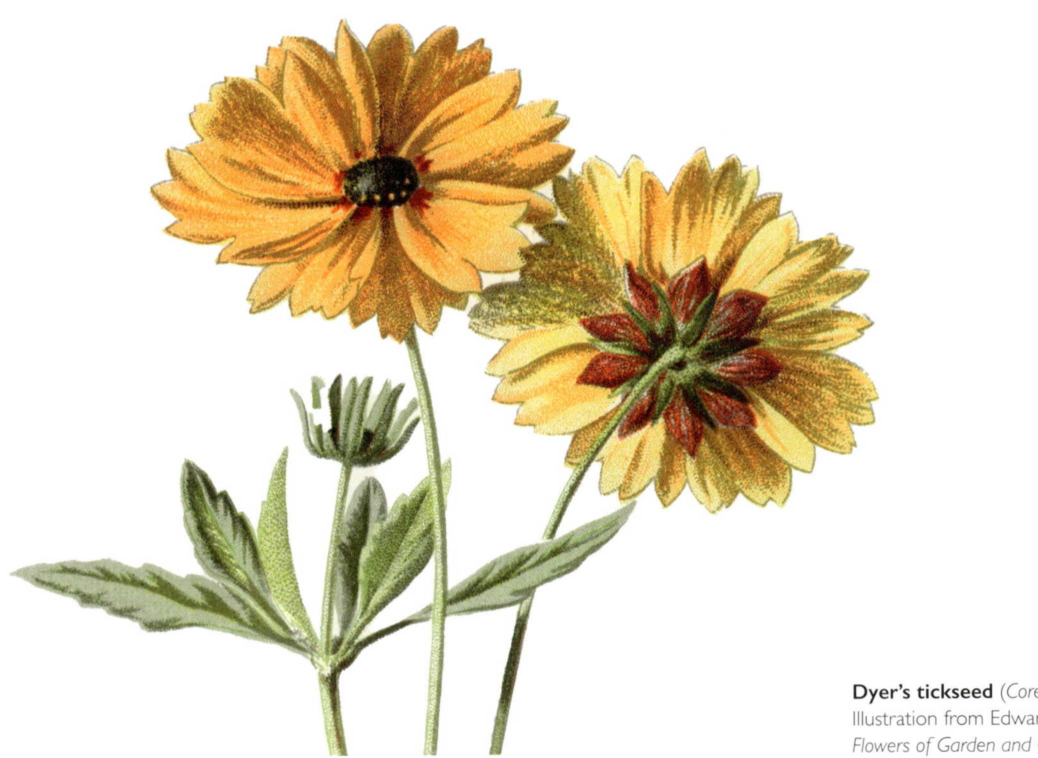

Dyer's tickseed (*Coreopsis tinctoria*). Illustration from Edward Step's *Favourite Flowers of Garden and Greenhouse* (1896).

Recipe 1: Making *Coreopsis* orange

Collect 25 *Coreopsis* flowers and heat them in 3 fluid ounces (90 milliliters) of water. Add half a teaspoon of baking soda to the dye and reduce by half. This produces an excellent orange color—the more layers you build up, the deeper the shade, eventually creating a deep reddish-orange color.

—Nabil's modern recipe

Recipe 2: Making terracotta paint

A terracotta pigment can be produced when dye made from dyer's tickseed is mixed with chalk and then tempered with gum arabic or egg white.

—Nabil's modern recipe

Cretan Hemp

Latin name: *Datisca cannabina*

Other names: False hemp

Family: *Datiscaceae*

Genus: *Datisca*

Native: Eastern Mediterranean to Central Himalayas

Type: Herbaceous perennial

There are two species in the *Datisca* genus: Cretan hemp (*Datisca cannabina*) and Durango root (*Datisca glomerata*), which is endemic to North America and Mexico. Cretan hemp is widely distributed across the world, including Siberia, the Middle East, northern India and the Indian Archipelago, Europe, and North America. As the name suggests, the plant was first discovered on the island of Crete in around 1594. It is a large herbaceous perennial with long leaves that resemble those of the hemp plant (*Cannabis sativa*).

Most *Datisca* foliage produces a strong yellow dye, which was used for religious Romanian velvets and embroidered epimanika (cuffs worn by clergy of the Eastern Orthodox Church) in the 15th to 18th centuries—these can be seen in the National Museum of Art of Romania. Other plants that can be used to produce yellow dye, as well as red and blue dyes, were used to dye the fibers of religious cloths. These include the smoke tree (*Cotinus coggygria*; see page 200), from which a yellow dye called young fustic is obtained; weld (*Reseda luteola*; see page 144); dyer's broom (*Genista tinctoria*; see page 116); dyer's savory (*Serratula tinctoria*); isparak (*Delphinium semibarbatum*); and buckthorn (*Rhamnus* species; see page 152).

Datisca dye, along with dye plants like madder (*Rubia tinctorum*; see page 100) and dyer's indigo (*Indigofera tinctoria*), was used by Turkish female weavers/dyers of the Ottoman Empire in the 16th century. These skillful weavers spun naturally dyed wool yarns by hand, ready to be made into brightly colored Usher carpets, with examples showcased in the Turkish and Islamic Art Museum in Istanbul. Rugs of this quality can also be seen in the work of the early 16th-century Hans Holbein the Younger, painted draped across tables and under the feet of royalty and sacred icons.

The 18th-century Swedish botanist Engelbertus Jörlin wrote a comprehensive index of dyeing materials called *Plantae Tinctoriae* in 1759. He mentions *Datisca* under the heading "Lutei" (yellow dye). Durango root was used by the Indigenous peoples of North America, including the Karok, who lived along the Klamath River in California, and the Wintoon and Costanoan in central California, for producing dyes. The plant as a source of yellow dye has its place in early modern history too, having been used across the globe and still in use today as a colorant by local dyers and artists.

Cretan hemp (*Datisca cannabina*). Illustration from John Sibthorp's *Flora Graeca* (Flora of Greece; 1840).

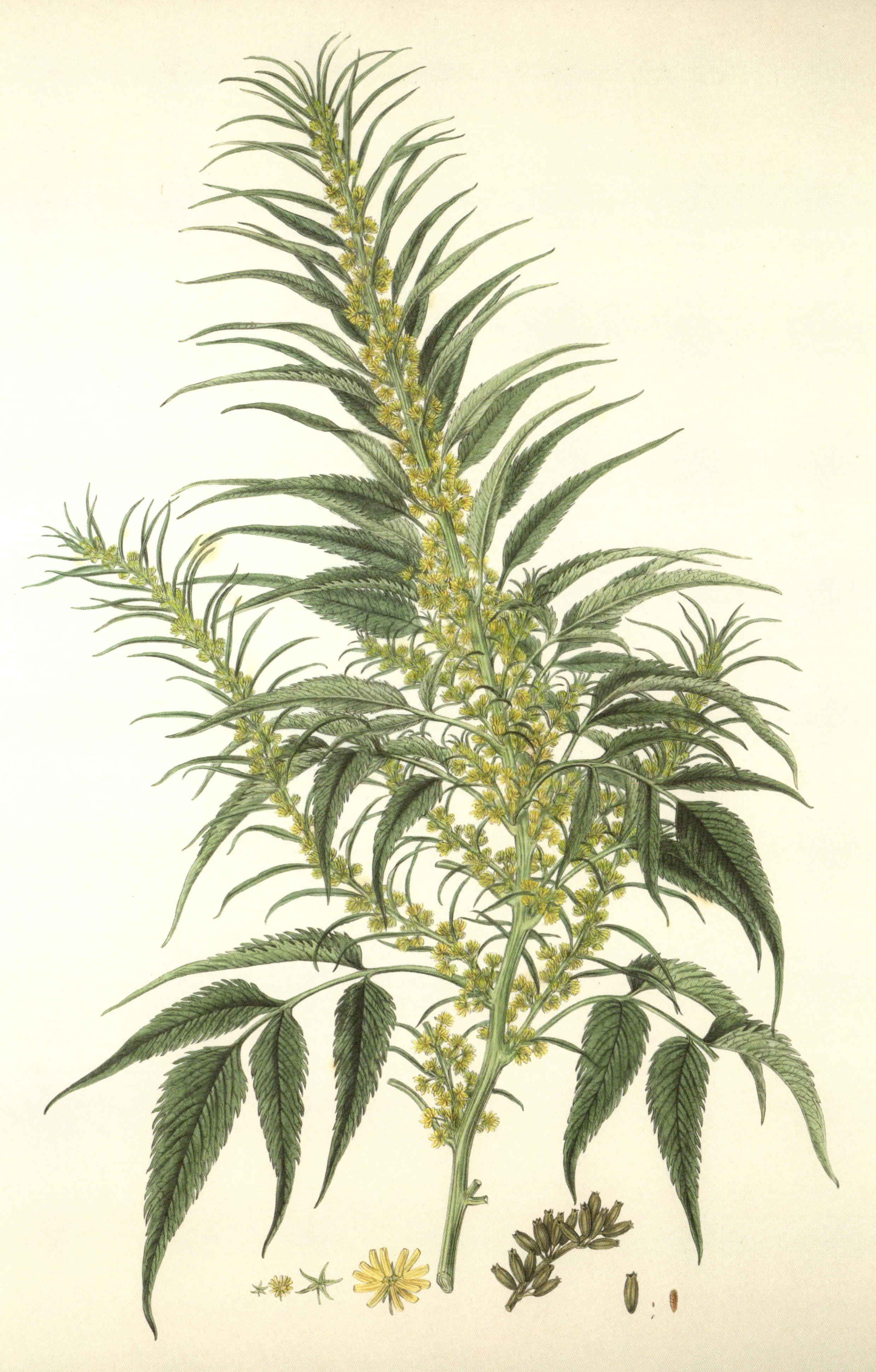

Recipes from Cretan Hemp

Few historical recipes document the making of dye from *Datisca*, although in the 16th century, the Venetian physician and botanist Prospero Alpini recognized the dyeing properties of plants in the genus in his publications. The dye may also have been used centuries earlier in ancient times by dyers to color wool, cotton, and silk for weaving into garments and carpets, but little archaeological evidence for this is currently available.

In the 19th century, French chemist and pharmacist Henri Braconnot noted that Cretan hemp leaves produce a strong, vivid yellow dye that is as good as that from weld or dyer's broom. Compared with weld, which has been used for thousands of years, Cretan hemp has served as a dye plant only relatively recently, with Indigenous people in Lahore, India (now in Pakistan), using the plant's roots to dye silks and other fabrics in the 19th century. The plant was also mentioned by colonial administrator and author Sir Walter R. Lawrence when describing his travels in Kashmir toward the end of the 19th century. He states that Cretan hemp was used by dyers and weavers in the Valley of Kashmir as a yellow colorant to dye cashmere woolen shawls.

Yellow dye can be made from most parts of the plant, including the leaves, stems, and roots. The roots should be cut to about 6–8 inches (15–20 cm) in length and ½–¾ inches (1–2 cm) thick, then bruised before being heated in water. This was the standard method used in Lahore and documented in around 1855 by the British chemist Dr. John Stenhouse in his examination of selected vegetable products from India.

Cretan hemp is well-established in the Systematic Beds at Cambridge University Botanic Garden. For my experiments, I collected a handful of the leaves early one spring morning and then processed them while they were still fresh to produce a yellow dye color (see *Recipe 1*). This proved a straightforward process with excellent results, especially when dyeing soft chamois leather which had been prepared with fish glue to coat the fibers, which raises the pH. Cutting just the top of the plant encourages more leaves to grow, which makes for a healthier plant. Although fresh leaves are preferable, color can still be made from the dried leaves if they are stored in an airtight container and kept out of sunlight.

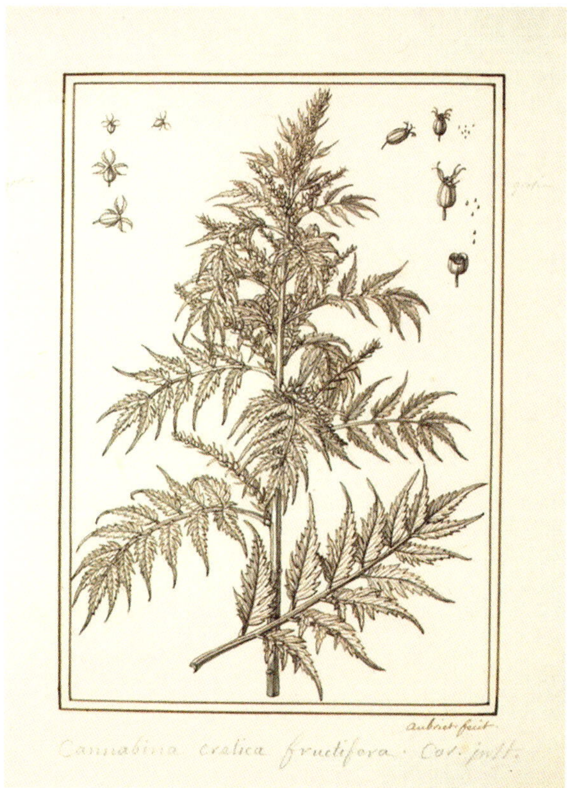

Cretan hemp (*Datisca cannabina*)—here labeled with an older name for the plant: *Cannabina cretica fructifera*. Pen-and-ink drawing on paper by Claude Aubriet (ca. 1700).

Recipe 1: Making yellow dye

Collect a handful of Cretan hemp leaves in spring or summer. Cut the fresh leaves, place them in a saucepan of water, and simmer for 1 hour. Strain the dye to remove the leaves. Alum some fabric the day before and place it in the dye for 30 minutes on a very low heat. You should then obtain a good yellow color.

—Nabil's modern recipe

Recipe 2: Making yellow paint

Extract yellow dye from Cretan hemp leaves (see *Recipe 1*) and brighten the dye with alum. Mix the dye with chalk or crushed cuttlefish, then leave to dry. Crush the resulting pigment with a pestle and mortar for a fine yellow pigment and add gum arabic to make a paint.

—Nabil's modern recipe

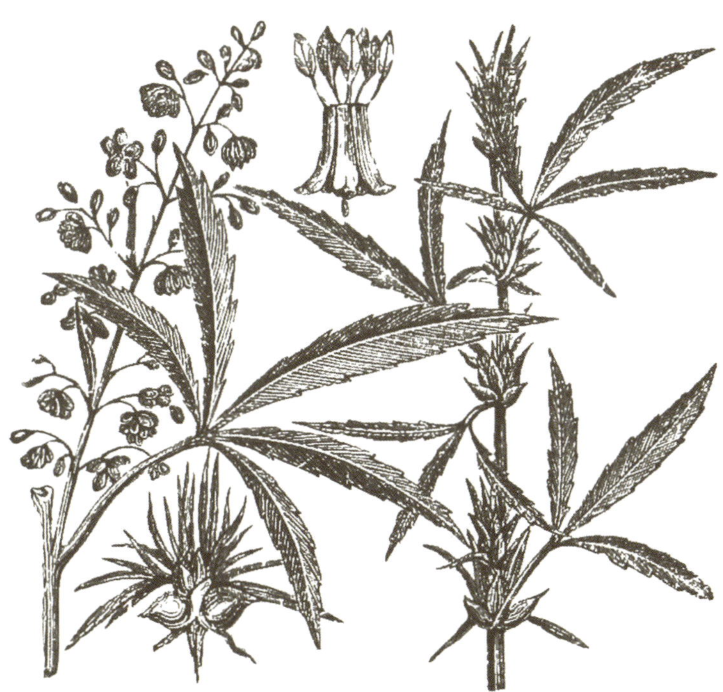

Cretan hemp (*Datisca cannabina*). Illustration from *An Illustrated Encyclopaedic Medical Dictionary* (1890).

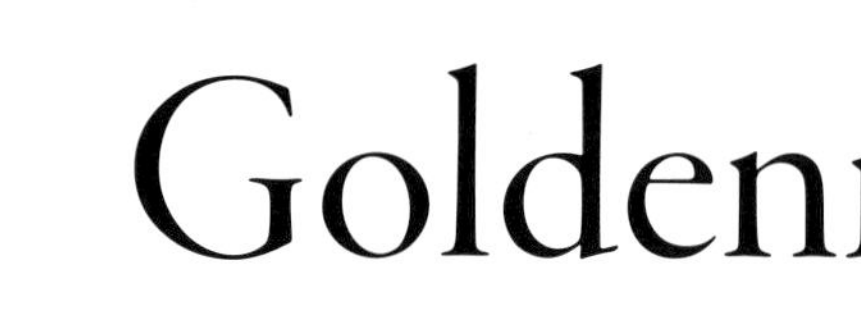

Goldenrod

Latin name: *Solidago virgaurea*

Other names: European goldenrod, Aaron's rod, woundwort

Family: *Asteraceae*

Genus: *Solidago*

Native: Western Europe to Central Siberia, the Philippines

Type: Perennial

There are over a hundred species of *Solidago*, including the European goldenrod (*Solidago virgaurea*) and the Canadian goldenrod (*Solidago canadensis*), which is native to North America. Although some *Solidago* species are classified as annoying and invasive "weeds," goldenrod will produce a good dye and the stems can be used to make woven baskets and mats.

The 16th-century German botanist Hieronymus Bock suggests in his 1539 *Kreuter Buch* (Book of Herbs) that ancient Germanic tribes used European goldenrod to heal wounds—hence one of the common names, woundwort. German botanist Jacobus Theodorus Tabernaemontanus in his *Neuw Kreuterbuch* (New Herbal), published between 1588 and 1591, also mentioned similar qualities. English herbalist John Gerard noted goldenrod's benefits in his 1597 *Herball, or Generall Historie of Plantes*. Later, Nicholas Culpeper cited the plant in his 1652 *Complete Herbal*. In around 1776, English botanist William Withering published his influential *British Flora: A Botanical Arrangement of All the Vegetables Naturally Growing in Great Britain*, describing European goldenrod in great detail.

Goldenrod is also embedded in North American history. Indigenous peoples, including the Meskwaki, Navaho, Goshute, and the Zuni of New Mexico, knew of goldenrod as a dye and medicinal plant. The Iroquois and Chippewa used it to treat cramps and internal hemorrhage and to counteract a love potion; the Chippewa called it *Gizisomukiki*, meaning "sun medicine." The American colonists also used goldenrod before the 18th century, having a good understanding of dyeing with European goldenrod, no doubt adapting their methods to American species. Goldenrod was popular among protestors in Boston, Massachussetts, as a "Liberty Tea" to replace the highly taxed Chinese black tea, and Blue Mountain tea made from sweet goldenrod (*Solidago odora*) was favored by people in the Mississippi area. Goldenrod is used today in herbal medicine, in America and Russia.

Goldenrod also features in the Tiffany & Co. *Magnolia Vase* (ca. 1893), in the Metropolitan Museum of Art in New York. Its foliage references regions of the US: pine for the North and East; magnolias, the South and West; and cacti, the Southwest. The goldenrod, made of gold from US mines, represents the country as a whole.

Goldenrod (*Solidago virgaurea* subsp. *virgaurea*). Illustration from O.W. Thomé's *Flora von Deutschland, Österreich und der Schweiz* (Flora of Germany, Austria and Switzerland; 1905).

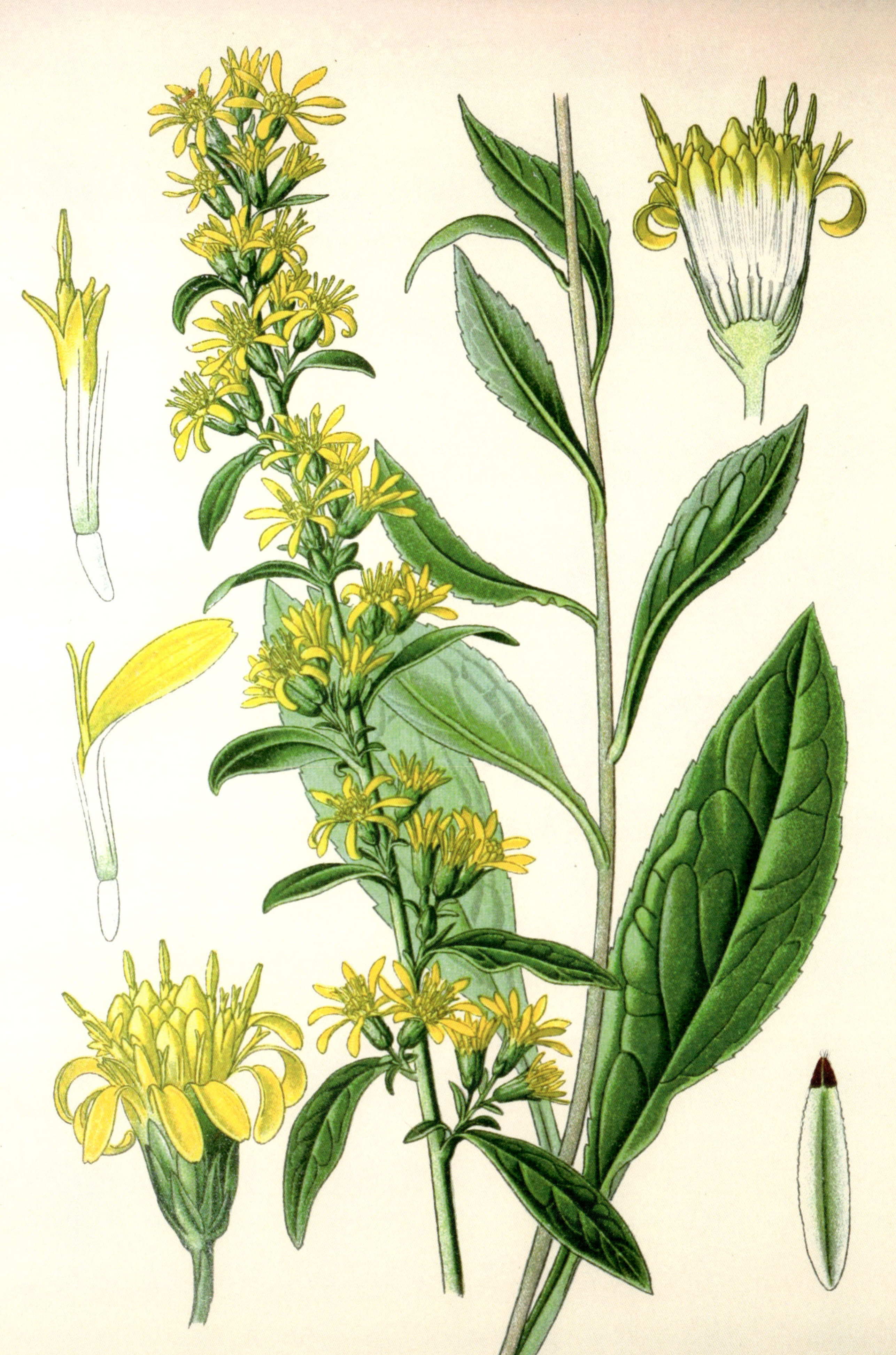

Recipes from Goldenrod

Historically, goldenrod has been used as a medicinal plant, with old herbals referencing its virtues, and as a dye plant by dyers and artists for hundreds of years. For my own experiments, I collected goldenrod from Cambridge University Botanic Garden in late summer to early fall and processed the leaves, stems, and flower heads to make a good yellow dye. Fresh flowers are best when making a colorant from the whole plant when it is in full bloom, but you can dry the flower heads to collect the seeds, then sow them to grow new plants or pass them on to fellow dyers and artists to grow. The plant can be invasive, so take care to prevent it setting seed when it flowers.

Combining several plants can produce stronger dye colors, especially when making a yellow dye. A traditional American combination is goldenrod with Queen Anne's lace (*Daucus carota*), which is also known as the wild carrot. When wild carrot comes into flower, cut the foliage from the base and use either fresh or after drying and storing, but ideally both plants should be used fresh. Care must be taken when handling *Daucus* species as the foliage can be a skin irritant if it is wet. Soak the wild carrot foliage overnight, then heat for about two hours in water and reduce by half. Separate the plant matter from the yellow dye, then mix it with some goldenrod dye, which is made using the same method when fresh.

Using only goldenrod flower heads produces clearer yellow dyes, while the dye will develop a more greenish tinge when the leaves are used at the same time. Introducing a tin mordant when preparing fabric for dyeing with goldenrod produces a vivid orange color and a reasonably permanent dye. An iron or copper mordant produces olive-greens if you use only the tops of the leaves. In this way, you can create a complementary color palette from just a few combinations of ingredients. The same rules apply when adding the dye to chalk or other inert substrates to make an organic paint.

The double-dyeing technique can also be used: first dye the fabric yellow using goldenrod or multiple yellow dye plants, then re-dye with madder root (*Rubia tinctorum*; see page 100) to produce orange colors on wool that has been pre-prepared using an alum–tartar mordant.

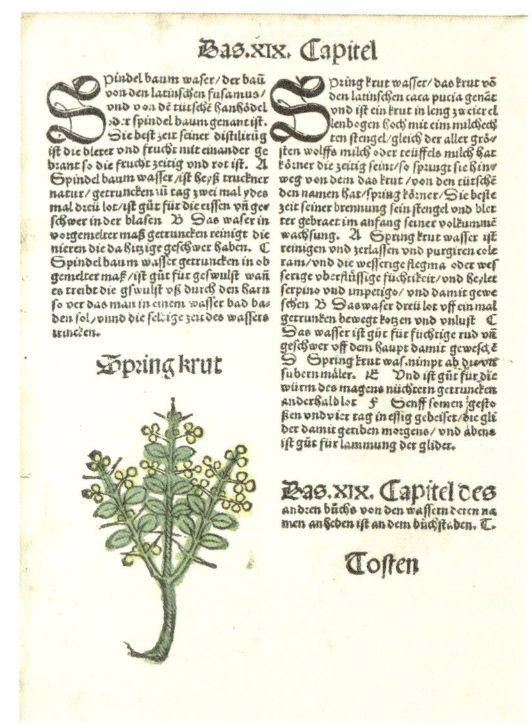

Das.XIX. Capitel

Spindel baum waſer/ der baū von den latiniſchen fuſamus/ vnd von dē tütſchē hanhödel oder ſpindel baum genant iſt. Die beſt zeit ſeiner diſtilierūg iſt die bleter vnd frucht mit einander ge brant ſo die frucht zeitig vnd rot iſt. A Spindel baum waſſer/iſt heyß truckner natur/ getruncken iii tag zwei mal ydes mal dreü lot/iſt güt für die eiſſen vñ geſchwer in der blaſen B Das waſer in vorgemelter maß getruncken reinigt die nieren die da hitzige geſchwer haben. C Spindel baum waſſer getruncken in ob gemelter maß/iſt güt für geſwulſt wañ es treibt die gſwulſt vß durch den harn ſo ver das man in einem waſſer bad baden ſol/vnnd die ſelbige zeit des waſſers trincken.

Spring krut

Spring krut waſſer/ das krut vō den latiniſchen caca pucia genāt vnd iſt ein krut in leng zweier ellenbogen hoch mit eim milchecht ten ſtengel/gleich der aller gröſten wolffs milch oder teüffels milch hat körner die zeitig ſeint/ſo ſpringt ſie hin weg von dem das krut/von den tütſchē den namen hat/ſpring körner/Die beſte zeit ſeiner brennung ſein ſtengel vnd bletter gebraet im anfang ſeiner volkummē wachſung. A Spring krut waſſer iſt reinigen vnd zerlaſſen vnd purgiren cole ram/vnd die weſſerige flegma oder weſſerige vberflüſſige füchtikeit/vnd heylet ſerpino vnd impetigo/ vnd damit geweſchen B Das waſer dreü lot vff ein mal getrunken bewegt kotzen vnd vnluſt C Das waſſer iſt güt für füchtige rud vñ geſchwer vff dem haupt damit geweſcht D Spring krut waſ.nimpt ab die vnſubern mäler. E Vnd iſt güt für die würm des magens nüchtern getruncken anderhalb lot F Senff ſomen geſtoſſen vnd vier tag in eſſig gebeiſet/die glider damit geriben morgens/ vnd abens iſt güt für lammung der glider.

Das.XIX. Capitel des andern büchs von den waſſern deren namen anheben iſt an dem büchſtaben. T.

Toſten

European goldenrod (*Solidago virgaurea*). Sample page from Leonard Fuch's *De Historia Stirpium* (1542).

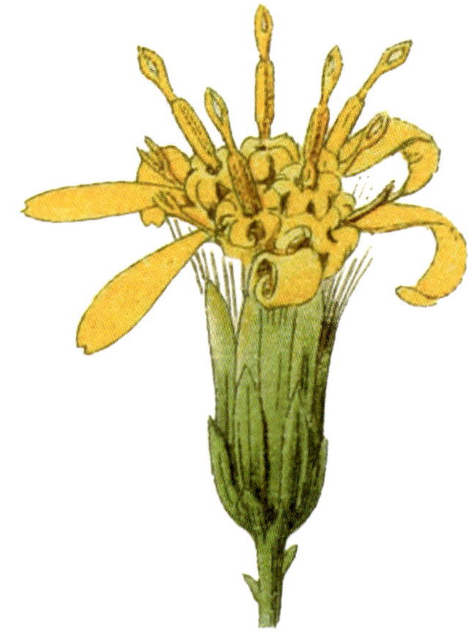

Detail of **goldenrod** (*Solidago virgaurea*) flower. Illustration from A. Mentz and C. H. Ostenfeld's *Billeder af Nordens Flora*, vol 1 (Pictures of the Flora of the North; 1901–03).

Recipe 1: Making yellowish-tan or gold dye

"*To prepare the dye*: Place one to one-and-a-half pecks [2.4 US gallons/9 liters] of goldenrod flowers in enough cold water to cover; bring to a boil and boil for an hour or longer to extract their color. Strain liquid into a bath for dye.

Use a mordant: For yellowish-tan, add alum; for old gold, add chrome."

—Etheljane Schetky, *Dye Plants and Dyeing: A Handboock*, Volume 20, No.3, Brooklyn Botanic Garden, ca.1964

Recipe 2: Making goldenrod yellow

Collect goldenrod flowers, leaves, and stems in late summer to early fall and heat the whole plant in water for an hour. Then:

To use as a dye: Mordant some fabric with alum in hot water and leave overnight. You should have a lemon-yellow color when you dip the fabric in the dye. If you want to make a vivid orange, use a tin mordant.

To make a paint: Add alum to the dye bath when heating the plant, strain, and then mix the dye with chalk, gesso, crushed cuttlefish, or kaolin (China clay) and leave to dry. Add gum arabic to the pigment to form a yellow paint.

—Nabil's modern recipes

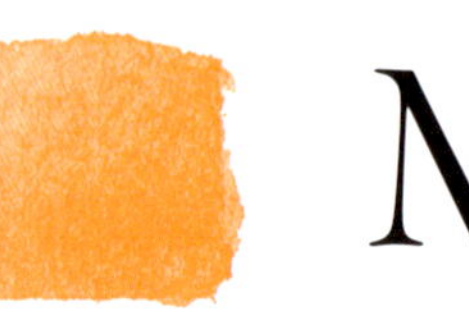

Marigold

Latin name: *Calendula officinalis* (pot marigold); *Tagetes erecta* (Mexican marigold)

Other names: *Calendula officinalis* (Mary's gold, goldins); *Tagetes erecta* (Aztec marigold, African marigold)

Family: *Asteraceae*

Genera: *Calendula* and *Tagetes*

Native: *Calendula officinalis* (Western Mediterranean); *Tagetes erecta* (Mexico to Guatemala)

Type: Perennial

The pot marigold (*Calendula officinalis*) was common in European medieval gardens alongside leeks and onions, for adding color to food and for its medicinal properties. The flowers were mentioned in *The Boke of Secretes*, by 13th-century German alchemist friar Albertus Magnus, noting that they could be wrapped in laurel or bay leaves with a wolf's tooth for the magical purpose of creating peaceful encounters.

Marigolds—in this case, the Mexican marigold (*Tagetes erecta*)—were known much earlier to Mesoamerican cultures, particularly the Aztecs, and were and still are associated with Mexico's *Día de los Muertos* (Day of the Dead) festival. It was believed that marigold flowers guided deceased souls back to the world of the living and were sacred to Tláloc (He who makes things grow), a pre-Aztec rain and fertility deity. In early images dating to the 3rd to 8th centuries, Tláloc is depicted as blue, wearing eye goggles, and with serpent-like fangs, bringing rain at the end of the dry season.

Mexican marigolds were cultivated by the Aztecs and are mentioned in an Aztec herbal called the *Libellus de Medicinalibus Indorum Herbis* (Little Book of the Medicinal Herbs of the Indians)—often known as the *Badianus Manuscript* or *Cruz-Badianus Codex*—held in Mexico's National Library of Anthropology and History. It was translated into Latin from the original Aztec in around 1552 in the College of Santa Cruz (near Mexico City) and states that the marigold was an important ritual plant for honoring the dead and for medical purposes; the Aztecs made a potion called *Copaliyacxinhtontli* from herbs, roots, leaves, and the marigold plant mixed together in water, used to treat abdominal pain and cramps.

The 16th-century court physician and naturalist Francisco Hernández de Toledo, who was sent by Phillip II of Spain on a scientific mission to Mexico in 1570, depicts *Tagetes* in his *Historia de las Plantas de Nueva España* (History of the Plants of New Spain), while botanist Leonhart Fuchs, founder of the first German botanical garden, describes plants in the genus as being of a pungent nature in his 1542 *De Historia Stirpium* (On the History of Plants). *The Florentine Codex* by the friar Bernardino de Sahagún—another 16th-century book, housed in the Laurentian Medici Library in Florence—documents Aztec culture and includes depictions of ritual dancers adorned with marigold flowers.

Pot marigold (*Calendula officinalis*). Illustration from *Köhler's Medizinal-Pflanzen* (Köhler's Medicinal Plants; vol.3, 1898).

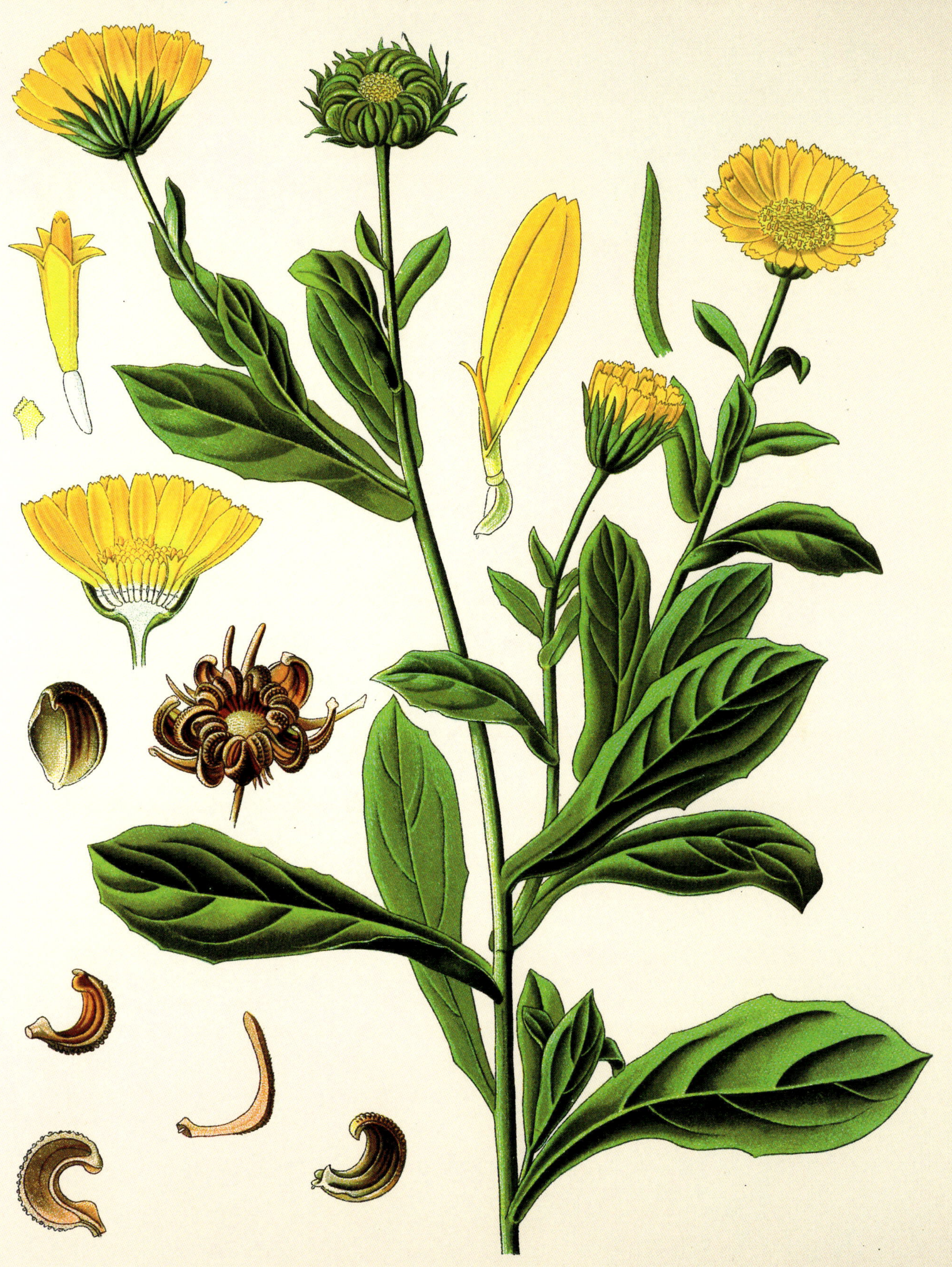

Recipes from Marigold

As *Tagetes* was introduced to Europe later than other trade plants, such as weld (*Reseda luteola*; see page 144), saffron (*Crocus sativa*; see page 30), and safflower (*Carthamus tinctorius*; see page 104), it has few recipes in Western medieval manuscripts. The *Badianus Manuscript* Aztec herbal documents marigolds being used as a dye and a medicine.

The *Dictonarium Polygraphicum, or The Whole Body of Arts, Regularly Digested* (1735) by mathematician and naval historian John Barrow includes a recipe for yellow dye to color thread by heating broom, Spanish Yellow (saffron), crab-tree rind, and corn marigolds (*Glebionis segetum*)—which are in the Asteraceae family like *Calendula* and *Tagetes*—with potash lye (see *Recipe 1*). He explains that "all yellows that are designed are to be dyed green—using [dyer's] indigo." Over a century later, Eliza Leslie's *The House Book: A Manual of Domestic Economy* (1840) gives a recipe for a "fine lemon-yellow dye" from French marigolds (*Tagetes patula*), boiling the fresh flowers in water with alum.

Indian artisans used *Tagetes* flowers as a cheap source of yellow and green dye, with India, South Africa, Zambia, South America, and Mexico being exporters of marigold colorants. There are two source of food colorant products: marigold meal from the dried powdered flowers, which is added to poultry feed to enhance the color of egg yolks and meat, and marigold extract, which is used as a food coloring to enhance the appearance of salad dressings, ice cream, and dairy products.

I grew many pot marigold plants from seeds and cultivated the flowers in my private garden under the watchful eye of a visiting hawk that perches on a neighboring tree. As the plant self-sows, there was an abundant supply of flowers in the flower beds for experimentation and for the local wildlife. I collected the flowers from spring to fall and dried them in my potting shed, then stored them for later use in an airtight container, which was kept out of sunlight. I made a yellow dye for paint and this can also be used to dye fabrics. Note that the lightfastness of marigold yellow, as with many natural yellow dyes, can be poor, although these dyes can be used for temporary color. Adding a few unripe green husks from black walnut (*Juglans nigra*; see page 140) to the dye will deepen the tone of yellow.

Pot marigold (*Calendula officinalis*). Illustration by German-Danish botanist Simon Paulli from *Flora Danica* (Flora of Denmark; 1648).

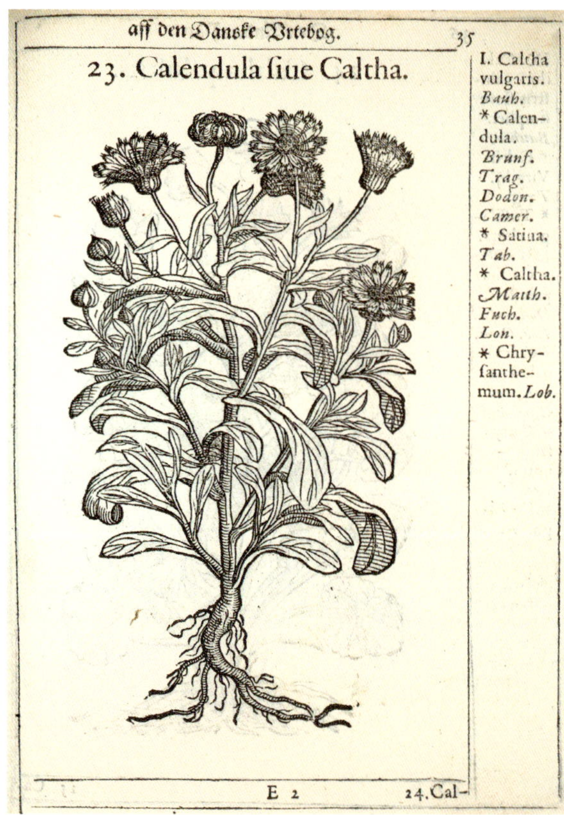

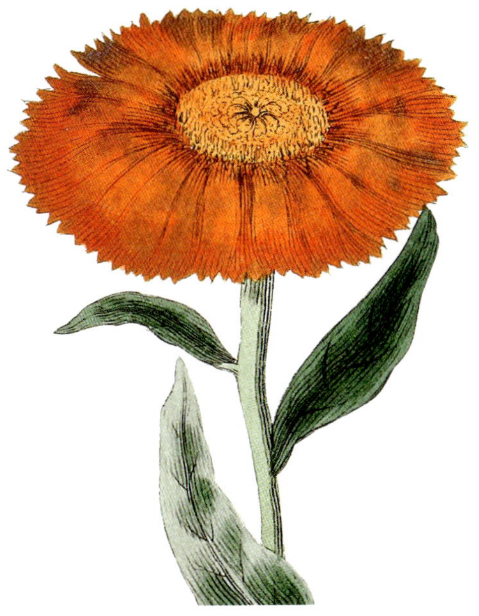

Pot marigold (*Calendula officinalis*). Illustration from Richard Duppa's *Elements of the Science of Botany, As Established by Linnaeus; With Examples to Illustrate the Classes and Orders of his System* (1812).

Recipe 1: Dyeing thread yellow

"Boil eight pound of broom, one pound of Spanish Yellow [saffron], one pound of crab-tree rind, and one pound of corn marigold in a kettle, with three quarts [around 3.5 liters] of sharp lye, and work the thread in the liquor three times successively, not suffering it to dry between whiles, and it will be of a beautiful and lasting color."

—John Barrow, *Dictonarium Polygraphicum*, Volume 2, page 552, ca. 1735

Recipe 2: Using marigold yellow

Collect a handful of orange and yellow marigold flowers, either dried or fresh, and soak them in water in a saucepan/dye bath overnight. Heat the flowers for less than an hour, then sieve through a fine-mesh strainer. To use as a dye, dip fabric pre-treated with alum into the dye to produce a beautiful yellow cloth.

To make a yellow paint, add alum mordant to the dye, then reheat and reduce by half (which is standard practice in my workshop). Add the dye to Italian gesso (at a ratio of 1:1), then mix well and leave to air-dry. Temper with gum arabic or egg white.

—Nabil's modern recipe

Mountain Pansy

Latin name: *Viola lutea*

Other names: Mountain pansy, yellow viola

Family: Violaceae

Genus: *Viola*

Native: Western and Central Europe

Type: Perennial

The mountain pansy (*Viola lutea*) grows widely throughout Europe and beyond and produces small, delicate yellow flowers. Hybridization of this species in the early 19th century led to the familiar cultivated garden pansy *Viola × wittrockiana*. It is described by apothecary and botanist William Hudson in his 1762 publication *Flora Anglica* (Flora of England).

Mountain pansy flowers were beautifully and accurately depicted in the margins of medieval Books of Hours, along with other *Viola* species and flowers. Pansies represent remembrance and thoughtful contemplation and were painted in late medieval illuminated prayer books. Although these illustrations do not suggest the mountain pansy could be used as a yellow colorant, the depiction of the flowers in such important books must have had symbolic significance—or they were perhaps simply appreciated for their beauty by observers and painters. The plant also appears in various medieval technical painting manuscripts, which implies that the dried flower heads were used to produce a dye resembling *sil atticus*—a yellow pigment called Attic Ocher that was used by the ancient Greeks and Romans.

Artists, notably since the late 19th century, have demonstrated a keen infinity with pansy flowers. For example, an 1874 oil painting by French painter Henri Fantin-Latour features deep purple, white, and yellow pansies growing in terracotta pots, along with some recently picked ripened apples, with green foliage against a dark background. In around 1903 Henri Matisse painted a simpler version of pansies in a glass jar entitled simply *Pansies*, while much later in 1967 American artist Joe Brainard produced a series of strikingly modern opaque watercolor and ink drawings of pansies. One of these, *Three Pansies*, depicts pink-, red-, and yellow-petalled pansies with green leaves and stems against a flat black background. The flowers remain timeless: small but everlasting, with evocative scents that etch themselves on the mind.

Violets also inspired various writers, such as American poet William Cullen Bryant, whose 1821 poem *The Yellow Violet* demonstrates the appeal of the downy yellow violet (*Viola pubescens*). The poem finishes with the following lines:

"And when again the genial hour / Awakes the painted tribes of light, / I'll not o'erlook the modest flower / That made the woods of April bright."

Mountain pansy (*Viola lutea*). Handcolored lithograph by Stroobant from Louis van Houtte and Charles Lemaire's *Flore des Serres et des Jardins de l'Europe* (Flowers of the Hothouses and Gardens of Europe; 1851).

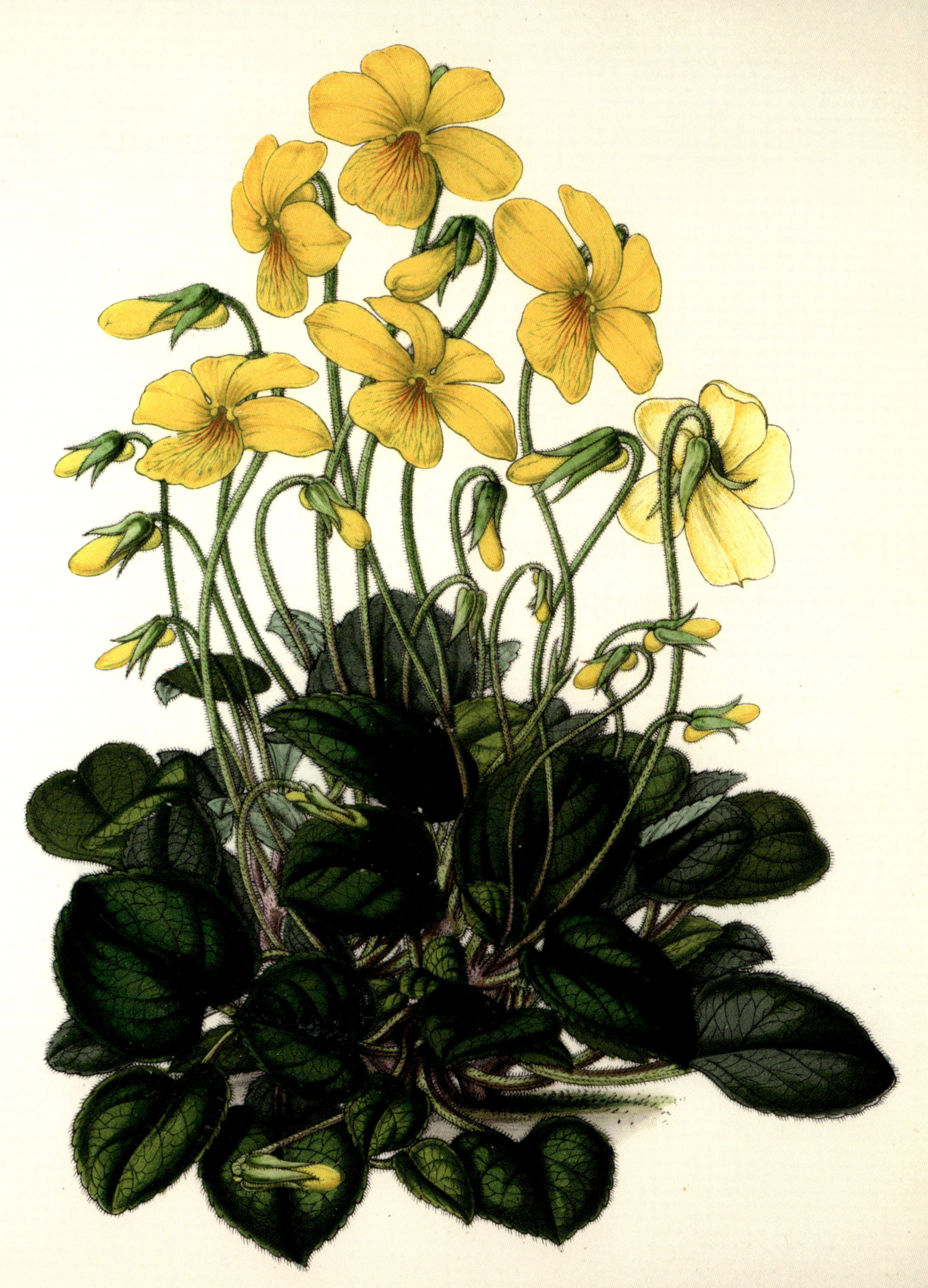

Recipes from Mountain Pansy

The 13th-century *Compendium Artis Picturae* (Compendium of the Art of Painting), held in Brussels Royal Library (MS 10152), features many recipes, such as Recipe 33 for making yellow from flowers. Similar recipes appear in Heraclius's 11th-century *De Coloribus et Artibus Romanorum* (On the Colors and Arts of the Romans) and the 14th-century *Liber Diversarum Arcium* (Book of Various Arts), which documents the use of dried violets (mountain pansies), as pointed out by Mary P. Merrifield in her 19th-century *Medieval and Renaissance Treatises on the Arts of Painting*.

Recipe 1.21.2 in the *Liber Diversarum Arcium* describes extracting color from dried mountain pansy flowers by heating them in water and fixing the colored lake to lime or a chalk substrate to make an yellow pigment called Attic Ocher. This process was documented in the 1st century BCE by Vitruvius (see *Recipe 1*) in *De Architectura Libri Decem* (Ten Books on Architecture). Similar processes are mentioned by Pliny the Elder in his *Naturalis Historia* when he describes staining pigeon droppings with dyer's indigo (*Indigofera tinctoria*) and coloring chalk with woad (*Isatis tinctoria*) to create a blue-gray pigment. This method was also mentioned in Heraclius's *De Coloribus et Artibus Romanorum*. Clearly, fixing using a chalk substrate was well-established long before medieval practice, and such techniques are still used by woad growers in East Anglia, England.

I once grew many mountain pansy plants in garden pots, collecting and drying the flower heads to create a yellow dye. This was made with fifty dried flower heads in 7 fluid ounces (200 milliliters) of water and 10 grams of alum to brighten the dye. I heated and reduced the mixture to ⅔ fluid ounce (20 milliliters) of liquid, then poured it through a fine-mesh cloth to separate the plant matter. You can make a paint sample by applying several layers of the dye, which deepens the hue on paper. Finally, add the dye to one-and-a-half teaspoons of chalk (calcium carbonate), then leave to dry. Once dried, crush into a fine pigment with a pestle and mortar to make an imitation Attic Ocher. I repeated this pigment lake process to deepen the colored chalk.

Mountain pansy (*Viola lutea*) (bottom left and right). Illustration from W. Winkler's *Sudetenflora: Eine Auswahl Charakteristischer Gebirgspflanzen* (Sudetenflora: A Selection of Characteristic Mountain Plants; 1900).

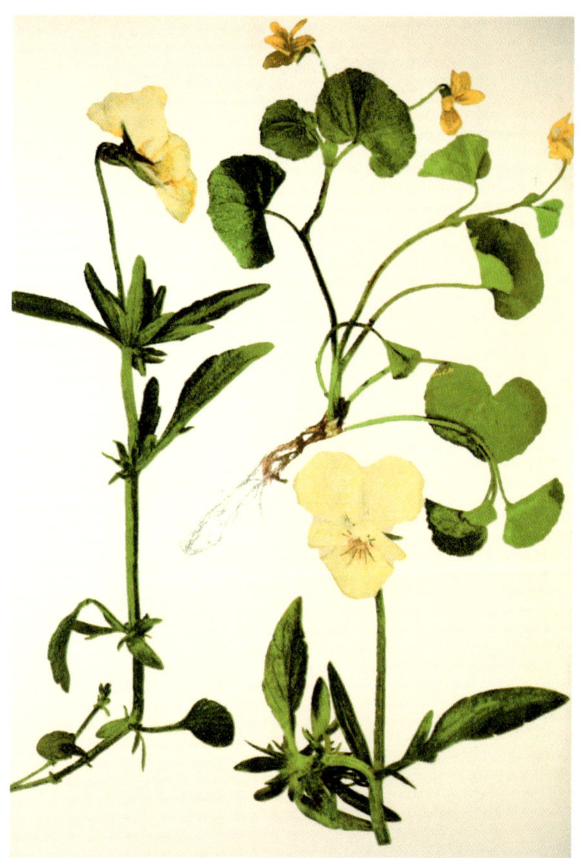

Recipe 1: Making Attic Ocher

"Thus if fresco painters should want to imitate Attic Ocher, they throw dried yellow violets into a vessel with water, boil it over a fire, and when it is ready pour it into linen. Wringing the linen with the hands, it soaks up the water, which has been colored by the violets, in mortars, and when they pour chalk into these and pound it creating the color of Attic Ocher."

—Vitruvius, *De Architectura Libri Decem*, 1st century BCE (Rowland & Howe, 1999, p.95)

Recipe 2: Making pansy-yellow paint

Collect 20 small yellow mountain pansy flowers in summer and use them either fresh or dried. Add the flowers to 3½ fluid ounces (100 milliliters) of water, then heat with 1 teaspoon of alum and reduce by half. Add the yellow dye to 4 teaspoons of chalk and mix well, then leave to dry. Crush the dry pigment with a pestle and mortar and add gum arabic to make into a paint.

—Nabil's modern recipe

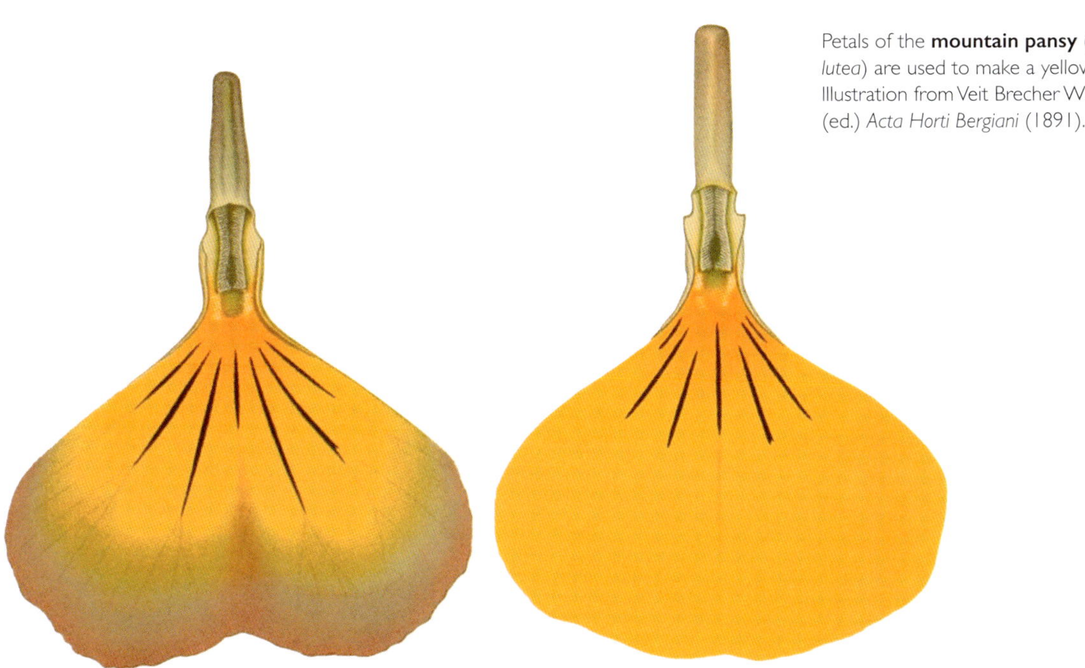

Petals of the **mountain pansy** (*Viola lutea*) are used to make a yellow dye. Illustration from Veit Brecher Wittrock's (ed.) *Acta Horti Bergiani* (1891).

Common Walnut

Latin name: *Juglans regia*

Other names: Persian walnut, Black Sea walnut

Family: *Juglandaceae*

Genus: *Juglans*

Native: Northeastern and eastern Turkey to Lebanon and the Western Himalayas

Type: Deciduous tree

The common walnut (*Juglans regia*) was introduced to Britain by the Romans. The genus name is derived from the Latin *Jovis*, the name of the Roman god Jupiter, also known as Jove, and *glans*, which means "nuts"—hence Jupiter's nuts. It is also associated with Bacchus, the Roman god of wine. The tree is valued for its good-quality wood, and its leaves and husks can be used as a natural dye. Pliny the Elder mentions walnuts in his *Naturalis Historia*, describing mixing dried mistletoe berries that had been left in water to rot with walnut oil to make hunter's birdlime to trap birds.

The Greek philosopher Plato, around the 4th century BCE, tried to pinpoint whether the Greek storyteller Aesop, who purportedly lived a few centuries before him, was real or fictitious, as little is known about him beyond his famous fables and the fact that he was a slave. Plato records Socrates using Aesop's analogies in mockery while awaiting execution in prison. One of Aesop's fables tells of a walnut tree being brutally struck with sticks and stones by people wanting the fruit, damaging the tree in the process. Ernest Henry Wilson noted in his 1913 book *A Naturalist in Western China* that in the Min Valley, China, the local people would hack the lower trunk of walnut trees to make them fruitful.

Walnut wood has been prized since antiquity, with the ancient craft of intarsia (a wood-inlay technique) practiced using various woods, especially walnut panels, to produce decorative wooden mosaics. Intarsia was developed in Italy from the 14th century; in the late 15th to 16th century, Fra Giovanni da Verona, an Italian Olivetan monk, improved the technique by using tinted oil stains and scorching the wood by dipping it in hot sand. Other craftsmen bleached the wood, using oil of sulfur, corrosive sublimation, or arsenic, but generally natural tones from different woods were favored. A detailed example of 15th-century intarsia can be seen in New York's Metropolitan Museum of Art. The *Studiolo from the Ducal Palace in Gubbio* (ca. 1478–82) shows the walls of a *studiolo* (study) carried out in intarsia for Duke Federico da Montefeltro. Highly realistic trompe l'oeils in wood inlay depict cabinets of objects that reflect the duke's artistic and scientific interests. These are created from beech, rosewood, oak, fruitwoods, and, of course, walnut wood.

Common walnut (*Juglans regia*). Illustration by German-Danish botanist Simon Paulli's from *Flora Danica* (Flora of Denmark; 1648).

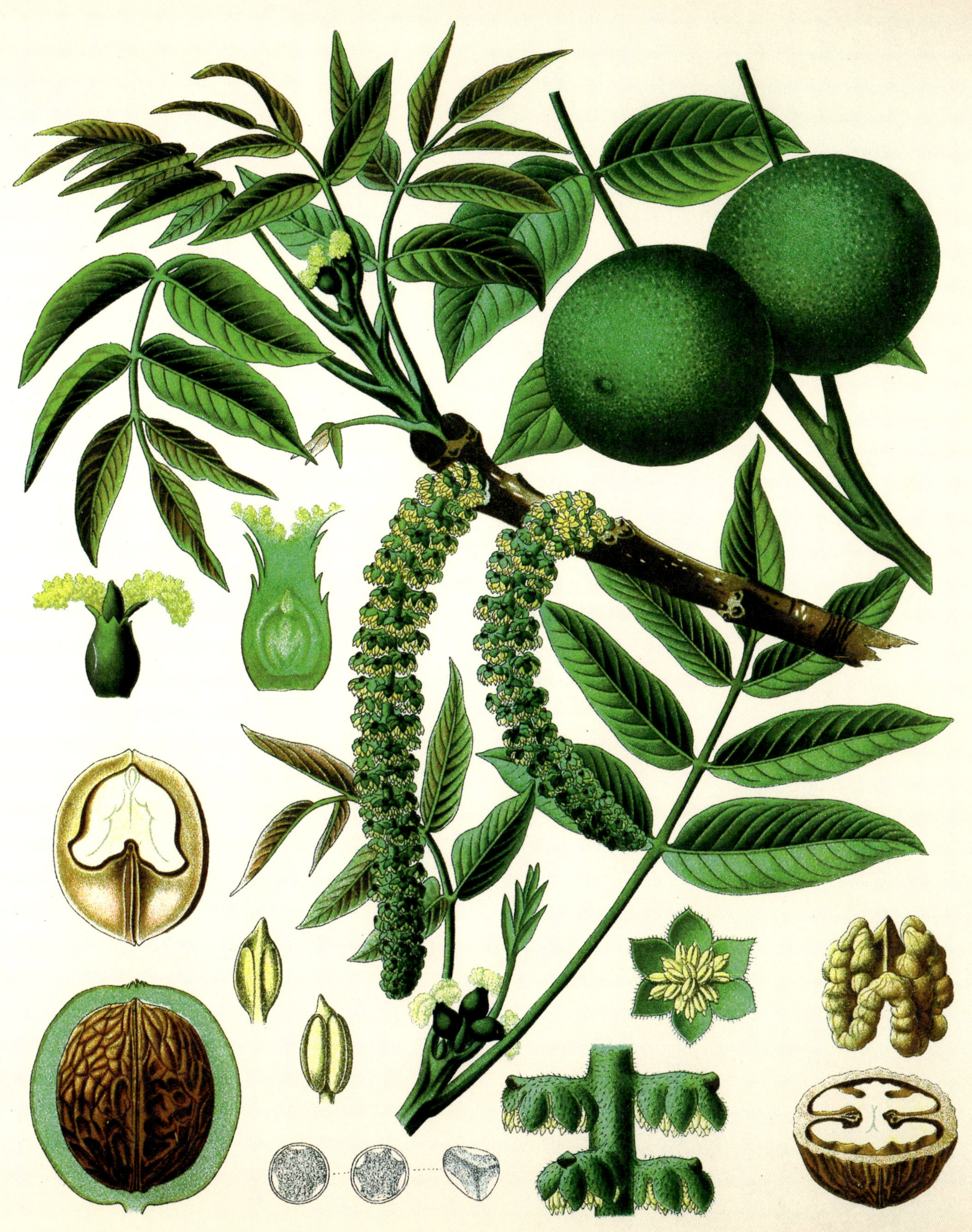

Recipes from Common Walnut

Walnut leaves, bark, twigs, catkins, and husks have been used for centuries to make yellow, brown, and black dyes. In the 1st century CE, Pliny the Elder mentions in his *Naturalis Historia* that walnut shells can be used to dye wool, and that young "green" walnuts can be made into a red hair dye, known from the stained hands of walnut pickers.

Later, the use of walnuts to enhance other colors is noted in Recipe 97 of the 3rd-century-CE *Leiden Papyrus X*, in the National Museum of Antiquities in the Netherlands. This recipe for making purple utilizes tannins in powdered walnuts (most likely the husks), which are added to clear vinegar with the roots of dyer's alkanet (*Alkanna tinctoria*) and some pomegranate bark. Alternatively, the petals of bear's breeches (*Acanthus mollis*) can be combined with a natural mineral salt called natron to make purple.

Walnut husks are mentioned many centuries later in the German *Innsbruck Manuscript* (MS. Cod. 355), held in the University Library in Vienna, Austria. This manuscript, from around 1330, was commonly referenced in 15th-century recipes for making green. A Middle English manuscript (MS Douce 45), held in the University of Oxford's Bodleian Library, documents using powdered walnut rinds and leaves or bark to make a green dye, which is then mixed with chalk and other pigments made from fustic (a yellow dye from dyer's mulberry, *Maclura tinctoria*), weld, a little brazilwood (*Biancaea sappan*; see page 84), and verdigris mixed with cider to make a green paint.

Centuries later, yellow dyes made from the white walnut (*Juglans cinerea*) were used by 17th-century settlers in America, both for fabrics and as a hair dye, notably in Pennsylvania and New Jersey. *The American Druggist and Pharmaceutical Record*, published in around 1899, includes a section on "Walnut Dye for the Hair." This describes taking the skins of fresh walnuts, beating them to a pulp, and preserving them in alcohol. When the dye is ready, the solution is applied directly to the hair for a few hours and exposed to air and light. The yellow dye was also used in the early 1800s as a semipermanent wood stain, made from both the bark and leaves.

Walnut tree (*Juglans regia*). Illustration by Persian artist Mirzā Bāqir for an 1889–90 edition of a 9th-century Persian translation of *Pedanius Dioscorides' De Materia Medica* (1st century CE).

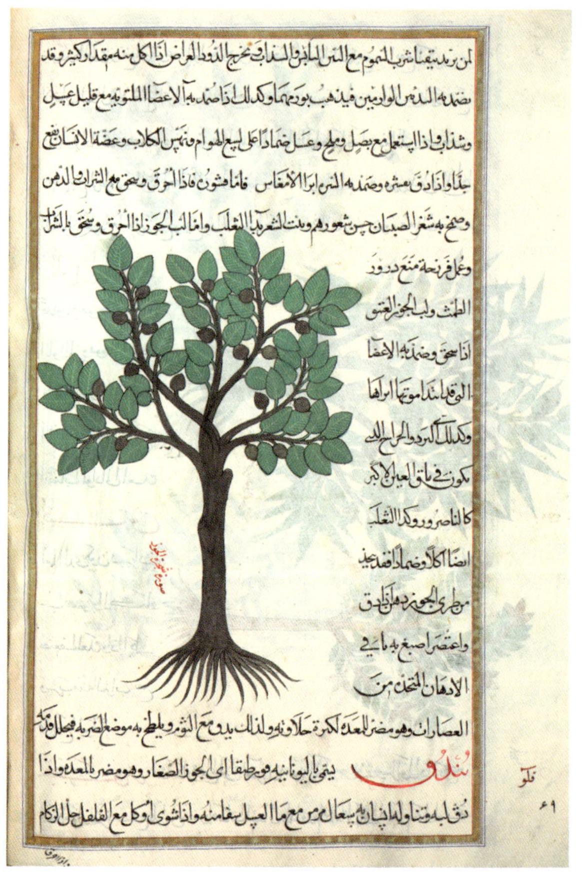

Recipe 1: Making yellow hair dye

"Take a potel [34 fluid ounces/1 liter] of small ale and stamp it together with three handful of walnut leaves and a quater [of a pound] of alum, and put them all together in a brass pan, and boil them together well. And when it is cold, put in your hair that you want to have yellow, until it is as dark as you want it, and then take it out."

—Recipe 2, MS 171, 15th century, Beinecke Rare Book & Manuscript Library, Yale University (Clarke, 2016. p.331)

Recipe 2: Making walnut-yellow

Collect a handful of black walnut (*Juglans nigra*) leaves in late spring or summer and let them soak in water for several days. Heat the leaves for an hour, then leave to cool. Strain the leaves and then add fabric pre-mordanted with alum to the dye and simmer for 30 minutes. You can make a yellow pigment by adding the alum directly to the dye, reducing by two-thirds, and then adding to chalk.

—Nabil's modern recipe

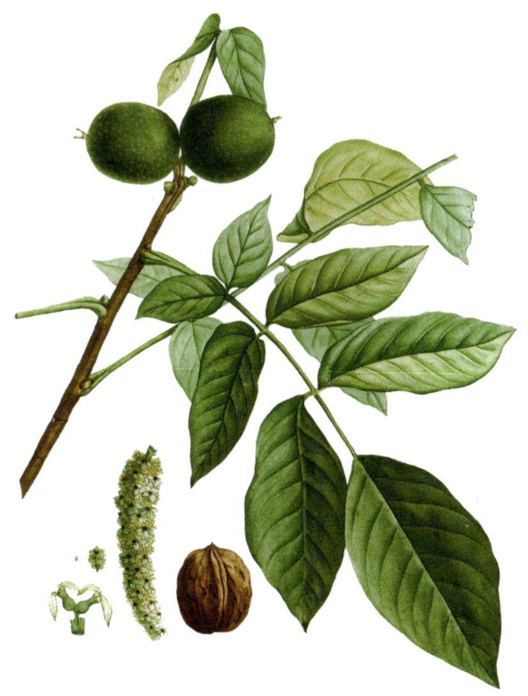

Walnut leaves and fruits (*Juglans regia*). Illustration by Pierre-Joseph Redouté from Henri-Louis Duhamel du Monceau's *Traité des Arbres et Arbustes que l'on Cultive en France en Plaine Terre* (Treatise on Trees and Shrubs Grown in France in Open Ground; 1801–19).

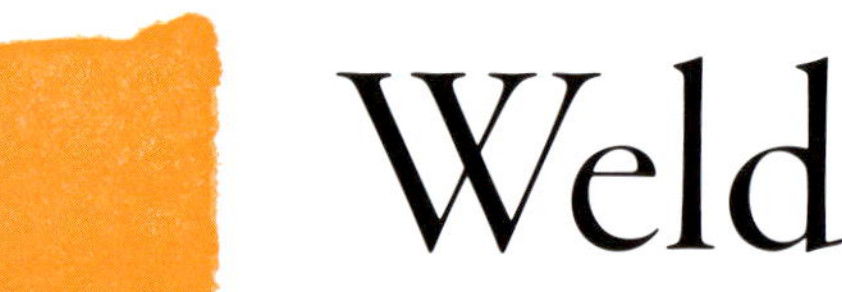

Weld

Latin name: *Reseda luteola*

Other names: Dyer's rocket, dyer's yellow weed, wild mignonette

Family: *Resedaceae*

Genus: *Reseda*

Native: Europe, North Africa, Central Asia, and Pakistan

Type: Biennial

Weld (*Reseda luteola*) is a common weed that has been used as a colorant for thousands of years. The spikes of small, yellow flowers produce thousands of seeds that germinate in disturbed ground and have been found during excavations of ancient Neolithic sites. In the 1st century CE, Pedanius Dioscorides writes about the corn mignonette (*Reseda phyteuma*)—which is in the same family as weld—in his *De Materia Medica*, noting that people would use the root in love potions. He also mentions the medicinal uses of bastard rocket (*Reseda alba*), which was ground up with hellebore and hydromel (a type of mead), mainly to encourage vomiting.

According to Dominique Cardon's *Natural Dyes* (2007), weld dye was found in Nubian textiles dating from the 4th century BCE to the 9th century CE (Nubia was located along the Nile River, mainly in today's Sudan and parts of Egypt). Weld dye was also popular centuries later in the Ottoman Empire for dyeing wool and silk and has been identified in 16th-century Turkish silks and large kilim rugs with floral designs. Although weld was commonly used as a dye for fabrics, artists and painters have successfully used it in paintings and contemporary dyeing.

In New York's Metropolitan Museum of Art, there is a fine example of a late 15th-century woven tapestry made in Paris. *The Unicorn Defends Himself* shows a chaotic hunting scene in a fruitful medieval garden, with flamboyantly dressed men holding spears, collared hunting dogs, wild birds among trees and plants, and a bucking unicorn piercing one of the hounds. The woolen yarns were dyed with the typical triad of dye plants: weld (*Reseda luteola*), madder (*Rubia tinctorum*; see page 100), and woad (*Isatis tinctoria*), which were also used to dye the woolen embroidery threads of the 11th-century Bayeux Tapestry.

In the 17th century, Dutch painter Aelbert Cuyp used weld as a yellow lake pigment mixed with other pigments to give green shades in his oil painting *A Distant View of Dordrecht, with a Milkmaid and Four Cows, and Other Figures*. His contemporary Johannes Vermeer also used weld mixed with indigo to create the deep green background in his painting *Girl with a Pearl Earring* (ca. 1665).

Weld (*Reseda luteola*). Illustration by Christiaan Sepps from Jan Kopp and F.W. van Eeden's *Flora Batava* (Flora of the Netherlands; 1872).

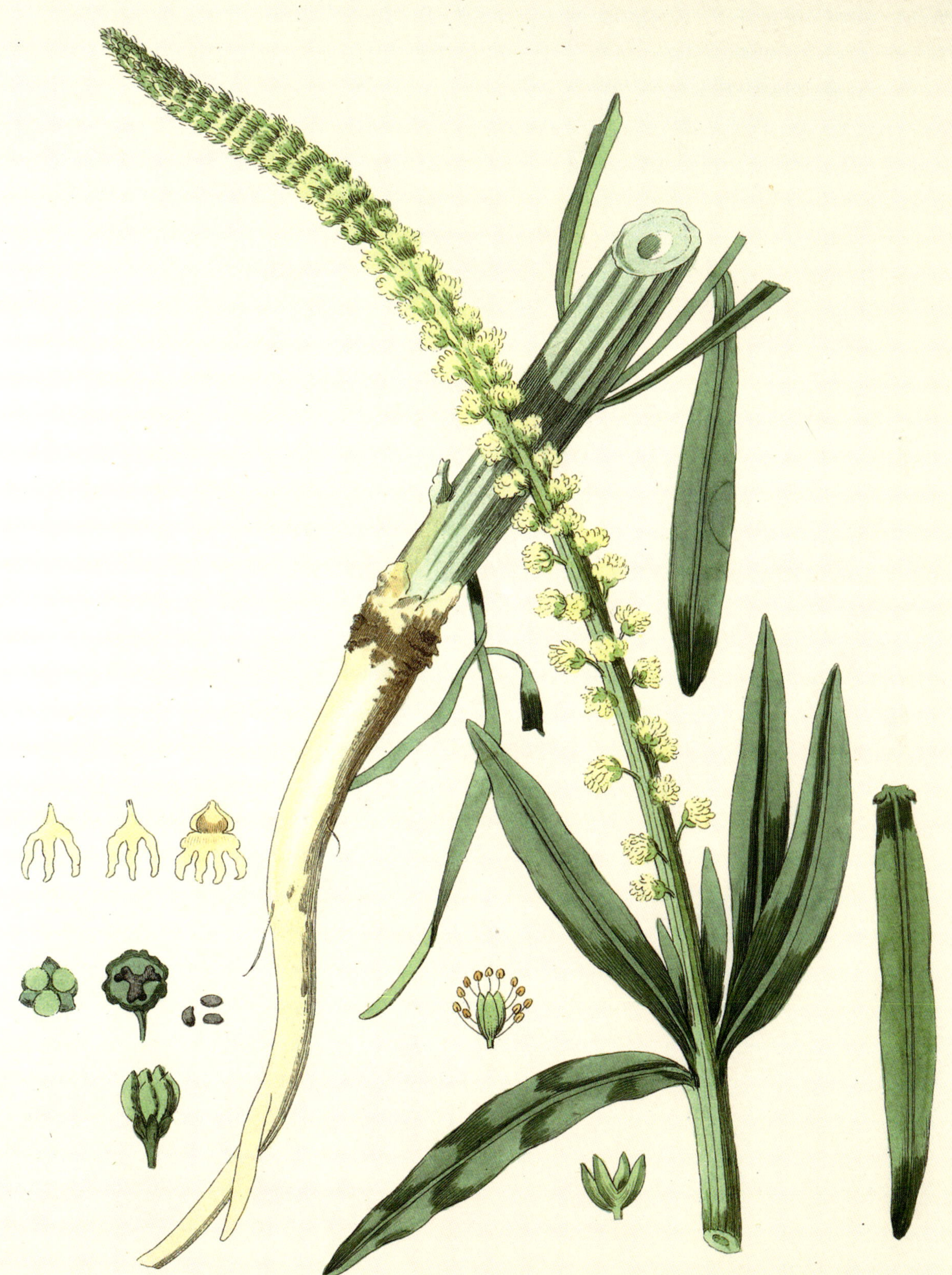

Recipes from Weld

Many recipes describe making yellow dye and other colors from weld. Indeed, the plant has been used as a dye source since prehistoric times and was popular with the Romans for dyeing bridal veils and robes. Vitruvius mentions in his *De Architectura Libri Decem* (Ten Books on Architecture) in the 1st century BCE that weld can be mixed with a blue pigment in fresco painting to create green. In fact, it was common practice to use yellow dye from weld to over-dye blue fabrics dyed with woad or indigo to produce a green fabric, a technique developed in ancient Egypt using woad and safflower (*Carthamus tinctorius*; see page 104). The 14th-century *Liber Diversarum Arcium* (Book of Various Arts) contains several dye recipes and mentions dyeing thread with weld before using a little indigo (from *Indigofera tinctoria*).

An East Anglian manuscript from around 1400 (MS.457/395), in Gonville & Caius College, Cambridge, contains a basic recipe for making yellow: "To make good yellow—take good weld and seethe it well and add alum to it." Similarly, Recipe 11 in a 15th-century manuscript (MS Hunter 110), held in the University of Glasgow Library, describes making yellow from either dyer's broom (*Genista tinctoria*; see page 116) or weld, which, after it is cut down, is placed in water overnight, then heated with alum—using the standard method of heating plants with alum to brighten the color. Alum is the most common mordant used when dyeing with the fresh tops of weld, but using a copper mordant produces the fastest shades of yellow, while a second mordant fixative bath, using just iron, creates beautiful olive-greens.

Several years ago, I noticed some weld growing among discarded bricks on a building site. Asking permission to venture onto the site, I collected the plants. On my way out, I picked up a brick and took both back to my studio. I dried the plants and prepared the brick to create a surface susceptible to color, using the same methods as a late-medieval easel painter: sizing the brick with rabbit-skin glue mixed with chalk collected by hand from chalk quarries in Bedfordshire. After making the yellow weld dye, I painted the brick with the dye for an art piece called *Dismantling the Walls of Differences*.

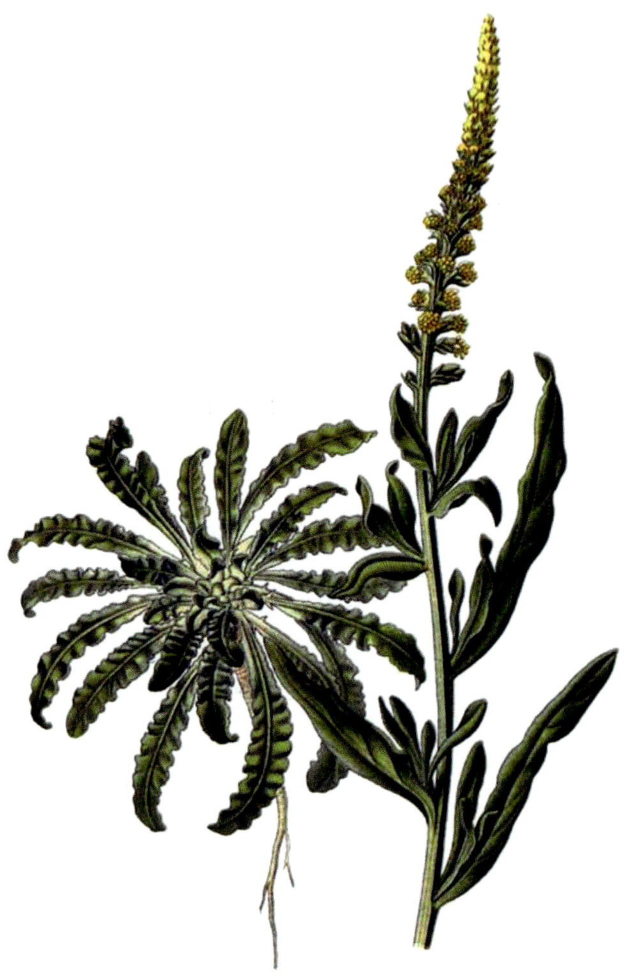

Weld (*Reseda luteola*). Illustration from Amédée Masclef's *Atlas des Plantes de France* (Atlas of Plants of France; 1891).

Recipe 1: Dyeing silk yellow

"On Colors. To dye all kinds of silk, you shall prepare like this before you dye any silk. The silk must first be washed in soap or [washing] soda, and soda is best of all. And after that take alum and gallnuts [oak galls], and seethe them together in water, but let the nuts and the alum be made into powder, and put in the water and let it seethe in the water, and so shall it best take in its color. Take weld and seethe it in there with lye, in an earthenware pot with a little alum, and so shall you make it yellow, with the preparation as said earlier."

—Recipe 1, Adv.MS. 23.7.11, 15th century, National Library of Scotland, Edinburgh (Clarke, 2016, pp.297–98)

Recipe 2: Making weld paint

Gather some wild weld flowers or the entire plant from a spot that has not been visited by dogs. Dry the whole plant in a paper bag hung from a hook for several days or more, so you can collect and resow the seeds. Chop the dried plant into small pieces and store in an airtight container for later use. Heat a handful of the dried weld in water (you can also use fresh flowers) with alum and reduce by half, then sieve through a fine-mesh strainer. Add the yellow dye to chalk, so it fizzes, mix well, and leave to dry. Once dry, mix the yellow pigment with some gum tragacanth to make a yellow paint.

To make other colors: Mix 50 percent madder dye with 50 percent weld dye to make orange, and 20 percent woad with 80 percent weld to make green (add more woad to make darker greens).

—Nabil's modern recipes

Colors from Nabil's Practice

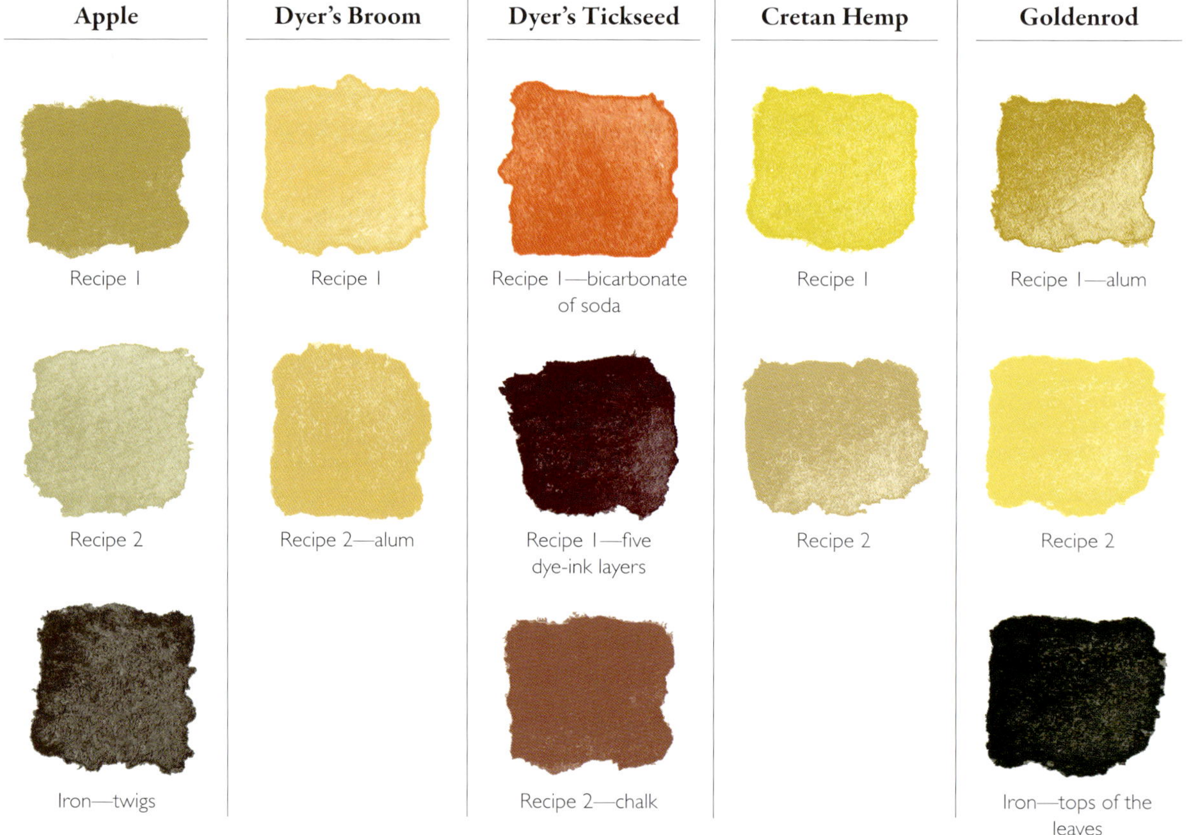

Yellow is the most common color produced from plants, with weld and broom being used since ancient times. These can easily be dried and stored for later use when the plants are out of the growing season. The primary mordant used to brighten the yellow color is alum, which can create a sparkle effect when light shines on dye paint applied to paper. Dyer's tickseed yellow is easy to make, with a very strong color. Cretan hemp, goldenrod, walnut, and apple also produce strong yellows. Growing mountain pansy plants in pots and harvesting the flowers as soon as they bloom will encourage more flowers to appear.

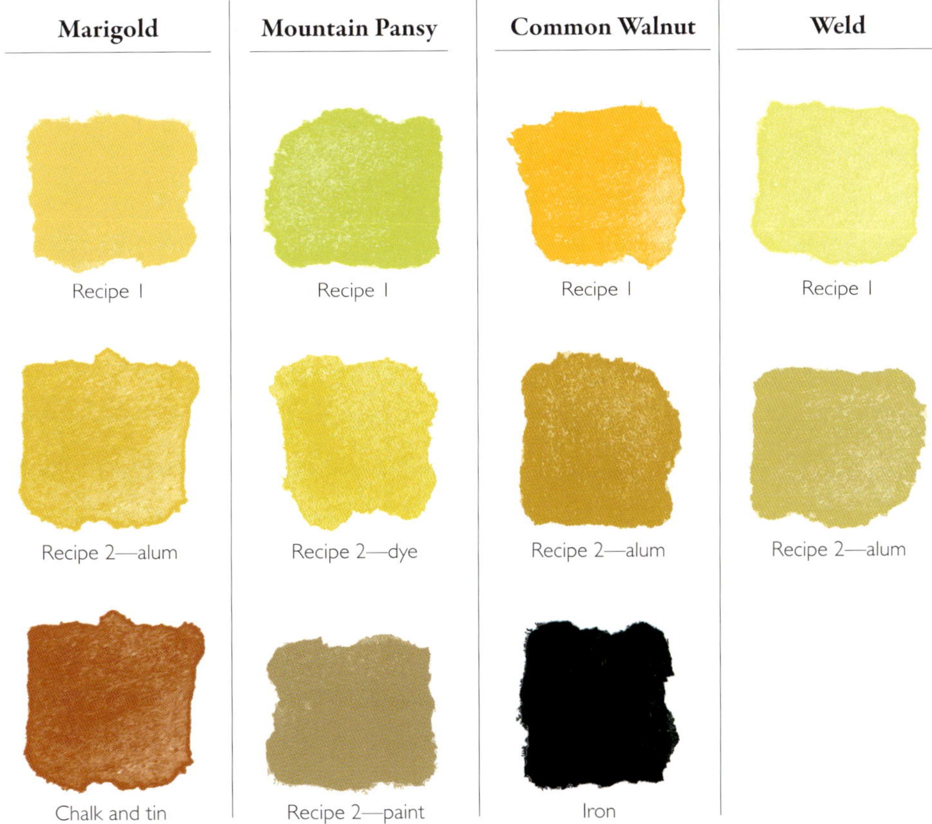

Chapter 5

Green

Common Buckthorn

Latin name: *Rhamnus cathartica*

Other names: European buckthorn, waythorn

Family: *Rhamnaceae*

Genus: *Rhamnus*

Native: Europe to Western Siberia, northwest China, northwest Africa

Type: Deciduous shrub or small tree

The berries of many buckthorn species were used around the world to produce various colors. The following are just a few of over 150 species (most of which are in the *Rhamnus* genus) with color potential: common buckthorn (*Rhamnus cathartica*), which is historically important in Europe; Mediterranean buckthorn (*Rhamnus lycioides*); Avignon or Persian berry (*Rhamnus saxatilis*), also called rock buckthorn or dyer's buckthorn; Italian buckthorn (*Rhamnus alaternus* subsp. *alaternus*); alder buckthorn (*Frangula alnus*); and Carolina buckthorn (*Frangula caroliniana*).

The bark of alder buckthorn was cut and used with other woods to make charcoal in Iron Age Britain, and is used artistically today for drawing and art pigments. A manuscript (Royal MS.7.F VIII) written by the 13th-century monk Roger Bacon, held in the British Library, London, gives formulas for making gunpowder from buckthorn, and in the late medieval period, lime wood (*Tilia* species) and poplar trees (*Populus* species) were used as a wood source.

Common buckthorn was used in ancient Greece and Rome, primarily as a medicine to treat shingles or erysipelas (a type of skin infection), with Pedanius Dioscorides describing it as "boxthorn" in his 1st-century-CE *De Materia Medica*, but it may also have been used as a green and yellow dye. In the 19th century, buckthorn was introduced to North America as an ornamental hedging plant and is reported to have first appeared in a physician's garden in Salem, Massachusetts. It would have been sold by apothecaries mainly for its medicinal properties rather than as a colorant. However, it was common for artists to buy organic materials like pigments and colorants from apothecaries frequented by doctors and herbalists. Therefore, the berries may have been used to make green or yellow dye as well as for medicine.

Several years ago, I grew some alder buckthorn in pots and waited for the berries to use for colorants. Everything was going well, until one morning I noticed the entire plant had been stripped of its leaves by an army of brimstone caterpillars, leaving behind only twigs! However, twigs can be used to produce yellow and brown dyes, so I decided to make charcoal sticks charred from my alder buckthorn for artwork and crushed the charcoal to make a black paint pigment.

Alder buckthorn (*Frangula alnus*). Illustration from *Köhler's Medizinal-Pflanzen* (Köhler's Medicinal Plants; vol. I, 1887).

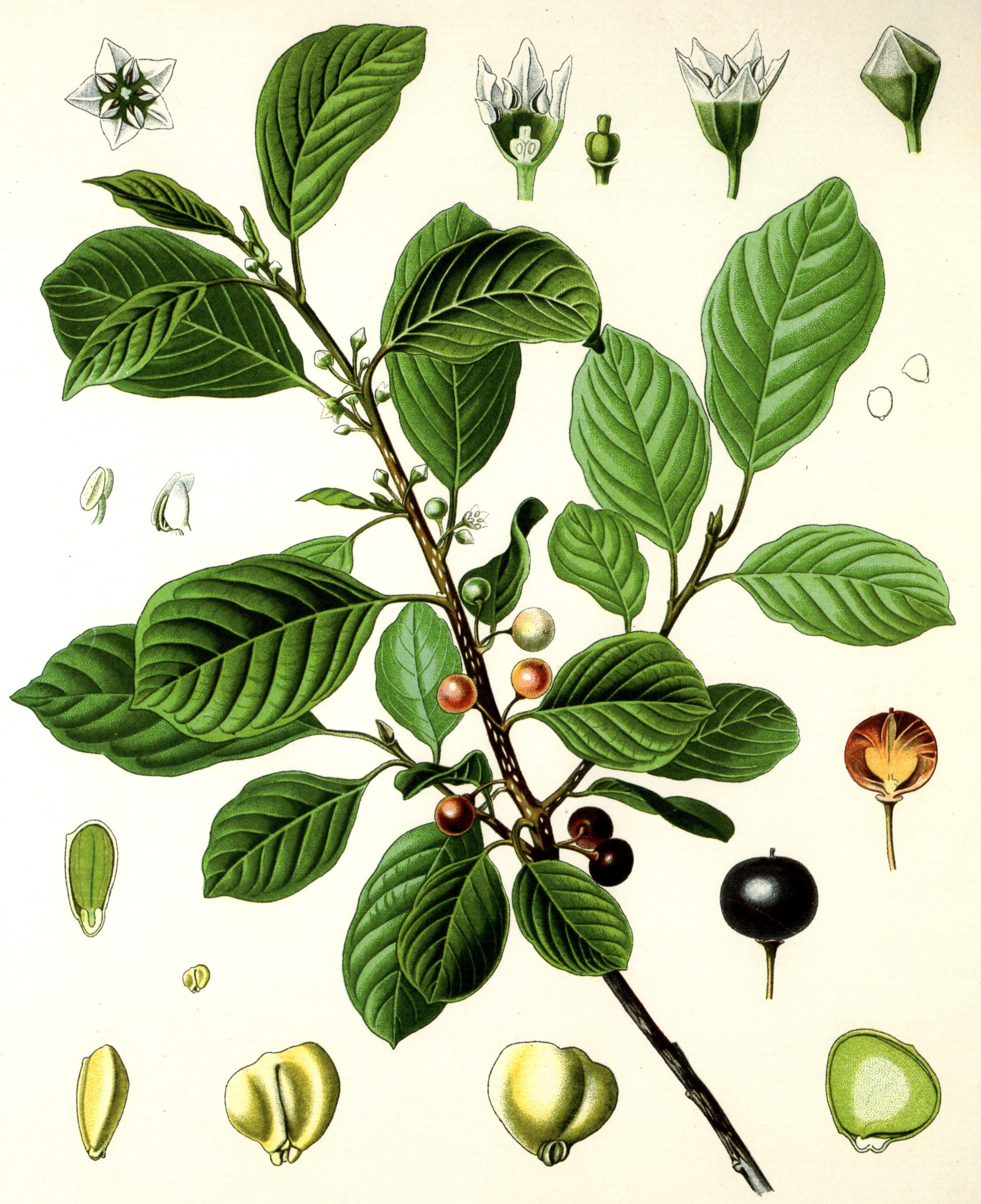

Recipes from Common Buckthorn

De Coloribus et Mixtionibus (Of Colors and Mixtures), a 12th-century Anglo-Norman manuscript in the British Library, London (MS Cotton Titus D. XXIV), includes a recipe for making a green dye with common buckthorn: "Take ripe berries of buckthorn and extract the juice and boil it, and place the yellow cloth in it and it will be green." From the same century, Theophilus Presbyter's *De Diversis Artibus* (On Diverse Arts) describes pigments for painting flesh undertones, including a colorant known as *sucus viridis*, a vegetable green made from ripe common buckthorn berries.

A green called sap green, made from buckthorn, has been found on artifacts by conservation expert Cheryl Porter, founder and director of the Montefiascone Conservation Project. She discovered samples of sap green on a 15th-century manuscript from St. Leonards church in Wollaton, Nottinghamshire, known as the *Antiphonal* (Latin MS 250, ca.1430), which is kept in Nottingham University Library. Analysis of several pieces of 15th-century alum-tawed bookbinding leather at the National Library of Science and Medicine in Copenhagen confirmed that buckthorn berry green was used. Tawing animal hide to make leather involves soaking it in a solution of alum and salt rather than the usual tannins.

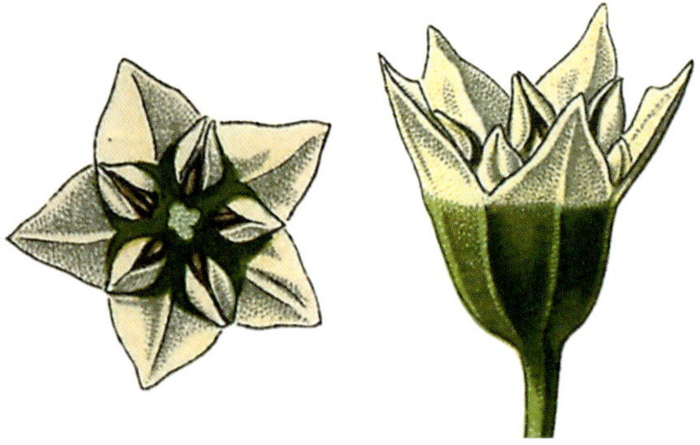

Traditionally, unripe common buckthorn berries were made into a yellow dye, also known as *Giallo Santo*, then mixed with azure (any blue colorant) to create a green dye, as described in a 15th-century manuscript (MS.Latin 6741.ff.43–51) held in the Bibliothéque Nationale de France, Paris. The berries were also documented in the 15th-century *Segreti per Colori* (Secrets of Color), also known as the *Bologna Manuscript*, in the University Library of Bologna (MS 2861). The *Strasbourg Manuscript*, also from the 15th century, provides similar techniques for producing sap green using the ripe berries (see *Recipe 1*).

Many other German recipe books, extensively explored by Sylvie Neven, a specialist in craft technical recipes, mention green made from buckthorn berries. These include two 15th-century manuscripts: the *Colmar Art Book* (150 MS.Helv.Hist.XII 45) in the City and University Library of Bern, Switzerland, and another manuscript (MS.Vad.429.ff.68v–70r) in the Cantonal Library Vadiana St. Gallen.

Contemporary artists such as myself, commercial paint producers, and scientists in art conservation still make sap green from buckthorn berries (see *Recipe 2*). I produced a beautiful green from common buckthorn berries collected in the Cambridge University Botanic Garden and used a basic dyeing recipe (Recipe 4.25.4) from the 14th-century *Liber Diversarum Arcium* (Book of Various Arts). Yet alder buckthorn berries from my own garden are suitable and produced green, purple, yellow, and a good blue colorant.

Detail of **alder buckthorn** (*Frangula alnus*) flowers. Illustration from *Köhler's Medizinal-Pflanzen* (Köhler's Medicinal Plants; vol.1, 1887).

Recipe 1: Making a green color

"If you want again to make another finer [and] greener color, take two parts of good light blue azure and mix less than the third of lead white into it. Pour sap green in it [so it becomes] neither too thick nor too fluid. Apply it for [painting] drapery or for trees, or for grass or on mountains. On that, one should shade with sap green or with Brazil [wood]or with deep rose color."

—Recipe 61, *The Strasbourg Manuscript*, 15th century (Neven, 2016, p.117)

Recipe 2: Making sap-green dye and pigment

To dye a piece of fabric green, collect 20 grams of ripe buckthorn berries (*Rhamnus* species or *Frangula alnus*) in late summer, then crush and leave in water overnight. Heat the water for 45 minutes and put fabric that has already been mordanted in alum or potash in the green dye. The unripe berries with alum as the mordant produce a good yellow, and a yellow pigment when added to chalk.

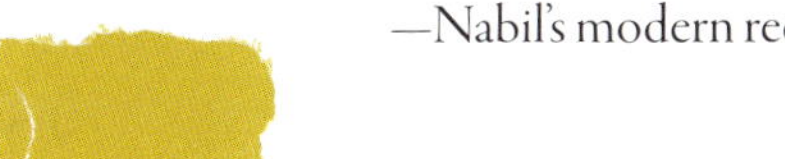

—Nabil's modern recipe

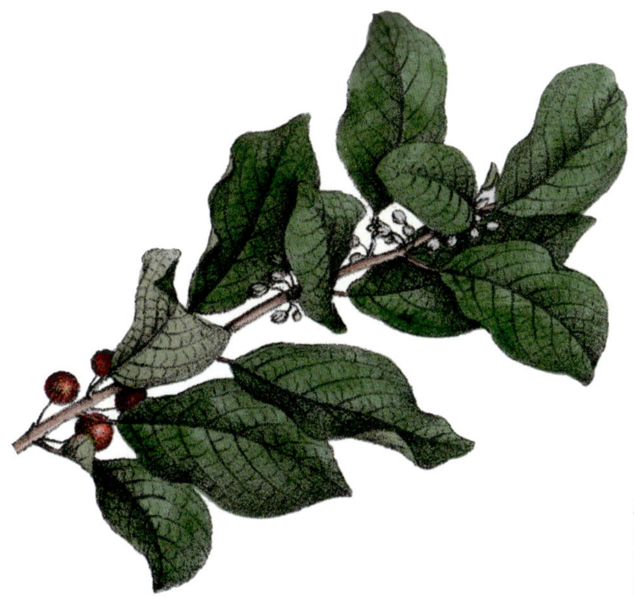

Alder buckthorn (*Frangula alnus*). Illustration from the *Dictionaire des Plantes Suisses* (Dictionary of Swiss Plants; 1853).

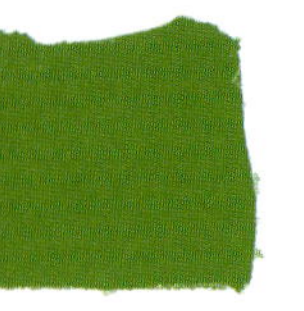

Common Columbine

Latin name: *Aquilegia vulgaris*

Other names: European crowfoot, granny's bonnets, culverwort

Family: *Ranunculaceae*

Genus: *Aquilegia*

Native: Europe

Type: Perennial

Common columbine (*Aquilegia vulgaris*) is associated with the Greek goddess of love, Aphrodite, with the flower spur seen as a phallic symbol, and, in Celtic mythology, with the realm of Tír na nÓg. The plant is slightly toxic, as every part contains cyanogenic glycoside. Despite this, it was recommended for medicines in Tudor herbals such as *The Grete Herball* (1526), printed by Peter Treveris, and *The Vertuose Boke of Distyllacyon of the Waters of All Maner of Herbes* by Hieronymus Brunschwig (1527).

Columbine is also depicted in medieval paintings, including Hieronymus Bosch's triptych *The Garden of Earthly Delights* (painted between 1490 and 1500), in which the central panel shows earthly pleasures such as oversized fruits, plants, birds, fish, and animals, and peculiar shapes morphed together to form eccentric structures. The right-hand panel depicts disturbing images of hell.

Columbine also became important in Christian symbolism. The three-part leaves were a symbol of the Trinity, and seven columbine flowers were seen to represent the Seven Gifts of the Holy Spirit: Wisdom, Intellect, Advice, Strength, Knowledge, Reverence, and Fear. Columbine appears in several 16th-century religious paintings, including *Mary with the Child, Saint Catherine, Saint Barbara and Angels* (ca. 1512) by Augsburg painter Jörg Breu the Elder, which was unfortunately destroyed in 1945. Breu worked alongside the German artist Albrecht Dürer in providing pen and ink marginalia for Emperor Maximilian I of Habsburg's Book of Hours (1514), often known as the Munich–Besançon Prayer Book. Dürer also produced a botanical painting called *The Piece of Turf with the Columbine* (1526), which indicates that the plant was of some importance in Renaissance Europe. Francesco Melzi, a friend of Leonardo da Vinci, painted *Flora* (ca. 1520), which depicts the Roman goddess of flowers delicately holding columbine flowers.

A later oil painting, *A Still Life of Lilies, Roses, Iris, Pansies, Columbine, Love-in-a-Mist, Larkspur and Other Flowers in a Glass Vase on a Table Top, Flanked by a Rose and a Carnation* (ca. 1610) by Flemish artist Clara Peeters, shows the color complexity of a range of flowers. The columbine, along with the other flowers depicted in Peeters' painting, can be made into organic paints.

Common columbine *(Aquilegia vulgaris)*. Illustration from Worthington G. Smith, John N. Fitch, and Walter H. Fitch's *The Floral Magazine* (1861–81).

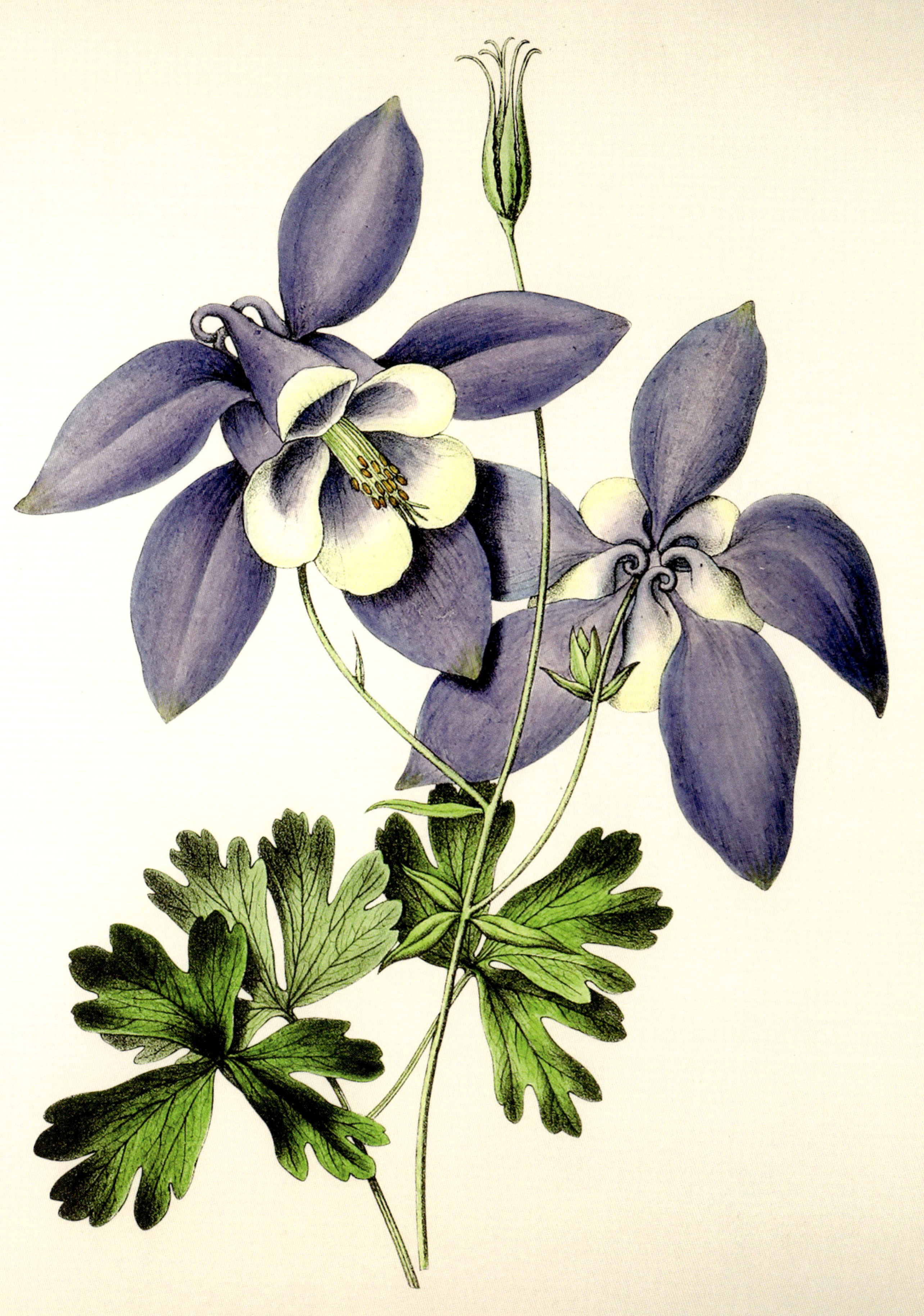

Recipes from Common Columbine

Several years ago, I noticed three or four purple flowers growing wild at the bottom of my garden. With a little research I discovered these were columbine flowers. I already knew of old recipes that use the flowers to make a green paint and decided to experiment. I collected the flowers from my garden to process into a colorant, following a standard method used when making green paint with purple flowers (see *Recipe 1*).

Only a few medieval recipes mention using columbine flowers as a colorant. One recipe in a 15th-century manuscript (MS Latin 6741) copied by Jehan Le Bégue, in the Bibliothéque Nationale de France, Paris, gives instructions for "How to make green without brass"—the verdigris-green crystals that grow on brass, copper, and bronze due to oxidation and in acidic environmental conditions. The recipe describes collecting "aquileia" flowers in late summer; according to Daniel Thompson in his *Materials and Techniques of Medieval Painting* (1936), the flowers are those of *Aquilegia vulgaris*. The petals are crushed with a pestle and mortar to extract the juice and then filtered through a fine linen cloth. The cloth can be dried and used later as a clothlet, a small piece of linen that acts as a portable pigment reservoir, from which the color is released when a little hot water with gum arabic is applied, and which is useful for prepreparing watercolors. The extracted columbine juice is then mixed with a green earth. This might be terre-verte or Verona Green, a marine clay colored by iron silicate and used by the Romans on the walls of Pompeii.

Common columbine (*Aquilegia vulgaris*) (top). Illustration from Georg Bocskay and Joris Hoefnagel's *Mira Calligraphiae Monumenta* (The Model Book of Calligraphy; 1561–96).

Many centuries after the Roman Empire, green earth was also used as an underpaint pigment by Italian masters when painting warm flesh tones, as described by Cennino Cennini in a recipe in his 15th-century *Il Libro dell'Arte* (The Craftsman's Handbook). For this, the columbine juice and green earth are mixed together and then left to bake in the sun until the paint becomes a dry, solid pigment. To use this color, water is first applied to soften the pigment, then egg glaire or gum arabic is added as a binder to formulate the paint. The recipe implies that the paint can be used for coloring wood, such as sculptures, panel paintings, or wooden objects in general; on walls in fresco paintings; or for illuminating artwork on parchment for books.

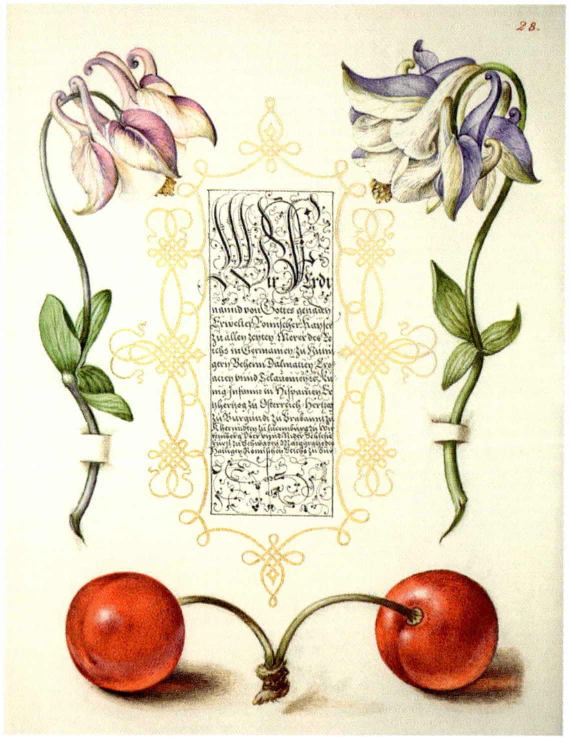

Recipe 1: *To mak a grene for lymnyng* (illuminating)

"Tak columbyn and stampe hem with a lityl alum, and þen wrynge it þrow a cloþe and let it stand til it be drye, and wan þu xalt wyrke þerwith temper it with water."

Translation: "Take columbine and stamp it with a little alum, and then wring it through a cloth, and let it stand until it is dry. And when you want to work with it, temper it with a little water."

— Recipe 10 (a Middle English paint recipe), MS 21.f.272r, 15th century, Pembroke College, Oxford (Clarke, 2016, p.272)

Recipe 2: Making columbine-green paint

Collect 20 purple columbine flowers, then heat the petals in 3 fluid ounces (80 milliliters) of water with 10 grams of alum and reduce the liquid to ⅔ fluid ounce (20 milliliters). Sieve the dye through a fine-mesh strainer to remove the petals and add 2 grams of tin mordant. Mix the dye with 1 or 2 grams of powdered chalk—depending on how light you want the green to be—and let it dry to form a pigment. Mix the pigment with a binder, such as gum arabic, egg glaire, crushed onion juice, or isinglass fish glue, to create a columbine-green paint. The more flowers you use, the deeper the shade will be.

—Nabil's modern recipe

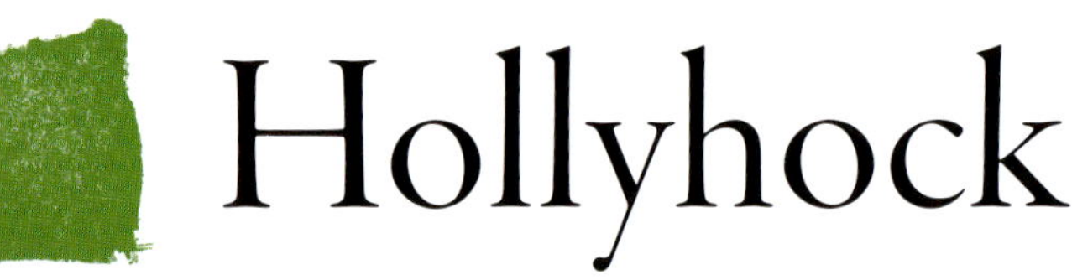

Hollyhock

Latin name: *Alcea rosea*

Other names: Common hollyhock

Family: *Malvaceae*

Genus: *Alcea*

Native: Turkey

Type: Biennial or short-lived perennial

On returning to England from the Eighth Crusade in 1270, Queen Eleanor of Castile, wife of Edward I, brought the common hollyhock (*Alcea rosea*) back from the Holy Land in Palestine. The word *hollyhock* comes from the Anglo-Saxon *holi*, meaning "holy," and *hoc*, a mallow plant.

The term *holyoke* was coined by the 14th-century English physician-herbalist William Turner to describe hollyhocks in *The Names of Herbes* (1548). The plant appears under the heading *Malua*—or jagged mallowe. In the 1881 reprint, botanist James Britten indicated that *holyoke* was, indeed, *Alcea rosea*. By the 16th century, the hollyhock was established in English gardens, with the seeds traveling across the Atlantic in the 1630s to the New England colony, where settler John Winthrop had established a garden on Governors Island in Boston Harbor, Massachusetts.

In the world of art, painting on a Chinese silk fan from the Ming Dynasty (ca. 1368–1644) depicts pink, white, and crimson hollyhock flowers alongside plump cats in the style of the 13th-century painter Luo Zonggui. Another artistic example can be found in New York's Metropolitan Museum of Art in the form of an early 17th-century Japanese piece called *Poem by Ōnakatomi Yoshinobu with Underpainting of Hollyhocks*, with calligraphy by Shōkadō Shōjō. The poem was originally written by the 10th-century poet Ōnakatomi no Yoshinobu. Shōjō commissioned painter Tawaraya Sōtatsu to provide an underpainting for the poem, which features hollyhock flowers. Sōtatsu was part of the Rinpa tradition (School of Korin—so named after painter Ogata Korin), a key art form in the Edo Period (1615–1868). Later still, the 20th-century Amerian artist Georgia O'Keeffe, renowned for her closeup paintings of flowers, produced an oil painting called *Black Hollyhock, Blue Larkspur* (1929). Incidentally, black or deep purple hollyhocks—*Alcea rosea* 'Nigra'—are ideal for processing into a paint (see page 163).

Claude Monet, famously created many paintings inspired by the wealth of colour in his gardens, both at his home in Giverny, Normandy, and elsewhere. *Hollyhocks Growing in the Garden at Argenteuil Admired by Madame Monet* (ca. 1876) depicts tall hollyhocks with other flowers in a large, circular garden.

Hollyhock (*Alcea rosea*). Print from hand-colored illustration by Walter Müller from *Köhler's Medizinal-Pflanzen* (Köhler's Medicinal Plants; vol. I, 1887).

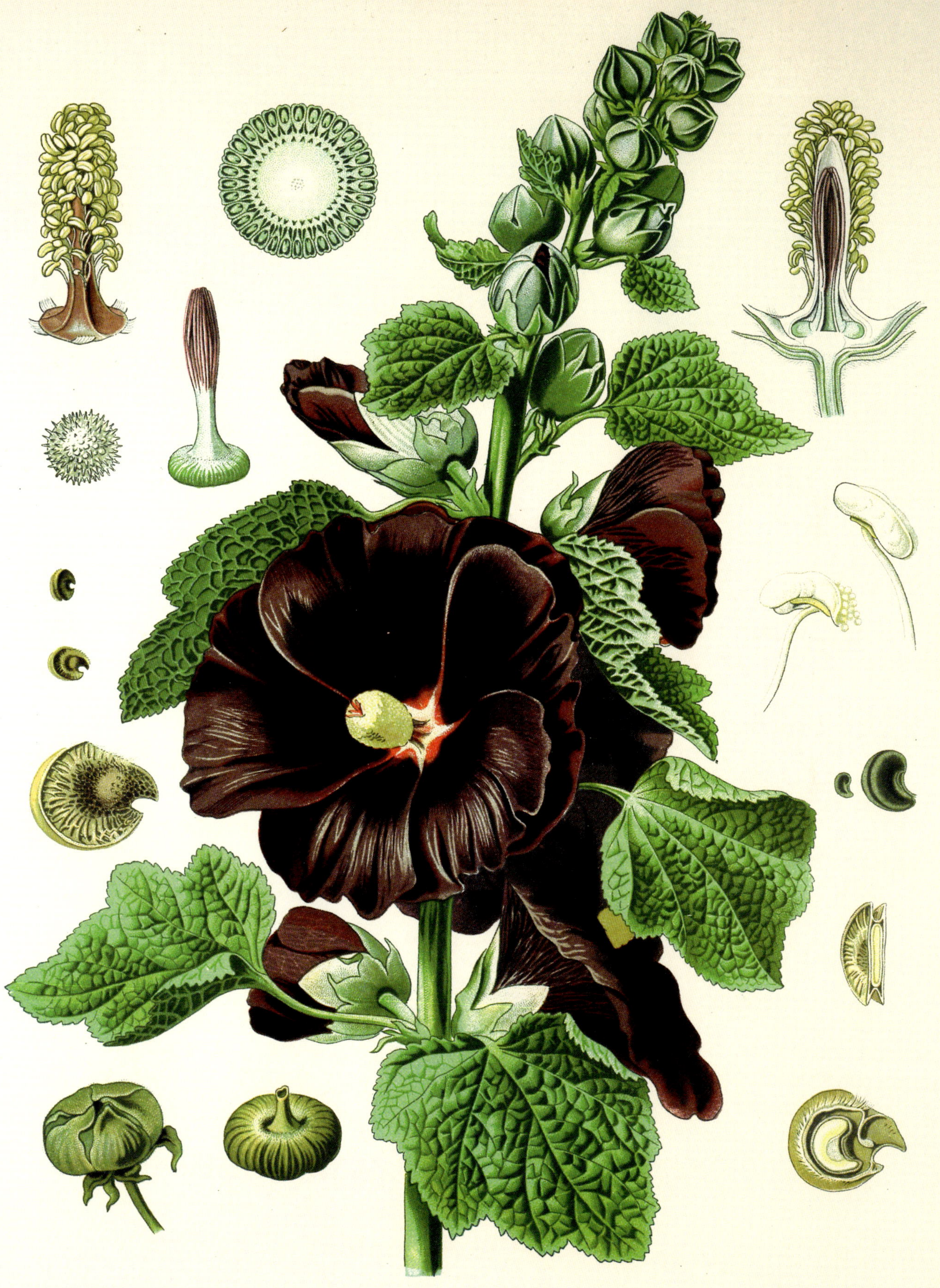

Recipes from Hollyhock

To my knowledge, there are few, if any, medieval recipes that indicate hollyhock was used as a dye or developed into a paint. Yet hollyhocks were grown as medicinal plants, probably among dye plants in monastic and other gardens. With their understanding of organic alchemy and colorants, the Arabs may have known hollyhocks as a dye plant. However, there is currently no evidence to suggest that this was the case.

Planting and growing hollyhocks is relatively easy, and I have cultivated them successfully in my garden for many years. Every time they bloom, I collect as many of the deep purple flower heads as possible and acquire flowers from friends' gardens too. I dry the flowers and then store them in a sealed bag or glass jar for processing at a later date.

For my own experiments, I collected a handful of purple hollyhock flowers (*Alcea rosea* 'Nigra') from my garden to see what colorants could be made. I heated the flowers in water and then reduced the liquid to make a concentrated purple dye. I added a tin mordant to the dye, then mixed this with chalk to make a blue pigment. I also made a deep purple paint by adding the concentrated dye to aluminum hydroxide with a small amount of tin mordant. Adding an iron mordant to the dye produced a nearly black colorant, dulling the shade. I like to adjust the quantities of these ingredients to achieve paints in a range of different shades in a process similar to cooking, with intuition and experience playing a key role, until I am satisfied with the results.

Another method for making paint with purple hollyhocks is to dry the flowers and grind them into a fine powder, first with an electric blender and then with a pestle and mortar—this may take some time, as this is a pure petal pigment rather than a dye. Then mix the petal pigment with gum tragacanth, which is prepared by adding gum powder to water and leaving it to swell into a sticky paste. Mixing the two should produce a blue watercolor paint. The color may vary depending on how many flowers were processed, the time of year they were collected, the location and soil in which the plants are growing, and also the water used—whether it is hard or soft, or rainwater collected from a garden rain barrel.

Hollyhock (*Alcea rosea*) (bottom). Illustration by Persian artist Mirzā Bāqir for an 1889–1890 edition of a 9th-century Persian translation of Pedanius Dioscorides' *De Materia Medica* (1st century CE).

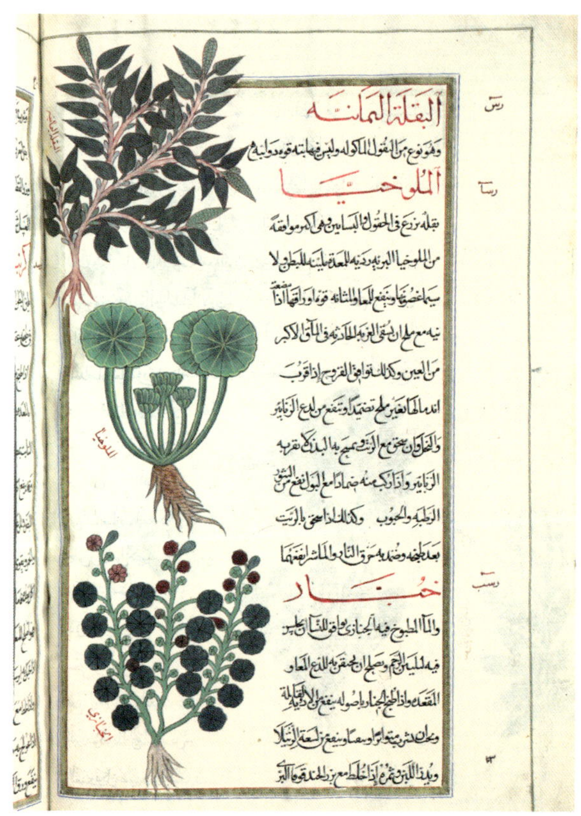

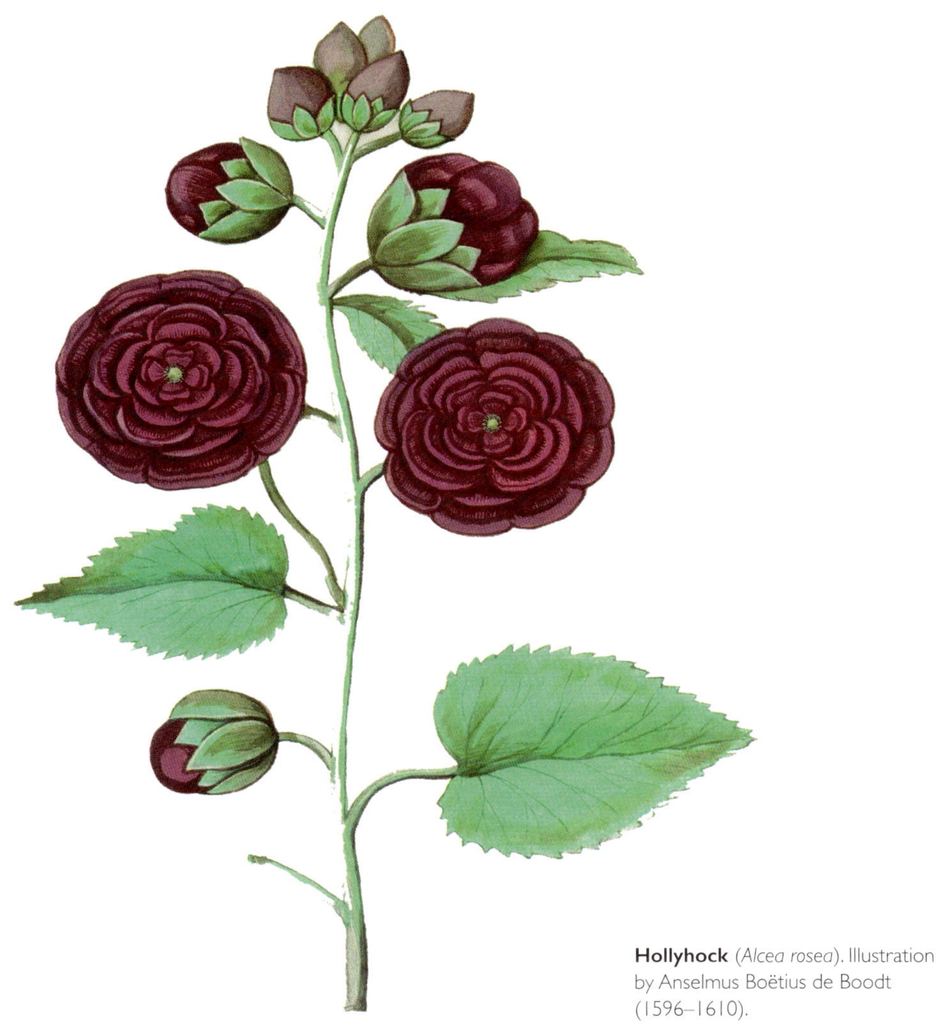

Hollyhock (*Alcea rosea*). Illustration by Anselmus Boëtius de Boodt (1596–1610).

Recipe 1: Making *holyoke*-green paint

Collect 10–20 deep purple hollyhock flowers (*Alcea rosea* 'Nigra') or any other deep purple hollyhocks. Then either dry the flowers for later use or use them fresh. Soak the petals in 3½ fluid ounces (100 milliliters) of water for a day. Heat the water for 30–60 minutes with 10 grams of alum and reduce the liquid to 1¾ fluid ounces (50 milliliters). Add the resulting concentrated dye to powdered eggshells and mix well, then leave to dry in a glass bowl. Once dry, scrape the dry pigment into a small mortar and grind with a pestle until fine. Mix the pigment with gum arabic to make a *holyoke*-green paint.

—Nabil's modern recipe

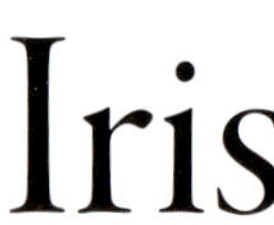

Iris

Latin name: *Iris × germanica*

Other names: Purple iris, bearded iris, German bearded iris, common German flag

Family: *Iridaceae*

Genus: *Iris*

Native: Northwest Balkan Peninsula

Type: Evergreen perennial

The *Iris* genus takes its name from the Greek goddess of the rainbow. This name reflects the wide variety of colors displayed by the many species in the genus and hints at the plant's potential for producing dyes. Some species are known as flag iris, from the Middle English *flagge*, meaning "rush" or "reed." There are some 310 species, making it the largest genus in the *Iridaceae* family, most found in the temperate Northern Hemisphere. Irises are perennials, growing from creeping rhizomes or bulbs, and have long, erect stems with symmetrical, six-lobed flowers.

The iris has a long tradition as a medicinal plant. The Greek philosopher Theophrastus, who lived in the 3rd to 4th century BCE, discusses the iris in his *Historia Plantarum* (Enquiry into Plants). In the 1st century CE, Pliny the Elder describes the "iris perfume of Corinth" in his *Naturalis Historia*, which was made from the dried rhizome, while the physician and botanist Pedanius Dioscorides refers to the flower having medicinal properties in his *De Materia Medica* from the same century. He describes how the iris can be used to treat coughs, induce sleep, and cure colic, spasms, and shivers, due to iridin and other chemicals in the plant.

Parts of Dioscorides' text were copied into a 6th-century herbal known as the *Juliana Anicia Codex* (Codex Vindobonensis med. gr. 1), or *Vienna Dioscurides*, held in Vienna's National Library. The codex includes a beautiful image of *Iris × germanica*, with roots, leaves, and deep purple flower. Several painted purple and yellow irises can be seen on the walls of Pompeii, painted five centuries earlier in the House of Paquius Proculus, where a water deity is shown with the plants.

The white lily, *Lilium candidum*, also painted on the walls of Pompeii, depicts the symbolic fleur-de-lis that we have come to know from the medieval period through to the present day. It is arguable whether this royal symbol used by European nobility signifies the iris flower, lily petals, or stylized acanthus leaves. Medieval artisans depicted the petals to represent the shape of each flower, with the yellow flag iris (*Iris pseudacorus*), for example, appearing on the shields of the 5th-century Frankish king Clovis I.

Purple iris (*Iris × germanica*). Watercolor illustration by Anselmus Boëtius de Boodt (1596–1610).

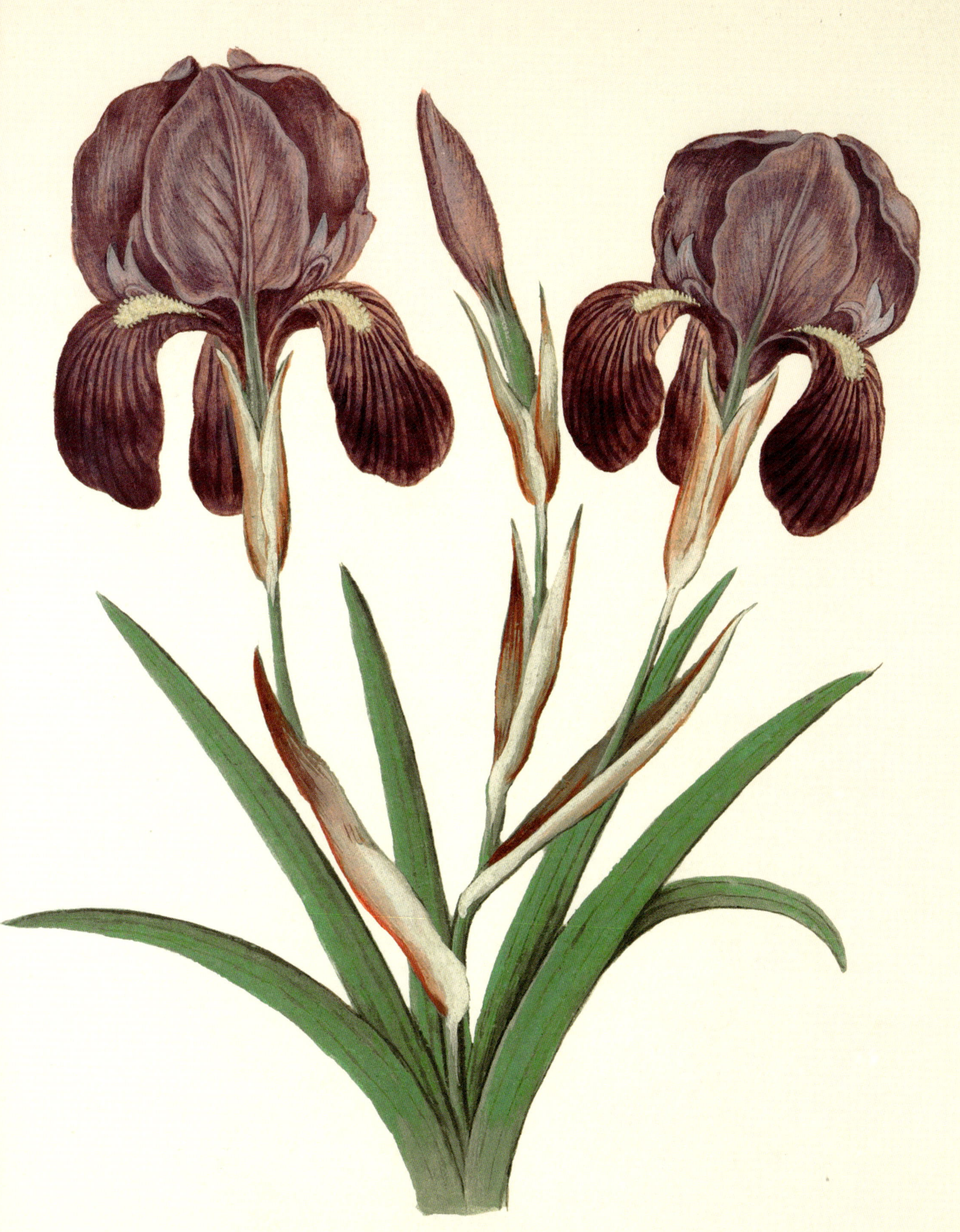

Recipes from Iris

Iris is mentioned in Greek-Latin technical manuscripts dating from the 9th to late 12th century, with some recipes reaching back even further to the 7th century. Iris flowers were cooked in urine along with powdered lily stamens, with the addition of Egyptian alum and lac producing a reddish-purple colorant called "ficarin."

The anonymous 14th-century treatise *De Arte Illuminandi* (On the Art of Illumination), held in the Biblioteca Nazionale Vittorio Emanuele II, Naples (MS Latin XII.E.27), translated from the Latin by Daniel V. Thompson and George H. Hamilton, describes how to make iris green: "[Take] fresh [iris] flowers [*Iris* × *germanica*] in springtime, pound them in a pestle and mortar, then squeeze the juice through a fine mesh, add alum, then soak the juice-dye onto a cloth. Leave the clothlet to dry and keep it between the pages of a book. It is a good green used on parchment." (Interestingly, when I made a dye-paint from *Iris* × *germanica* petals, I found it released a pleasing scent after it had dried in a glass dish.)

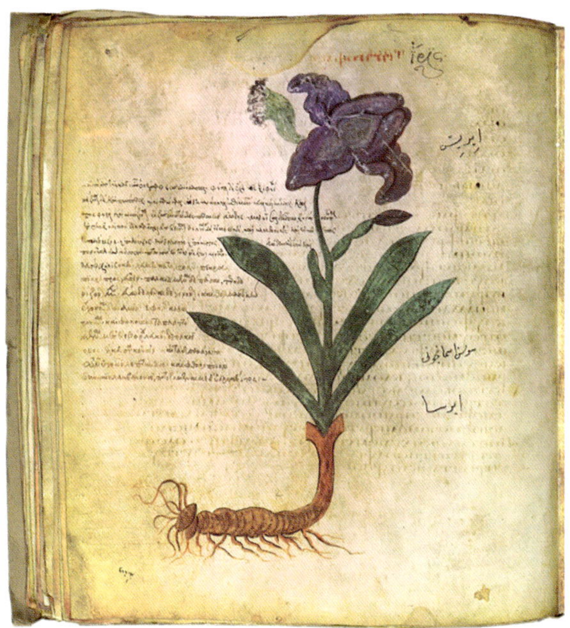

Over the years I have grown a crop of *Iris* × *germanica* in my garden and harvested about eighty plants every season purely for making colors. The dried flowers can be stored for several years if they are kept dry and out of sunlight. One year, I reached out to iris societies around the world and the American Iris Society (AIS) was keen to participate. I asked the Master Iris Judge, Chad Harris of the AIS, who grows a national display of iris, if I could acquire some purple iris flowers from his field crop. When he asked why, I explained that I wanted to use medieval iris recipes to make paint. Chad not only collected the flower heads for me, but also made a mesh rack and dried the flowers. He then posted a big bag of dried purple Japanese water iris (*Iris ensata*) from the West Coast of America across the Atlantic to Essex, England (see *Recipe 2*).

It is interesting to note that a black dye can be produced by heating the fresh or dried roots of yellow flag iris (*Iris pseudacorus*) in water while adding iron sulfate to the vat, making the water turn black. This technique can be repeated in late winter or early spring using the roots of the giant rhubarb (*Gunnera manicata*; see page 196) and the translucent dye from the leaves of the smoke tree (*Cotinus coggygria*; see page 200). Both plants produce a solid black dye or ink when gum arabic—resin from the acacia tree—is added to the solution, which is then heated until it has reduced down to a third.

Purple iris (*Iris × germanica*). Illustration from the herbal manuscript known as the *Juliana Anicia Codex*, or *Vienna Dioscurides* (6th century).

Recipe 1: Making green for illuminating

"To make a good green for illuminating. Take fleur-de-lys [probably *Iris × germanica*] and stamp it with a little alum, and then wring it through a cloth and let it stand until it is dried. And then when you want to work with it, temper it with a little water."

—MS 21 (folio 272r), 15th century, Pembroke College, Oxford

Recipe 2: Making iris-green pigment

Collect some purple iris flowers—the Japanese water iris (*Iris ensata*) or any other purple or deep reddish irises—then dry and store in an airtight container. When you want to make a green color, heat the flowers in water with alum and reduce by half. Sieve the flowers through a fine-mesh strainer, then mix the purple juice with powdered chalk in a small glass dish and allow to dry into a green pigment. To make a paint, mix some gum arabic or egg white into the pigment to suit your needs.

–Nabil's modern recipe

Recipe 3: Making iris blue

Heating freshly picked, purple-colored irises in alum water can produce a blue dye or a blue pigment when the dye is added to chalk. You can also heat a handful of dried *Iris × germanica* or *Iris* 'Sable' flowers in alumed vinegar to turn the liquid purple. Adding a very small amount of chalk—about the size of a garden pea, so as not to lighten the color too much—transforms the concentrated dye into a dark blue color that resembles indigo.

—Nabil's modern recipe

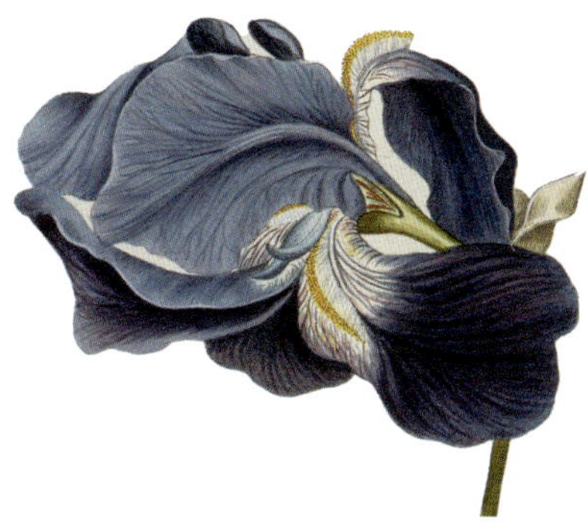

Flower head of the **purple iris** (*Iris × germanica*).

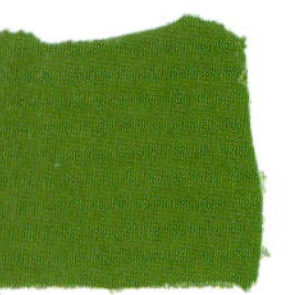

Common Elderberry

Latin name: *Sambucus nigra*	
Other names: Black elder, black elderberry	
Family: *Adoxaceae*	
Genus: *Sambucus*	
Native: Europe to Western Iran, the Azores	
Type: Deciduous shrub or small tree	

The healing properties of the common elderberry (*Sambucus nigra*) were noted by the Greek physician Hippocrates, regarded as the Father of Medicine, who lived around the 5th century BCE. The use of elderberry leaves to treat boils, ulcers, bruising, and chilblains was recorded centuries later by Theophrastus in his *Historia Plantarum* (Enquiry into Plants).

In the 1st century CE, Pedanius Dioscorides suggested using elderberry in his *De Materia Medica* to treat phlegm and bile, with the root boiled in wine and given as part of a diet to help those bitten by vipers. He also mentions that the berries can be applied to the hair as a blackish (purple) hair dye, a claim echoed by Pliny the Elder. The 6th-century CE *Juliana Anicia Codex* (Codex Vindobonensis med. gr. 1), which contains part of Dioscorides' *De Materia Medica* and is held in the Austrian National Library, Vienna, features a botanical drawing of the elderberry plant. The plant is also depicted in the late 6th- to early 7th-century copy of Dioscorides' Greek text written on vellum called the *Naples Dioscurides* (MS. gr. 1), which can be seen in the Biblioteca Nazionale Vittorio Emanuele II, Naples.

The elderberry tree has been rooted in magic and folklore for many centuries. In medieval Europe, Walpurgis Night (April 30) was celebrated before the Christian feast day of St. Walpurga, an 8th-century English nun, on May 1—the same date as the old pagan Beltane fire festival. Walpurgis Night became associated with witches' sabbaths. Elderberry wood was worn as a charm to protect against evil and ghostly powers. In antiquity, the pith from elderberry wood was removed to make flutes and panpipes, played at joyous celebrations where elderberry wine was consumed.

Some years ago, an elderberry grew at my front gate. It attracted insects and produced an abundance of fruit each season. I collected the berries and carefully air-dried them in a fine net bag suspended from the ceiling by fishing wire. The berries took weeks to dry. A little later, a local women cautiously stood by my gate and stated that I had a witch's tree in my garden—"Are you a witch?" she asked. I ground the berries using an electric herb grinder, then sifted the powder to create a very fine pigment for painting.

Common elderberry (*Sambucus nigra*) is associated with witchcraft and wine, and was also used as a herbal remedy. Illustration from *Köhler's Medizinal-Pflanzen* (Köhler's Medicinal Plants; vol. 1, 1887).

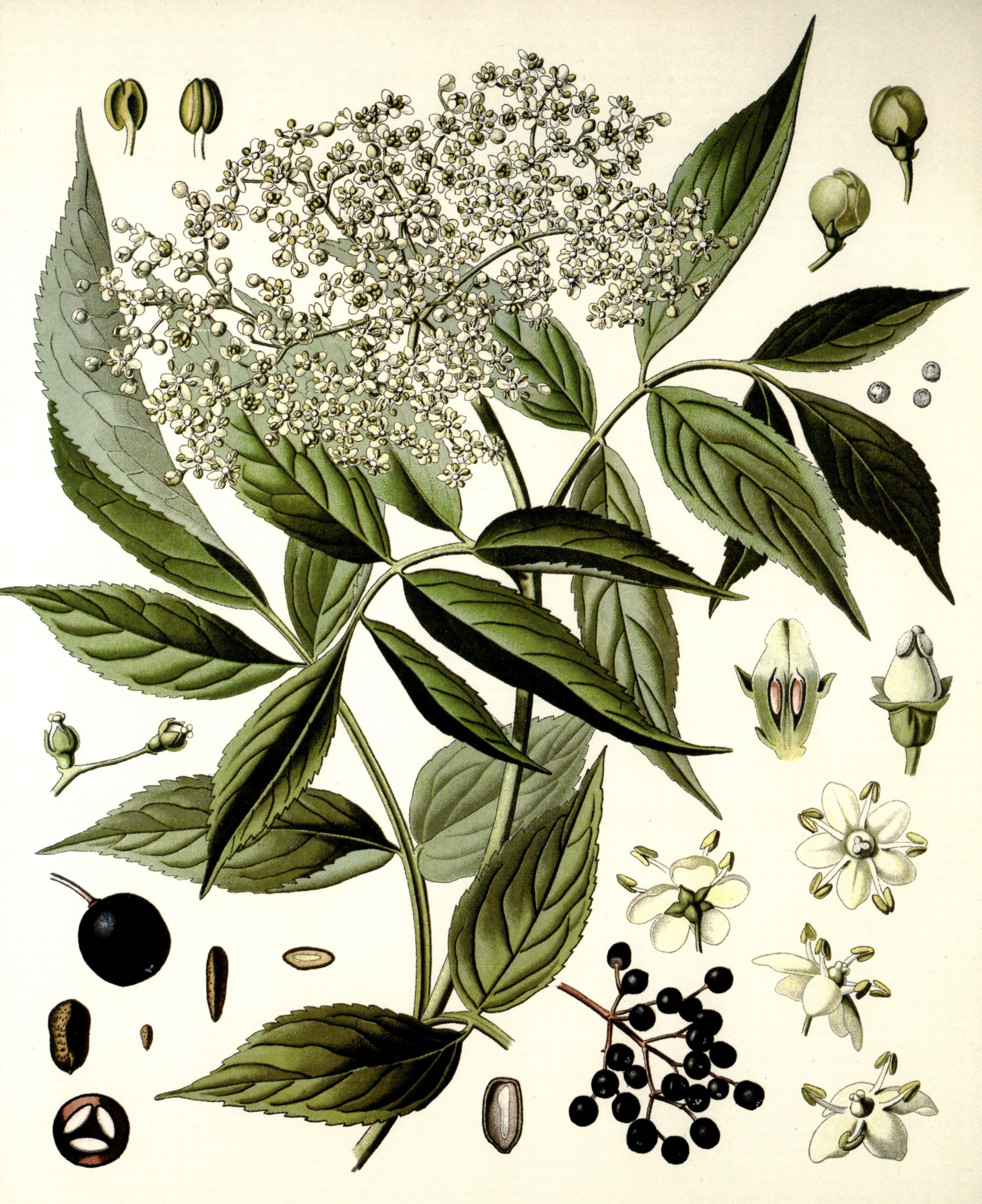

Recipes from Common Elderberry

Many medieval recipes mention elderberry as a colorant, used on its own or as an additive to create different colors. Theophilus Presbyter noted in his 12th-century treatise *De Diversis Artibus* (On Diverse Arts) that elderberries can be used to make green. A stable green paint, good for more than ten years, can be made by mixing elderberry juice, with alum to brighten the color before adding chalk.

A 6th-century-CE Byzantine illuminated manuscript known as the *Codex Purpureus Rossanensis*—or *Rossano Gospels*, housed in Italy's Museum of the Diocese—was made with a range of pigments. Analysis has revealed indigo, red and white lead, gold, silver goethite, and many more pigments, with the vellum pages dyed purple with elderberry. A painting recipe for depicting clothes (Recipe 1.28.27), which features in the 14th-century *Liber Diversarum Arcium* (Book of Various Arts), explains how to mix orpiment or yellow ocher with indigo (*Indigofera tinctoria*) to make green, or with the freshly extracted juice of elderberry to paint purple lines or shadows. This technique for painting purple shadows on green robes had already been introduced by Theophilus in his *De Diversis Artibus*.

A 15th-century manuscript (MS Rawlinson C.506), in the Bodleian Library, Oxford University, gives several recipes using elderberry to produce purple, the midway stage for making green (see *Recipe 1*), and turnsole, another term for blue (see *Recipe 2*, page 57). Medieval turnsole recipes were common, with a late 14th- to early 15th-century recipe in a manuscript (MS Dd.v.76) held in Cambridge University Library explaining that you should crush elderberries, add alum, heat the juice with water, and evaporate the dye by half. Dip strips of cloth in the dye several times, then leave to dry.

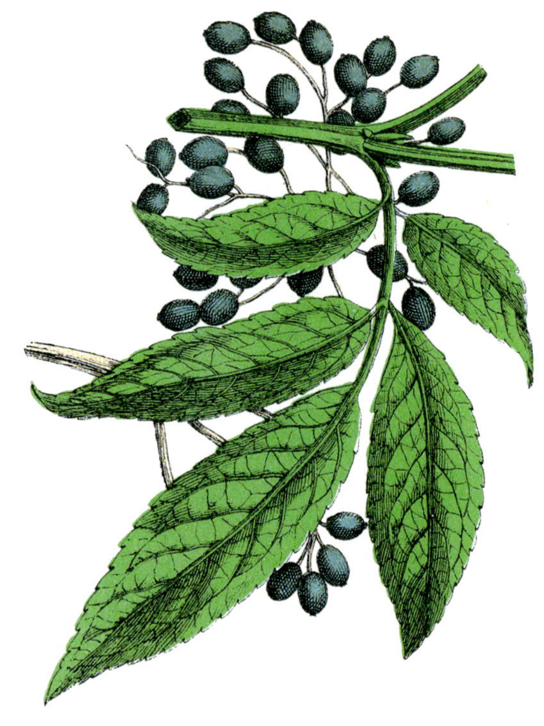

Common elderberry (*Sambucus nigra*). Illustration from Gotthilf Heinrich von Schubert's *Naturgeschichte des Pflanzenreichs* (Natural History of Plants; 1887).

I also made a good blue from elderberry juice. I added 2 grams of the juice to 1 gram of tin mordant (stannous chloride) and mixed them with a pestle and mortar. I then added 2 grams of finely crushed eggshells with 2 milliliters of clear vinegar and mixed until a blue paint formed. Yellow can also be made from the leaves by pouring boiling water over them and leaving overnight. Then, next day, simmer the leaves with alum for an hour to make yellow. Using iron instead of alum produces olive-greens. Black can be made from elderberry bark and roots due to their gallic acid and tannin content, which react with iron sulfate.

Recipe 1: Making elderberry green

Collect ripe elderberries and air-dry them for several weeks (or dry on a low heat in the oven). Once dried, crush the berries to make a fine pigment. To produce a good green, mix the pigment with ½ fluid ounce (16 milliliters) of egg white and ¼ fluid ounce (8 milliliters) of liquid gum arabic, then add this to 1¾ fluid ounces (50 milliliters) of white wine and 5 grams of alum. You can lighten the color by mixing in powdered chalk or gesso.

—Nabil's modern recipe

Common elderberry (*Sambucus nigra*). Illustration by German-Danish botanist Simon Paulli from *Flora Danica* (Flora of Denmark; 1648).

Two varieties of **common elderberry** (*Sambucus nigra*). Illustration by Persian artist Mirzā Bāqir for an 1889–90 edition of a 9th-century Persian translation of Pedanius Dioscorides' *De Materia Medica* (1st century CE).

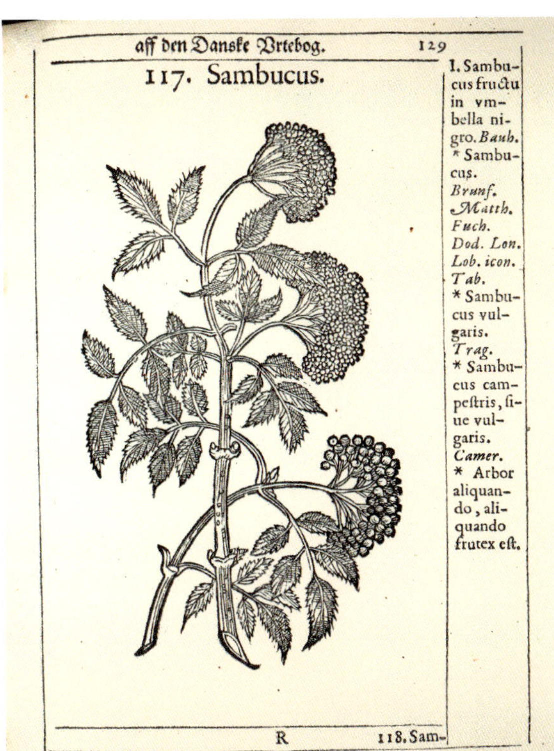

aff den Danske Urtebog. 129

117. Sambucus.

I. Sambucus fructu in umbella nigro. *Bauh.* * Sambucus. *Brunf. Matth. Fuch. Dod. Lon. Lob. icon. Tab.* * Sambucus vulgaris. *Trag.* * Sambucus campeſtris, ſiue vulgaris. *Camer.* * Arbor aliquando, aliquando frutex eſt.

R 118. Sam-

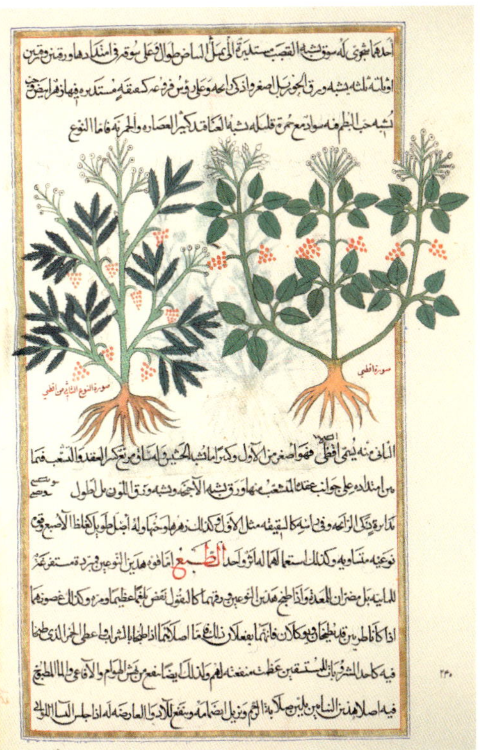

Common Rue

Latin name: *Ruta graveolens*

Other names: Rue, herb of grace, herb of repentance

Family: *Rutaceae*

Genus: *Ruta*

Native: North Balkan Peninsula to the Crimean Peninsula

Type: Evergreen shrub

Common rue (*Ruta graveolens*) is a toxic plant, but it was used in antiquity for various purposes. It was introduced to Britain by the Romans, with Pedanius Dioscorides describing it in his 1st-century-CE *De Materia Medica* as an antidote against reptilian poisons. He also notes that concentrated rue juice can be an irritant that will "redden and puff up the skin with itching and with a violent rash." Similarly, a spell in the 5th-century-CE *Papyri Demoticae Magicae* (Demotic Magical Papyri; PDM. XIV. 563–74) for extracting venom from the heart of a man suggests drinking fresh rue in wine before a meal as a purgative. From the same set of demotic magical texts, the *Papyri Graecae Magicae* (Greek Magical Papyri; PGM CXXIII) describes making magical ink from rue, onion juice, and frankincense for writing spells on tin.

Like many plants in the past, rue also found its way into heraldry, and in around 1181 Frederick I Barbarossa, Duke of Saxony, was given the right to bear a chaplet of rue on his coat of arms. Six centuries later, in about 1807, the Order of the Crown of Rue was established by Fredrick Augustus I of Saxony.

Also known as the herb of grace, rue was generally restricted to monastic infirmary gardens such as the lost 7th-century abbey garden on the Isle of Ely, in Cambridgeshire, where the East Anglian princess Etheldreda collected healing herbs, or the forgotten plant beds within the ruins of the Gothic Rievaulx Abbey in North Yorkshire of the 12th and 13th centuries. Rue was used as a medicinal plant—having sedative, antispasmodic, and emmenagogic (promoting menstruation) properties—to ward off the plague and witches, and to treat fevers. A 12th-century illustrated medieval herbal (Sloane MS 1975), in the British Library, London, shows rue painted in a graphic style, enclosed within a frame, alongside a variety of the ailments it could treat. Rue would also have been grown in the gardens of the monastery at Ourscamps, in Picardy, France, which once owned this manuscript. Rue is, in fact, depicted in many medieval herbals, having being valued since antiquity, and is one of the more common medieval European organic greens used as an additive to strengthen and modify the green color of verdigris.

Common rue (*Ruta graveolens*). Illustration from John Stephenson and James Churchill's *Medical Botany* (1836).

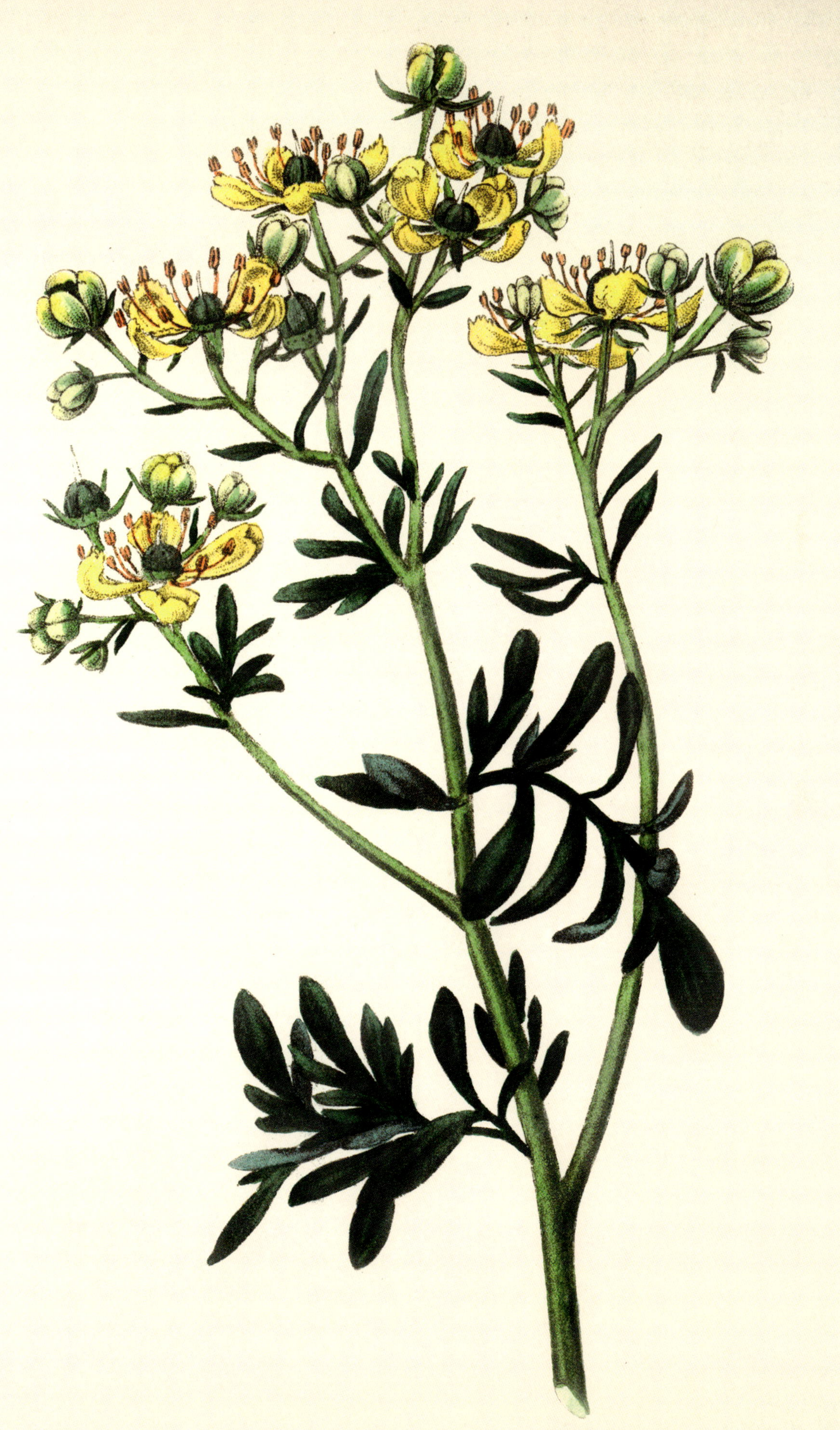

Recipes from Common Rue

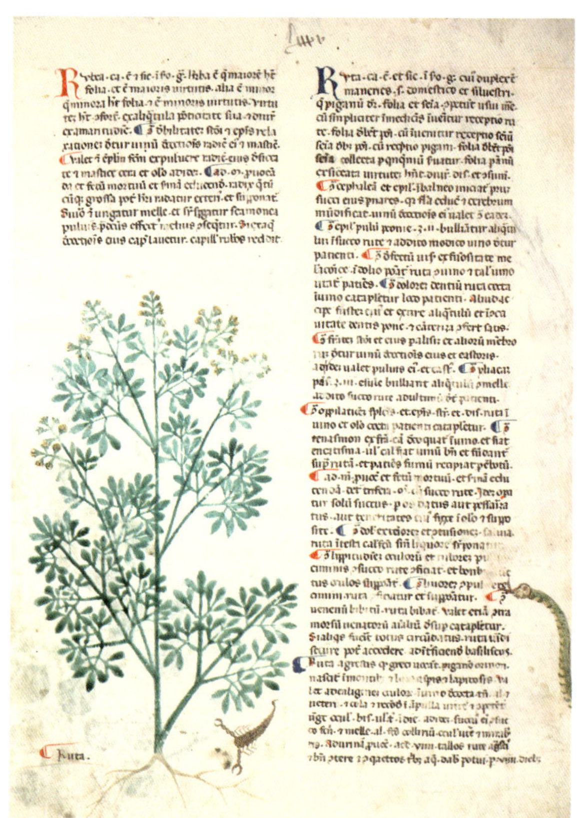

Common rue (*Ruta graveolens*). Illustrated page from *Bartholomaei Mini de Senis Tractatus de Herbis* (Bartholomaeus Mini de Senis's Treatise on Herbs; ca. 1300).

For centuries rue juice has been mixed with verdigris to produce a bottle-green paint. Many illuminated manuscripts used this color, with numerous medieval recipes explaining the process. The first step in making bottle-green paint is to "grow" some verdigris on flat copper strips—copper plant labels work well. Drill a hole in the strip and in the lid of a large glass jar. Thread some fishing wire through the hole to suspend the strip in the jar, so it does not touch the sides. Pour a small amount of clear vinegar into the jar, so the copper is not touching the liquid. Seal the jar and leave the crystals to grow for at least a week, then scrape off the green verdigris. Let the verdigris dry out before grinding it into a fine pigment with a pestle and mortar. Gather some rue leaves with stalks and crush them to extract the juice, then mix with the verdigris pigment. For a richer green paint, you can add a binder like gum arabic, if you wish.

A 16th-century recipe in Middle English for making a "fine green dye"—Recipe A19 in *The Art of Making Colours for to Lymme for Books* (MS Sloane 122), which is housed in the British Library, London—explains that you should "take rue [juice] and the yolk of an egg, both mixed together, [then] blend it with vinegar and dry it quickly as soon as it is ready." The ingredients are easily obtainable, and the process is similar to other rue recipes and produces a satisfying color.

An earlier 15th-century recipe, Recipe 7 in a manuscript (MS 21) in Pembroke College, Oxford, describes making a smooth green color by mixing rue juice with verdigris and then adding saffron dye to create a warmer green than if only rue and verdigris are used. A century later, the *Liber Diversarum Arcium* (Book of Various Arts) includes a section on making green with gum arabic, white wine (vinegar would also work), and rue juice. It also gives other plant juice options, such as cabbage leaves, elderberry leaves, or any other green plant, which would modify the color. In the past, I have extracted raw juice from the leaves of woad (*Isatis tinctoria*), which also produces a good green stain.

In my first herb garden, I grew rue from seed and watched it develop into a mature shrub. Several years later, I began to understand the plant's qualities and uses as a medieval colorant, and I still make a green paint from the leaves.

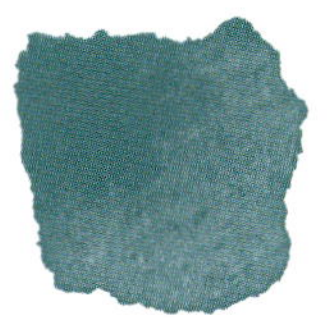

Recipe 1: Making verdigris

"You must prepare verdigris this way: after pouring very sharp vinegar into a cask or into another similar container, cover it with an inverted copper vessel; it is good if the cover were concave, but if not, it may be flat. It should also be clean and tight-fitting. After ten days, removing the cover, scrape off the verdigris that coats it. Or, after making a thin blade of the same copper, hung it over the container so as not to touch the vinegar and, after the same number of days, scrape it."

—Dioscorides, *De Materia Medica*, 1st century CE (Beck, 2011, p.364)

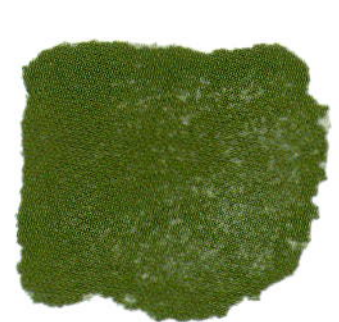

Recipe 2: Making "verruta" green paint

Collect 50 grams of fresh rue leaves and crush them with a pestle and mortar, then place inside a fine-mesh cloth and squeeze the juice from the leaves. Wear protective gloves because the sap contains furanocoumarins, which can blister the skin when exposed to natural UV. Grind 2 teaspoons of verdigris with 1 teaspoon of clear vinegar and add 1 teaspoon of gum arabic and 2 teaspoons of rue juice to form a green paint I call verruta (verdigris and *Ruta*).

—Nabil's modern recipe

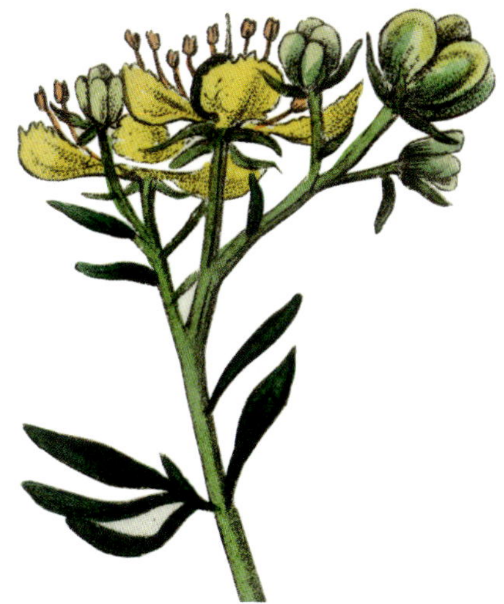

Detail of **common rue** (*Ruta graveolens*) flower. Illustration from John Stephenson and James Churchill's *Medical Botany* (1836).

Colors from Nabil's Practice

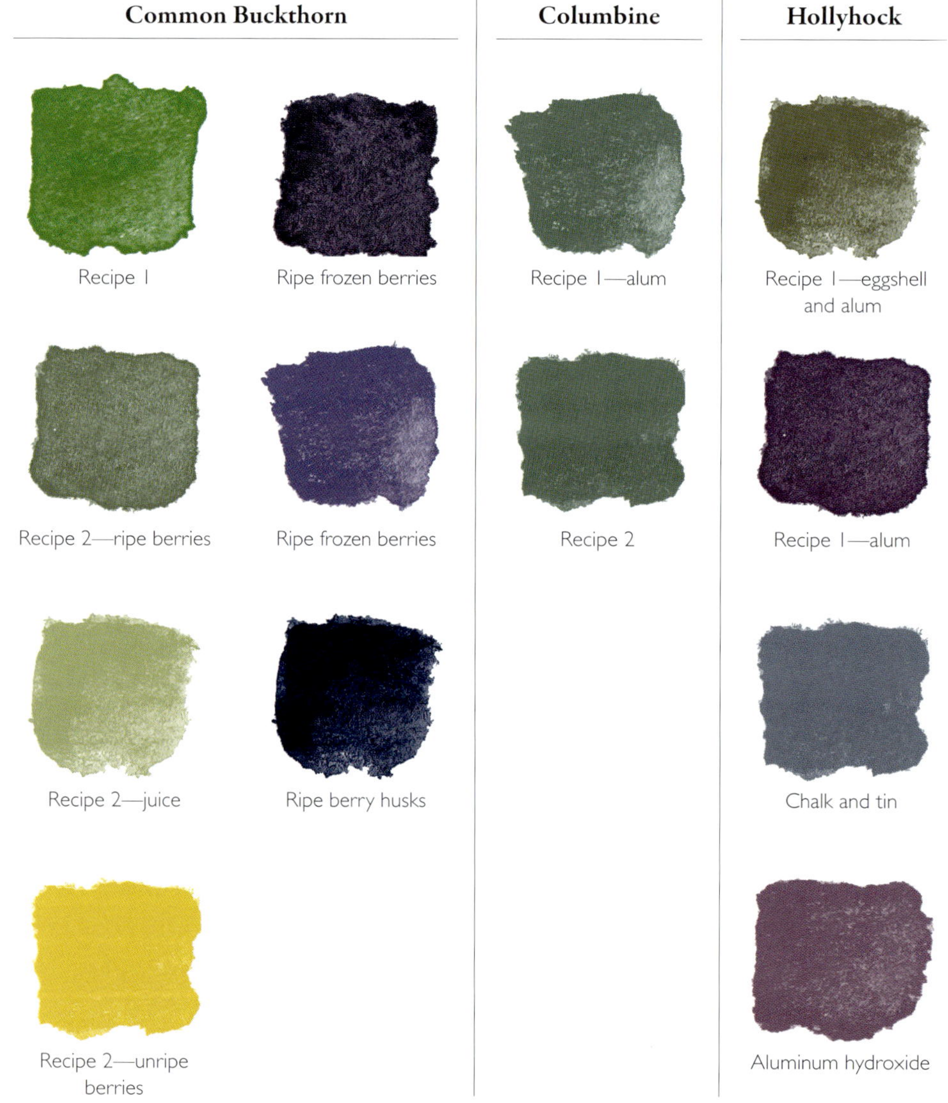

Green colorants are fairly stable and have a long history, with buckthorn and rue being mentioned in medieval technical manuscripts for illuminators. Even though rue juice can be an irritant to the skin, the fresh juice, when mixed with verdigris, needs no binder to stick to a substrate, making it a useful color process. Columbine, hollyhock, iris, and elderberry can be overlooked as green colorants, but when the purple juice is mixed with alum and applied to chalk, stable greens are produced.

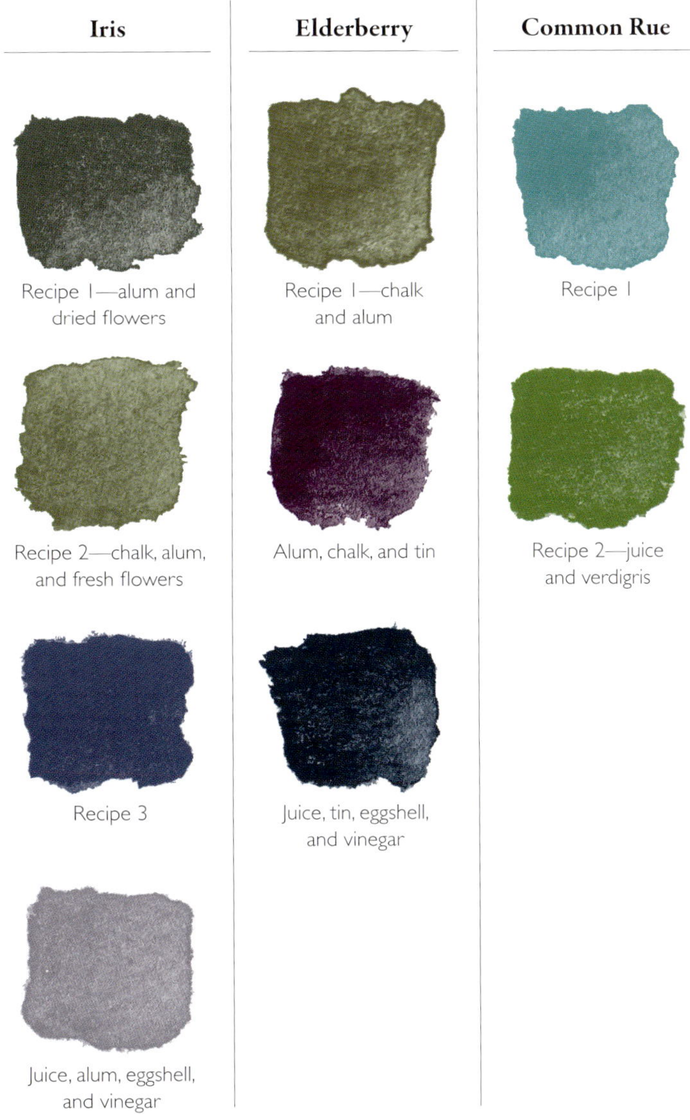

Chapter 6

Black & Gray

Common Hazel

Latin name: *Corylus avellana*

Other names: Common hazel, common filbert, cobnut

Family: *Betulaceae*

Genus: *Corylus*

Native: Europe to Western Asia

Type: Deciduous tree or small tree

The *Shijing*, also known as the *Book of Odes*, was written during the Chinese Zhou Dynasty (1046–256 BCE) and records the use of hazelnuts as food and offerings in ancestral worship. Archaeological evidence suggests that hazelnuts were cultivated and charred in shallow pits on the Scottish island of Colonsay as early as 6000 BCE. In the 3rd century BCE, cultivated common hazel (*Corylus avellana*) was documented by Theophrastus in his *Historia Plantarum* (Enquiry into Plants), where he describes the tree's catkins and mountainous habitat. It is also one of the species noted by Pedanius Dioscorides in *De Materia Medica* in the 1st century CE and by Pliny the Elder as a measurement equating to "the size of a hazelnut." Similar phrases appear many centuries later, including the "size of a chick pea" in the 11th-century Arabic *Book of the Staff of the Scribe* by al-Muʿizz ibn Bādīs and the "greatness of a bean" in a 14th-century manuscript (MS Bodley 177) held in the Bodleian Library, University of Oxford.

Hazel is embedded in wichcraft, folktales, and superstition. In Celtic Irish lore it is associated with the "tree of knowledge" and the great salmon that fed on nine hazelnuts that fell into the water—the "nine hazels of poetic art." When hero Fionn mac Cumhaill accidentally tasted the salmon, he gained legendary knowledge.

The hazel was also honored by King Mac Cuill or Mac Cool (Son of Hazel) of the Tuatha Dé Danann, a supernatural group said to be descended from a Bronze Age people called the Pelasgians, from which the name Hermes (the Egyptian Thoth or Hermes-Trismegistus) is said to be derived, according to 5th-century-BCE Greek historian Herodotus. One meaning behind the winged wand of Hermes with two serpents entwined—known as the caduceus—is the dualism of healing, with the hazel rod symbolizing communication, reconciliation, and commerce.

Herodotus also mentions divination rods in his writings, noting that they can be used to foretell the future. The art of divination with forked rods, or witch sticks, made from hazel and other trees like rowan or willow, is known as dowsing. It has been practiced for centuries and is still used by some people today. Dowsing was very much a practical tool, especially for locating metal ore in 16th-century European mines, where belief in underground spirits was prevalent.

Common hazel (*Corylus avellana*). Illustration by Pierre-Joseph Redouté from Henri-Louis Duhamel du Monceau's *Traité des Arbres et Arbustes que l'on Cultive en France en Pleine Terre* (Treatise on Trees and Shrubs Grown in France in Open Ground; 1801–19).

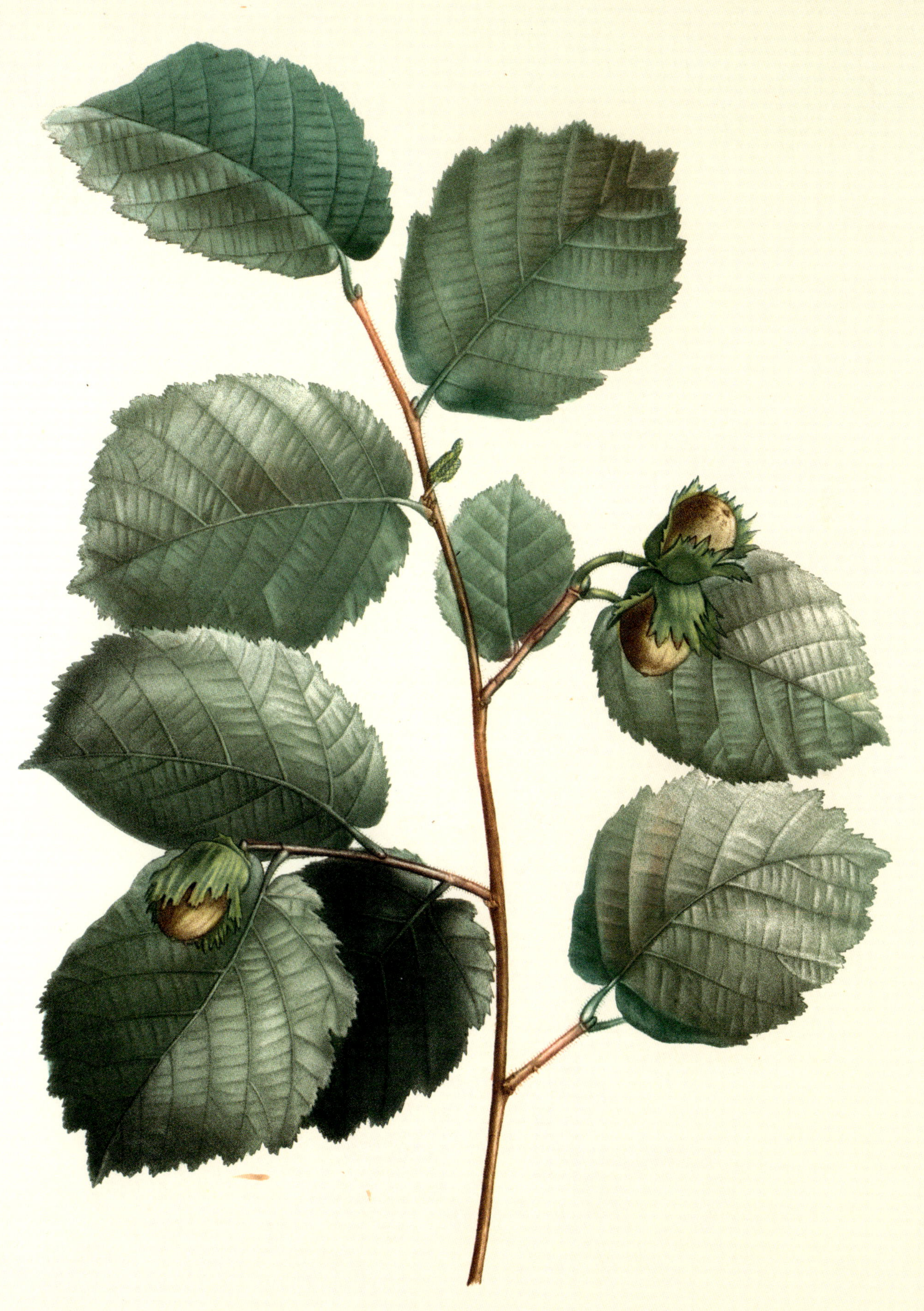

Recipes from Common Hazel

Dioscorides describes how to make a medicinal ointment from hazel husks. To do this, burn the empty husks in a small tin box with holes in the lid. Place the box on a fire for an hour until the husks have charred, then crush with a pestle and mortar. The resulting carbon black pigment can be mixed with oil to make the ointment. Dioscorides' method can also be used to make a black carbon oil paint (see *Recipe 1*), while the pigment can be developed into a black ink by adding gum arabic.

Although most *Corylus* species are native to Europe and Asia, *Corylus americana* originates from the United States where the husks and inner bark were used by many Indigenous American peoples to make a dye. For example, the Chippewa used hazel husks and the inner bark of the white walnut (*Juglans cinerea*) to make a black dye, which was used to stain rushes. The hazelnut and white walnut both contain tannic acid and adding a small amount of iron mordant helps the oxidation process when making a black dye.

British Columbian tribes produced a blue dye from the roots of *Corylus cornuta* subsp. *californica*, which can be combined with other *Corylus* species to produce a deeper blue-black. Black earth (charred wood, soot, or crushed and burned animal bones), the American black alder (*Alnus incana*), red osier dogwood (*Cornus sericea*), and bur oak (*Quercus macrocarpa*) can be added separately with hazelnut and the inner bark and root of the white walnut to make a black dye (see *Recipe 2*).

Other colors such as yellow can be made from fresh *Corylus* leaves, which can be stored in the freezer for later use. For my own experiments, I collected *Corylus avellana* leaves in late summer and early fall, which I found near Cory Lodge in Cambridge University Botanic Garden, and used them to make a warm yellow dye. Dyeing fabric pre-mordanted with alum results in a bright yellow; a tin mordant makes a brighter yellow; and an iron mordant produces a brown color. The fresh leaves can be soaked for up to a week in water before processing. This can make the dye bath acidic, so add baking soda to raise the pH level to neutral before heating the leaves for 30 minutes or up to two hours. Over-dyeing the yellow with blue from *Corylus avellana* roots will make a good green.

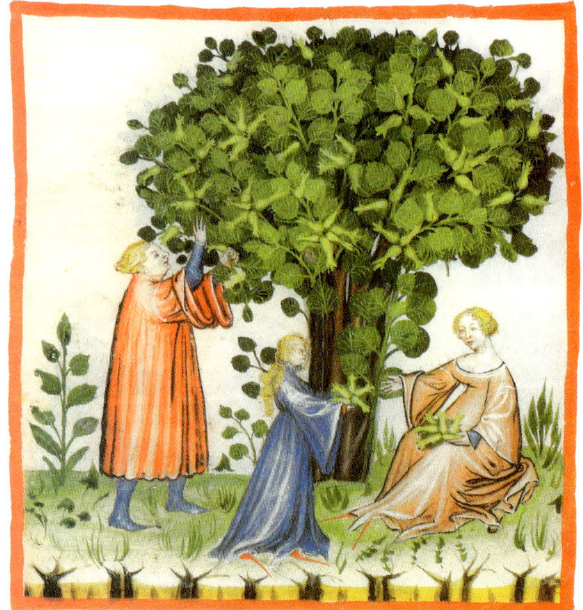

Common hazel (*Corylus avellana*). Illustration from the *Tacuinum-Sanitatis-Manuskript* (14th century).

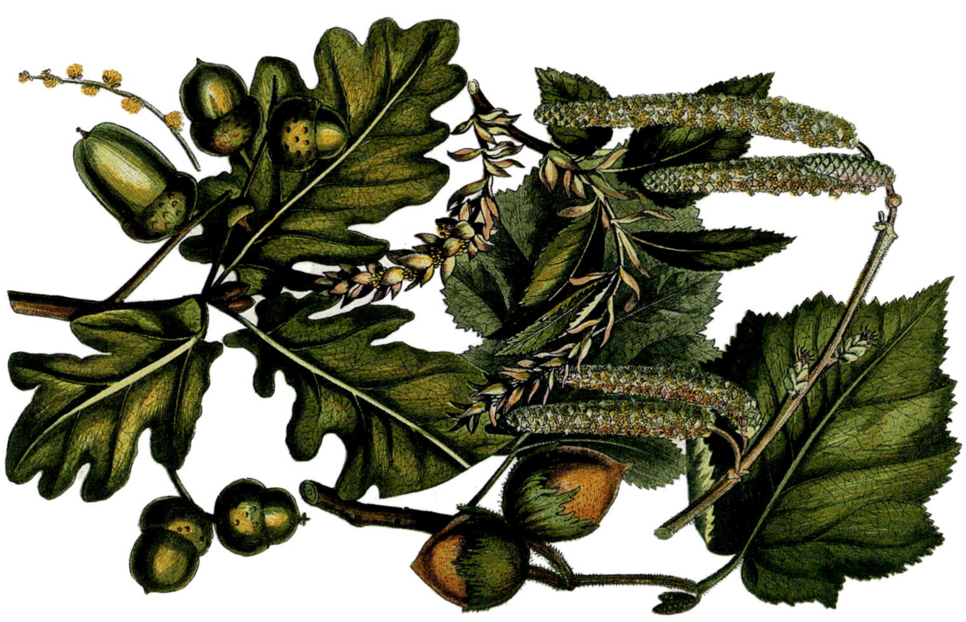

Fruiting and flowering twigs of **common hazel** (*Corylus avellana*), English oak (*Quercus robur*), and common hornbeam (*Carpinus betulus*). Chromolithograph by W. Dickes & Co. (ca. 1855).

Recipe 1: Making *Corylus*-black paint

Eat as many hazelnuts as possible in a day and place the empty shells inside a small tin box with holes in the lid. Place the box on a fire or over hot coals and leave the shells to char for an hour. Remove the box from the heat, place the charred shells in a mortar, and crush with a pestle into a very fine pigment. Mix the pigment with flaxseed (linseed) or walnut oil until you have a buttery black oil paint.

—Nabil's modern recipe

Recipe 2: Making American black dye

To make a black dye, boil some *Corylus americana* husks with white walnut (*Juglans cinerea*) and add a little iron mordant.

—Nabil's modern recipe

Black Mulberry

Latin name: *Morus nigra*

Other names: Common mulberry, Persian mulberry

Family: *Moraceae*

Genus: *Morus*

Native: Iran

Type: Deciduous tree

The mulberry tree has many applications. It can be used to make colorants and dyes, to make paint for creative artistry and inks for writing, to promote general health, and in cookery (both white and black fruits can be used to make mulberry pie). Its leaves are still fed to silkworms in silk production, with the preferred species being the white mulberry (*Morus nigra*). This species was cultivated commercially to feed the larvae of the domestic silk moth (*Bombyx mori*), which is native to China. The moth has been farmed for thousands of years for its silk, as a result of which the insects have underdeveloped flying ability and perception of environmental odors, and depend on humans for survival. The ancient Chinese aristocracy guarded the secrets of domestic silk production, and illegally exporting silkworm cocoons with the pupa inside and white mulberry seeds was punishable by death.

In the 1st century CE, Dioscorides mentions in his *De Materia Medica* that black mulberry (*Morus nigra*) was well known for health benefits relating to the stomach. He also notes that mixing the sap from the leaves with honey can relieve inflamed tonsils and mouth ulcers. The sap could also treat bites from venomous spiders and burns. A decoction of the bark and leaves was used as a mouthwash and for toothache. When added to grapevine and the leaves of black fig (*Ficus carica*), which both contain tannins that oxidize to black, the bark could be used as a hair dye.

The black mulberry was introduced to Britain by the Romans, and seeds have been found mineralized in an early Roman town settlement near the River Thames at Silchester, in Hampshire, dating to around 50–70 CE. Over the centuries, the tree became established across Europe in medieval monastery and abbey gardens and was introduced to North Africa and Spain by the Arabs and Moors.

The mulberry has inspired various artists over the centuries. Vincent van Gogh painted *The Mulberry Tree* (1889), now in the Norton Simon Museum, Pasadena, California, while at an asylum in Saint-Rémy-de-Provence. The painting shows flaming orange and yellow fall leaves that explode against a vivid blue sky and imply death on the horizon with the approach of winter. The mulberry also features in Ovid's tragic story of Pyramus and Thisbe who die beneath a mulberry tree.

Black mulberry (*Morus nigra*). Copperplate engraving by Elizabeth Blackwell from Joseph Miller's *A Curious Herbal* (vol.1, 1751).

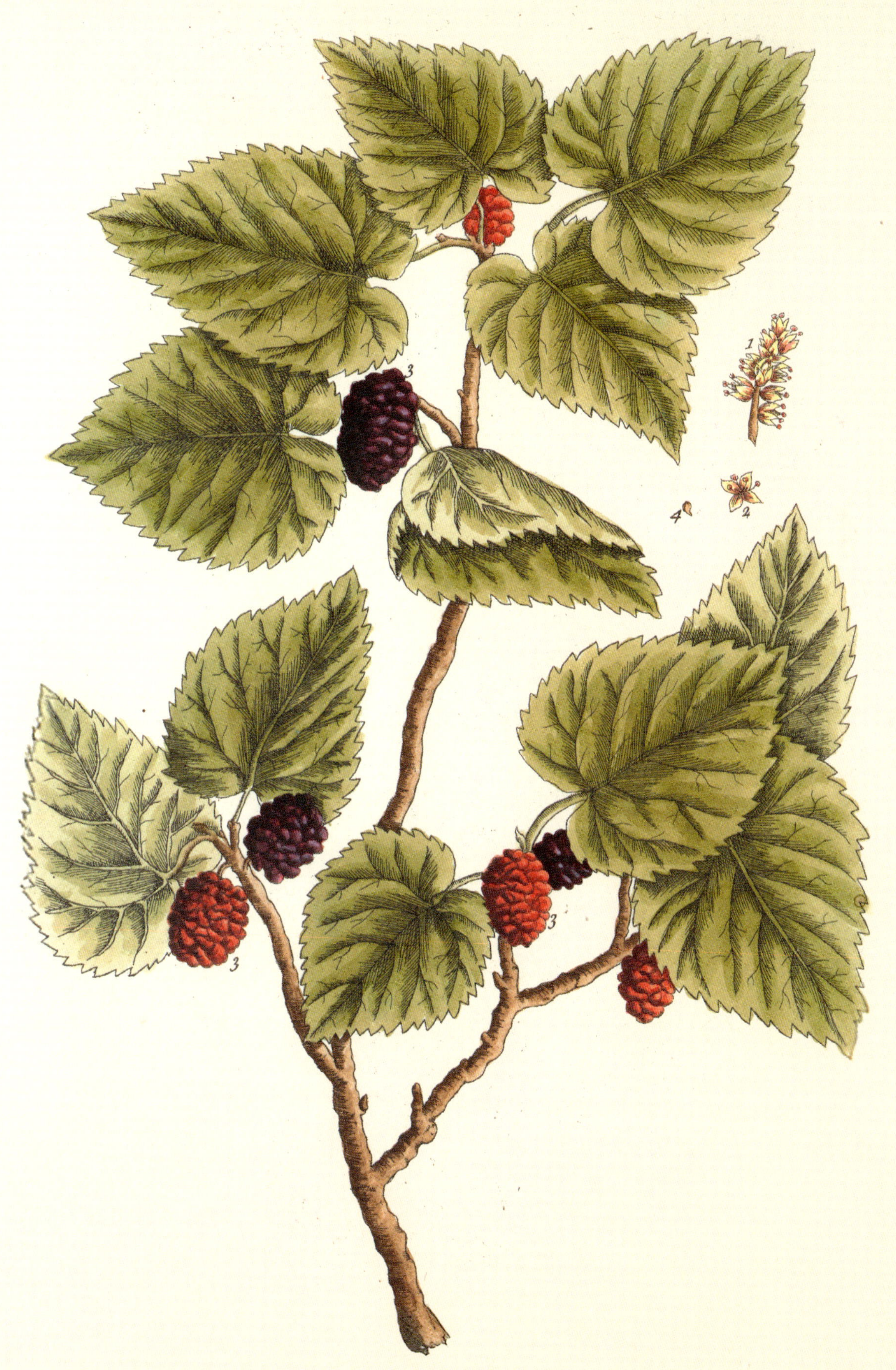
1
4
2
3
3
3
3

Recipes from Black Mulberry

Mulberry has been used since antiquity. The 5th-century-CE *Papyri Graecae Magicae* (Greek Magical Papyri; PGM VII.222–49) contain instructions for making an ink for dream oracles using red ocher, blood from a crow and a white dove, lumps of incense and myrrh, black writing ink, cinnabar, rainwater, the juice of a single stem of wormwood (*Artemisia absinthium*), vetch (*Vicia* species), and sap from a mulberry tree. Combining a variety of plants in a recipe in this way was common and exploited the qualities of each ingredient. An example is found in the *Papyrus Graecus Holmiensis*, also known as the *Stockholm Papyrus* (ca. 300 CE), in which mulberry root is used as a mordant to improve the lightfastness of dyer's alkanet root (*Alkanna tinctoria*) when producing purple and can be dissolved with the root of henbane (*Hyoscyamus niger*).

Any purple dye or colorant can be darkened with black mulberry ink made from indigo (*Indigofera tinctoria*) pigment, producing a blackish, blue-purple color, or any other shade depending on the quantity of ink and pigment used (see *Recipe 2*). The 11th-century *Book of the Staff of the Scribe* by al-Muʿizz ibn Bādīs describes how to make a black ink using mulberry leaves and iron sulfate (see *Recipe 1*).

Other colors can be made from mulberries, including reds and pinks from the berries, as explained by an anonymous 15th-century Persian author in a treatise called *Resāleh dar Bayān-e Kāğhaz Morakkab va Ḥall-e Alvān* (Treatise on Paper, Inks, and Dyes), in which *Shāhtūt* berries (*Morus nigra*), are boiled in water to make a red dye for dyeing paper. Further mulberry recipes were noted down by Simī Nishapuri, a master calligrapher and librarian from Mashhad, in his treatise on paper, color, and ink called *Jawhar-i Simī* (Simi's Jewel), including one for making red from the berry juice.

I collected a bag of mulberry leaves (wearing gloves to protect my hands from the sap) from a black mulberry tree growing in the Cambridge University Botanic Garden. I made a yellow dye from the leaves using an alum mordant; green, brown, and black colors with iron sulfate as the mordant (see *Recipe 2*); and a salmon-pink with a potash mordant. Ripe mulberry berries can be used to make vivid pink colors by heating a handful in water and simmering for an hour with alum. The resulting red dye can be used as an ink when gum arabic or gel from *Aloe* species is added as a thickener. A paint can be made if the red dye is mixed with a small amount of kaolin (China clay), making a lighter pink.

"Here We Go Round the Mulberry Bush" by artist Walter Crane and wood engraver Edmund Evans from *The Baby's Opera* (1800–1923).

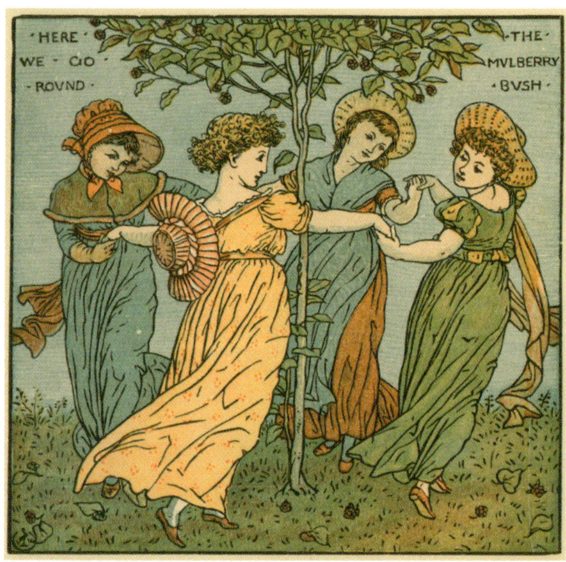

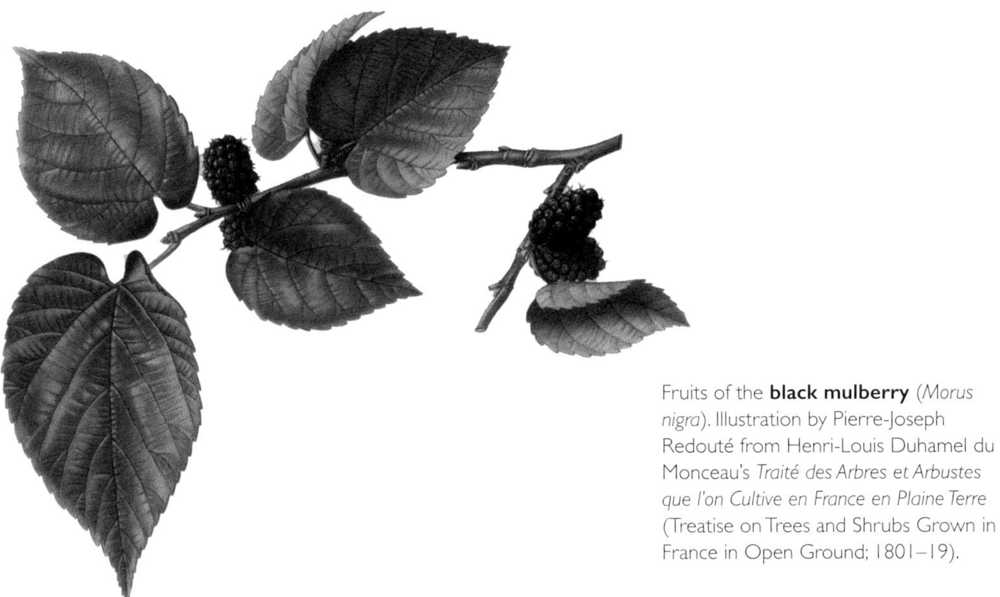

Fruits of the **black mulberry** (*Morus nigra*). Illustration by Pierre-Joseph Redouté from Henri-Louis Duhamel du Monceau's *Traité des Arbres et Arbustes que l'on Cultive en France en Plaine Terre* (Treatise on Trees and Shrubs Grown in France in Open Ground; 1801–19).

Recipe 1: Making black Arabic ink

"Preparation of ink. The water of the black ripened Syrian mulberry [black mulberry] is taken in the amount of a ratl [500 grams]. With it are put ten dirhams [30 grams] of pulverized sieved gum arabic. A little vitriol [iron sulfate] is added to it. It is put into a [sealed] pot in the sun [for] forty days."

—Al-Muʿizz ibn Bādīs, *Book of the Staff of the Scribe*, 11th century

Recipe 2: Making black mulberry ink and paint

Making mulberry ink: Heat the leaves in water or white wine and add iron sulfate. If you want a glossy ink, add some gum arabic, then evaporate the ink by half. Strain off the waste and the ink can be used immediately and kept in a sealed ink jar. Alternatively, let the ink dry naturally in a small glass dish and, once dried, place a brush in hot water then apply it to the dried ink.

Making mulberry paint: You can make a blue-black paint by mixing a small amount of indigo pigment with gum arabic and a little kaolin (China clay). Then mix with mulberry ink to form a dark paint.

Note: Wear gloves when collecting mulberry leaves, as the milky sap is an irritant.

—Nabil's modern recipe

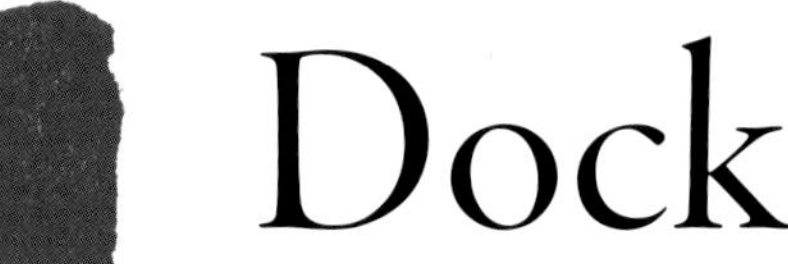

Dock

Latin name: *Rumex* species

Other names: Sorrel

Family: *Polygonaceae*

Genus: *Rumex*

Native: Europe, West and Far East Asia

Type: Herbaceous perennial

Pedanius Dioscorides mentions various *Rumex* species in his 1st-century-CE *De Materia Medica*, including dock sorrel (*Rumex aquaticus*) and oxylapathon or monk's rhubarb (*Rumex alpinus*). He outlines several other dock species and recommends drinking the seeds with wine to settle the stomach and using the roots to treat leprosy, as well as other beneficial effects. Today, broad-leaved dock (*Rumex obtusifolius*) and other *Rumex* species are believed to have anti-cancer properties, though research is still in the exploratory stage.

Bald's Leechbook (Royal MS 12 D XVII), in the British Library, London, is a 9th-century Anglo-Saxon manuscript of medical recipes, diagnostic guides, and charms. The manuscript features dock leaves used to treat water-elf disease and possibly measles or chicken pox—there was a superstitious belief that these ailments were caused by witchcraft. The 13th-century alchemist friar Albertus Magnus notes in *The Boke of Secretes* that the Greek physician and the philosopher Galen, who lived around the 2nd century BCE, mentions common sorrel (*Rumex acetosa*) as a remedy to "lessen the belly" and to treat pox.

Many medicinal plants in ancient texts can also be used to make colorants and dyes, using similar processing techniques of boiling the plant in water with additives like mordants and thickeners. Sorrel plants contain oxalic acid, used in textile mills to bleach fabric before dyeing, and also as a mordant, a reducing agent in developing photographic film, and a grinding agent for polishing marble. William Withering mentions *Rumex hydrolapathum* in *An Arrangement of British Plants* (1776), noting that "The powdered root is one of the best things for cleaning the teeth."

When I was young, stinging nettles (*Urtica dioica*) grew at the bottom of our garden, among gooseberries, rhubarb, and various weeds. I used to watch slow worms (*Anguis fragilis*) slithering across the compost heap and common frogs (*Rana temporaria*) leaping into a neighbor's pond, which was full of great crested newts (*Triturus cristatus*). Occasionally, I was stung by the nettles and looked for broad-leaved dock to rub on my skin. Dock leaf sap has anti-inflammatory and antibacterial properties, while the roots contain tannins and gallic acid, making them ideal for producing a black dye-ink.

Red water dock
(Rumex sanguineus var. *sanguineus*). Illustration from *The Works of Jacques Le Moyne de Morgues*, a book about the 16th-century artist published in 1977.

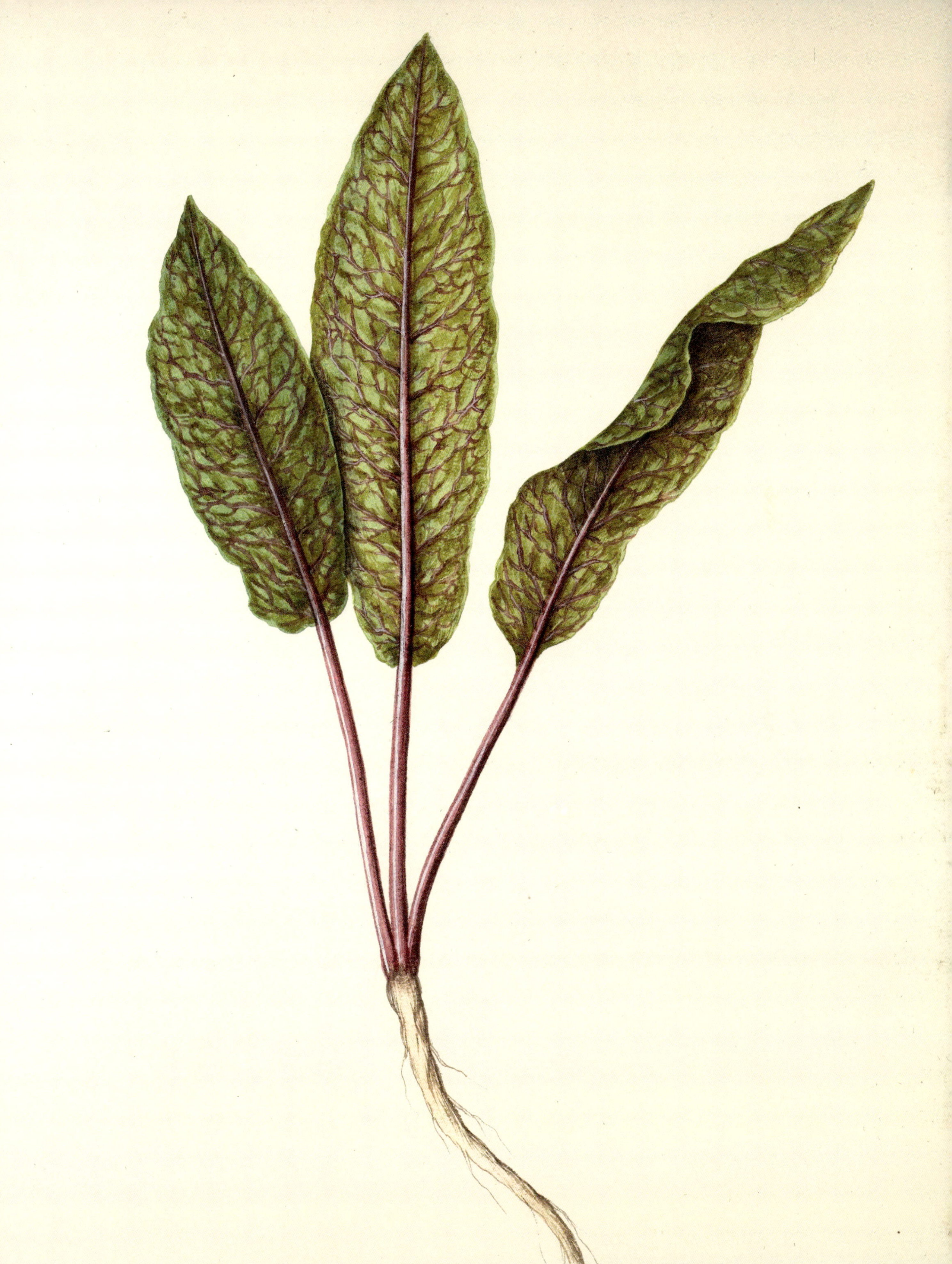

Recipes from Dock

Many cultures have used dock to make dyes. The root of the broad-leaved dock (*Rumex obtusifolius*) was gathered by Scottish clans to dye their kilt yarns black, while in Tibet, dyers used various *Rumex* species to produce a yellowish-green. The canaigre or Arizona dock (*Rumex hymenosepalus*) was used by Indigenous American weavers and dyers for coloring fabrics and yarns for hundreds of years, with the Hopi and Papagos turning to it as a food ingredient and to make a black dye from the roots, a technique obtained from them by the Navajo to color fibers. Cheyenne dyers used the roots and stems of curly dock (*Rumex crispus*) for yellows and the veiny dock or winged dock (*Rumex venosus*) for dyeing feathers red.

The roots of common sorrel (*Rumex acetosa*) produce a yellow dye and a red-brown when applied with sodium carbonate or a potash mordant, which is highly alkaline. Using an iron mordant with the roots will create a solid black dye. The species with the highest tannin content, ideal for making black dyes and ink, is the canaigre dock, followed by the patience dock (*Rumex patientia*).

You can dye fabric green by simmering dock leaves for an hour in water before straining off the leaves. Mordant the fabric first in alum, then heat in the green dye bath for about 45 minutes until the dye takes to the fibers. To achieve olive-green colors, soak the dried fabric in a clean water pan to which iron sulfate has already been added. Dyeing wool a red to coral color is also possible with the fall seeds and stems. To do this, mordant the wool first in alum, then simmer with the seeds and stems in water for several hours until the water turns deep red.

For other investigations, I dug up and processed the roots of the curled dock and broad-leaved dock from the Systematic Beds located in Cambridge University Botanic Garden. I was primarily seeking to make a black dye using old samples kept in the Cambridge University Herbarium and following the experiments shown in W. L. Lindsay's 19th-century *Illustrations of the Economical Applications of Common Weeds* (see *Recipe 1*). Once the black dye was made from the roots using iron and white wine, I added gum arabic to make a black ink.

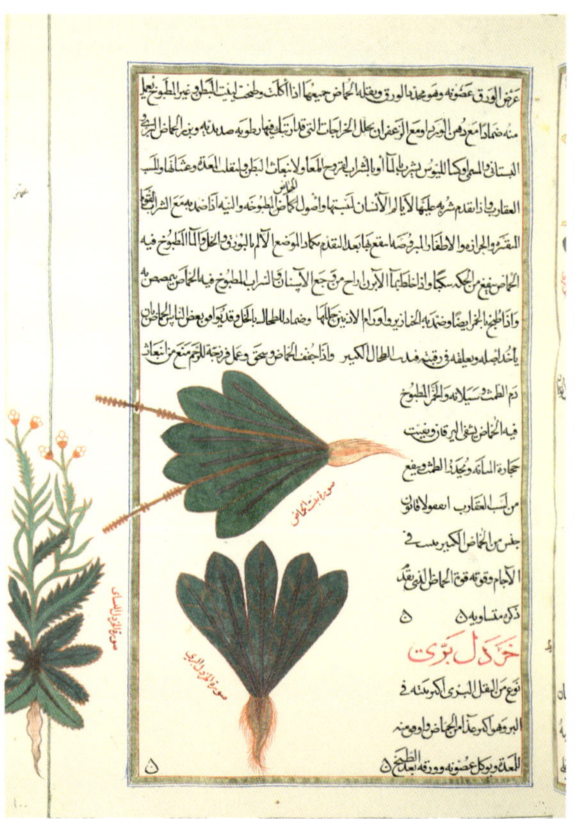

Garden dock or monk's rhubarb (*Rumex patientia*) (top). Illustration by Persian artist Mirzā Bāqir for an 1889–90 edition of a 9th-century Persian translation of Pedanius Dioscorides' *De Materia Medica* (1st century CE).

Recipe 1: Making Lindsay's black dye

"Yarn was boiled for a few hours with [sorrel/*Rumex acetosa*] and a little copperas [ferrous sulfate]."

—W. L. Lindsay, *Illustrations of the Economical Applications of Common Weeds*, Cambridge University Herbarium (M0000747), 19th century

Recipe 2: Making *Rumex* colors

Use a weed-puller to dig up a dock root and cut it into small pieces when fresh, or dry the root and grind to a powder with a pestle and mortar. Place the root in rainwater and simmer for over an hour. Adding an alum or tin mordant will give yellow-tan colors, while potash produces a reddish-brown. To make a gray to black color, use an iron mordant on chamois leather or mix concentrated black dye with chalk or gesso to make a gray pigment, then add a binder if you want a gray paint.

—Nabil's modern recipe

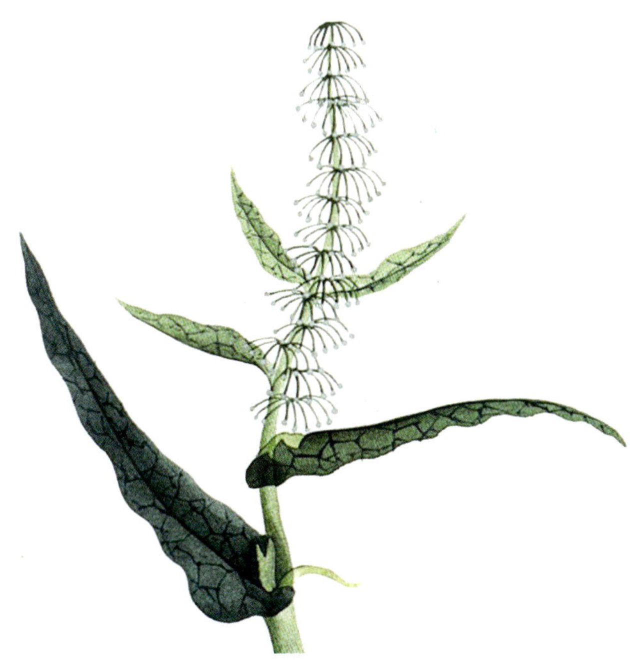

Rumex japonicus. Watercolor (ca. 1817–1823) by Japanese artist Ishizaki Yūshi.

Elephant Ears

Latin name: *Bergenia crassifolia*

Other names: Korean elephant's ears, pigsqueak, Siberian tea, badan

Family: *Saxifragaceae*

Genus: *Bergenia*

Native: Siberia to North Korea

Type: Evergreen perennial

Bergenia crassifolia is a hardy, evergreen, clump-forming perennial with large, leathery leaves. Growing in woods and on shady rock ledges, it was used as a tea and as a medicinal plant in Siberia, Russia. It was described by the 18th-century German naturalist and Russian naval officer Johann Georg Gmelin, who received a fellowship from the Russian Academy of Sciences, whose former headquarters are now home to Russia's oldest museum, the Kunstkamera, on the Neva River in St. Petersburg.

The name *Bergenia* was given to the genus in honor of the 18th-century German botanist and physician Karl August von Bergen, and the Swedish naturalist Daniel Solander introduced the plant to Britain in around 1765. The plant arrived in America in the same year, introduced by the American explorer William Bartram, a gifted artist, especially in botanical illustrations. His work can be seen in the Fothergill Album at the Natural History Museum in London.

Elephant ears became very popular in the 19th century, with artist and horticulturist Gertrude Jekyll teaching the value of common plants, especially *Bergenia crassifolia*, in her design schemes. It was one of her favorite edging plants and can be seen in her famous garden at Munstead Wood, Surrey, and in other gardens she designed or influenced such as RHS Wisley in Surrey, West Dean Gardens in West Sussex, and Barrington Court in Somerset.

The plant's roots and rhizomes are used in natural folk medicine in Mongolia and have a variety of benefits, including antimicrobial, anti-inflammatory, and astringent properties. For example, in Tibet the leaves are made into a paste that acts as a sun lotion and is applied to exposed skin to protect it from harmful ultraviolet radiation. The leaves are also widely used in Chinese medicine.

Bergenia drawings are seen from the 18th century and include a handcolored etching by the 18th-century French physician, artist, and naturalist Pierre-Joseph Buc'hoz in the *Collection Précieuse et Enluminée des Fleurs Les Plus Belles et Les Plus Curieuse* (Precious and Illuminated Collection of the Most Beautiful and Curious Flowers; 1776). A later illustration of the plant's flowers, with just the outline of the leaf, can be seen in *The Botanical Magazine* (Plate 196, Volume 6, 1793).

Elephant ears (*Bergenia crassifolia*) Handcolored etching from Pierre-Joseph Buc'hoz's *Collection Précieuse et Enluminée des Fleurs les Plus Belles et les Plus Curieuses* (Precious and Illuminated Collection of the Most Beautiful and Curious Flowers; 1776).

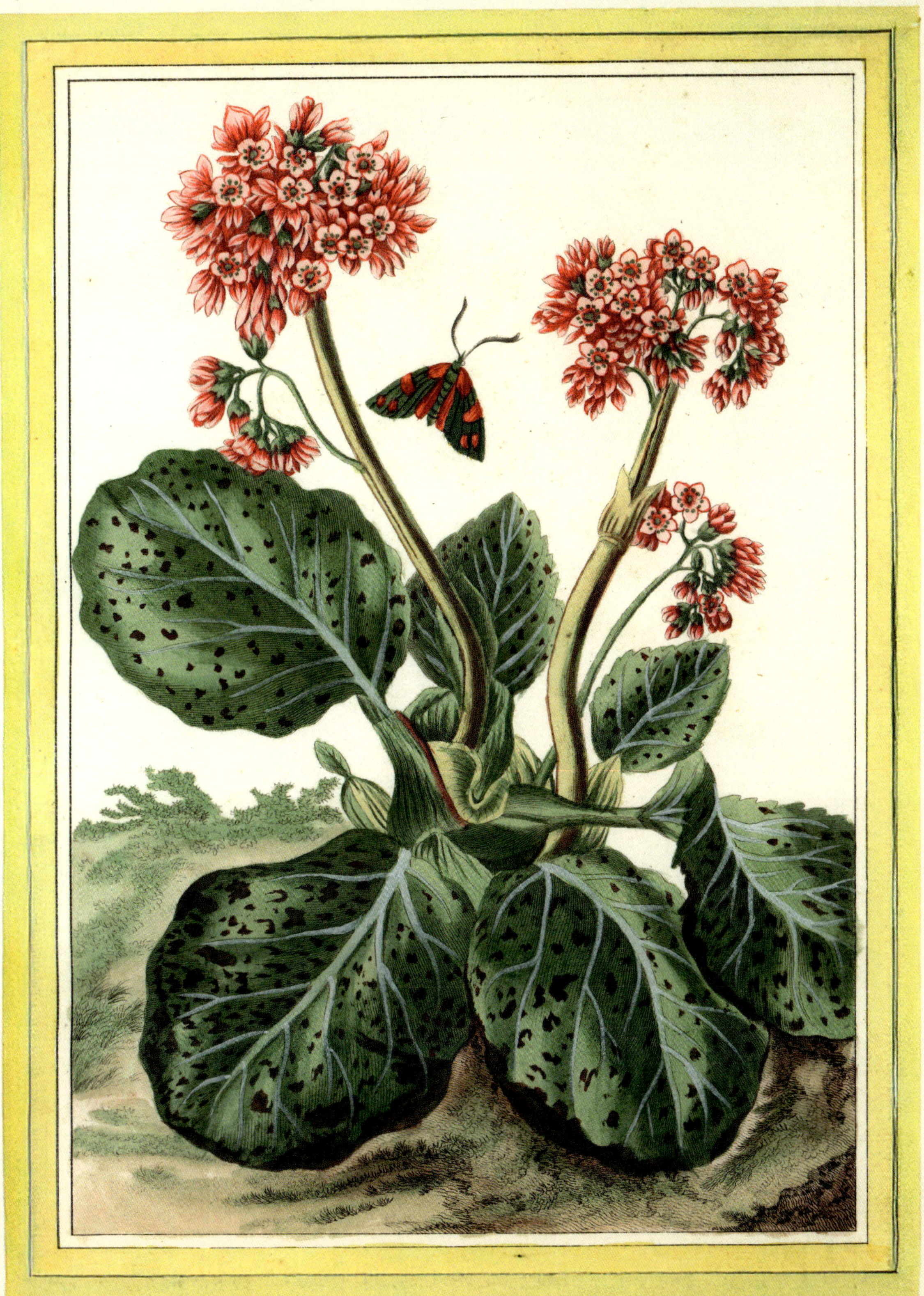

Recipes from Elephant Ears

Bergenia crassifolia is abundant in Cambridge University Botanic Garden, and I conducted several experiments using different mordants to see how they behaved with the tannin in the plant's leaves. There are few, if any, historical recipes for using *Bergenia* as a colorant, but it was used to dye Russian leather black. The following techniques and methods are experimental, but produced fairly successful results.

I collected some of the plant's leaves, which were growing in the borders of Brookside Lawn in the botanic garden, threaded them with some jute garden string, and hung them across the garden shed until they dried. A helpful attribute of the dried leaves is that they crumble into small pieces with ease and can be ground into a fine pigment with a pestle and mortar. I heated the powdered leaves for an hour in water with an iron mordant. This produced a black dye due to the gallic acid content of the plant, which oxidized chemically. When the dried leaves were heated with an alum mordant, the wool I was dyeing became more of a gray color. The fresh young leaves hold higher levels of gallic acid, and if used with a tin mordant produce a very satisfying golden yellow ocher on wool and a light mustard shade on chamois leather. A strong black can be made when an iron mordant is used on leather and cotton, yet the color still remains gray on wool, resembling the color of pewter or tarnished silver.

Recipe 2 is the result of an experiment involving animal bones, made possible when I found a pigeon in my garden that had been killed by a hawk. There were feathers and blood everywhere and I received a quick glance from the hawk before it flew elegantly into the sky. I quickly dug a hole and buried the dead pigeon, then left nature to do what it does best.

Interestingly, the 14th-century *Liber Diversarum Arcium* (Book of Various Arts) includes a recipe for using calcined (cremated) bones from a cow or other animal. It is used to make a good white for wooden drawing tables, and when added to a colored pigment like orpiment, cinnabar, indigo (*Indigofera tinctoria*), or, in my case, a black, which I made from *Bergenia crassifolia*.

Elephant ears (*Bergenia crassifolia*). Color lithograph from *Vilmorin's Blumengärtnerei* (Vilmorin's Flower Nursery; 1896).

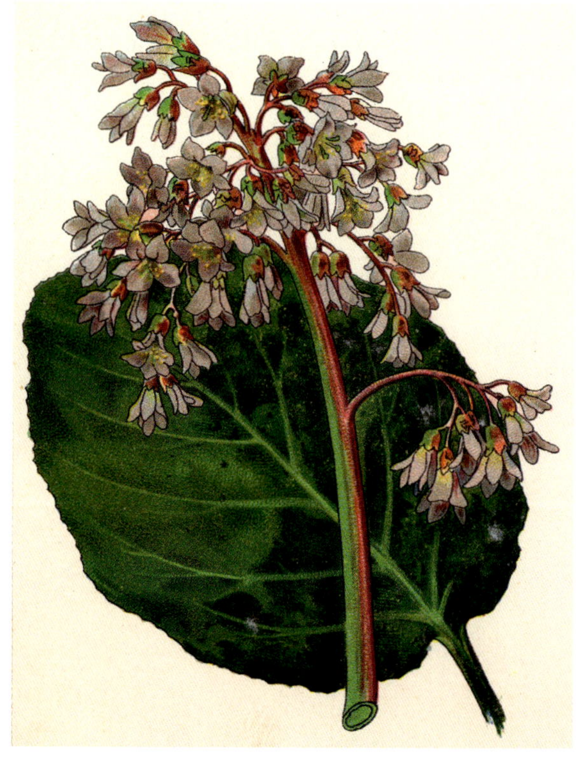

Recipe 1: Making *Bergenia* ink

Take four fresh *Bergenia crassifolia* leaves and cut them into small pieces, then heat in water with iron sulfate and gum arabic. Simmer for an hour and reduce the water by half, then strain off the waste. Let the ink stand for a day or so and it will be ready to use for brushwork. If the ink is too watery, reheat the ink until it flows beautifully.

—Nabil's modern recipe

Recipe 2: Making blacks and grays from elephant ears and pigeons

To make a black dye: Collect some elephant ears at any time of year, chop into small pieces, and heat for an hour with iron sulfate to make a black dye. Strain off the waste (reserve this if you want to make a black paint) and let the leaves dry. Place the dye back on the heat, then reduce by half.

To make a black paint: Once the elephant ear waste has dried, crush it with a pestle and mortar until you have a fine pigment, then temper (mix) with gum arabic or egg white to produce a black paint.

To make a gray paint: Summon a winged hawk (which is sacred to the Saracens), so it may feed on the body of a dead pigeon—or look for a carcass on your lawn or in a nearby field (you can, of course, use a discarded chicken carcass or similar). Bury the pigeon under the earth for four full moons, then dig up the remains. Wash the bones, then crush with a pestle and mortar until you have a fine white pigment, then add the black elephant-ear dye to the bone-white pigment to form a pleasing gray paint.

—Nabil's modern recipes

Giant Rhubarb

Latin name: *Gunnera tinctoria*

Other names: Chilean rhubarb, prickly rhubarb

Family: *Gunneraceae*

Genus: *Gunnera*

Native: Central and Southern Chile to Southwest Argentina

Type: Herbaceous perennial

The genus *Gunnera* was named by the 18th-century Swedish botanist and father of taxonomy Carl Linnaeus in honor of the Norwegian bishop and botanist Johan E. Gunnerus. As the name suggests, giant rhubarb (*Gunnera tinctoria*) is one of the largest species and was introduced to Britain in the 19th century. The species is now listed as an invasive plant in England because of its vigorous growth and potential to spread, which led to it being banned in England in 2023.

Growing to a height of 6½ feet (2 meters) or more and sending up large, heavily veined, sharply toothed leaves on prickly stems, the giant rhubarb is an outstanding architectural plant that grows in moist conditions, which encourage larger leaves. *Gunnera* × *cryptica*, a hybrid between the Chilean *Gunnera tinctoria* and the Brazilian giant rhubarb (*Gunnera manicata*), does not spread as vigorously but is also banned. Not all *Gunnera* have enormous leaves, however, with the dwarf gunnera (*Gunnera magellanica*), native to parts of South America and the Falkland Islands, having much smaller leaves no bigger than your hand.

A black dye can be made from the roots and leaves of the Brazilian giant rhubarb. Several large *Gunnera* specimens grow in the beds of the Stream Garden at Cambridge University Botanic Garden, which is fed from Hobson's Conduit and was built in 1610. The *Gunnera tinctoria*, also known as prickly rhubarb, showcases an exotic display of gigantic foliage that dominates the bed where plume thistle (*Cirsium rivulare*), Culver's root (*Veronicastrum virginicum*), bleeding heart (*Dicentra spectabilis*), and great horsetail (*Equisetum telmateia*) all grow.

The roots of *Gunnera* can be sensitive to the cool climate in Cambridge, but Simon and his team of skilled horticulturalists at the botanic garden use the leaves and stems to create a natural tepee to protect the roots from winter frosts. The leaves are cut back to the ground buds and separated from the stems. The smaller leaves are placed at the base, with the larger leaves on top creating a natural blanket. The leaves are weighed down with the stems and left until the following spring, when new green shoots start to sprout. The natural tepees look spectacular, especially when the frost settles on the leaves, creating an art garden installation.

Giant rhubarb (*Gunnera tinctoria*). Handcolored lithograph from Louis van Houtte and Charles Lemaire's *Flore des Serres et des Jardins de l'Europe* (Flowers of the Hothouses and Gardens of Europe; 1870).

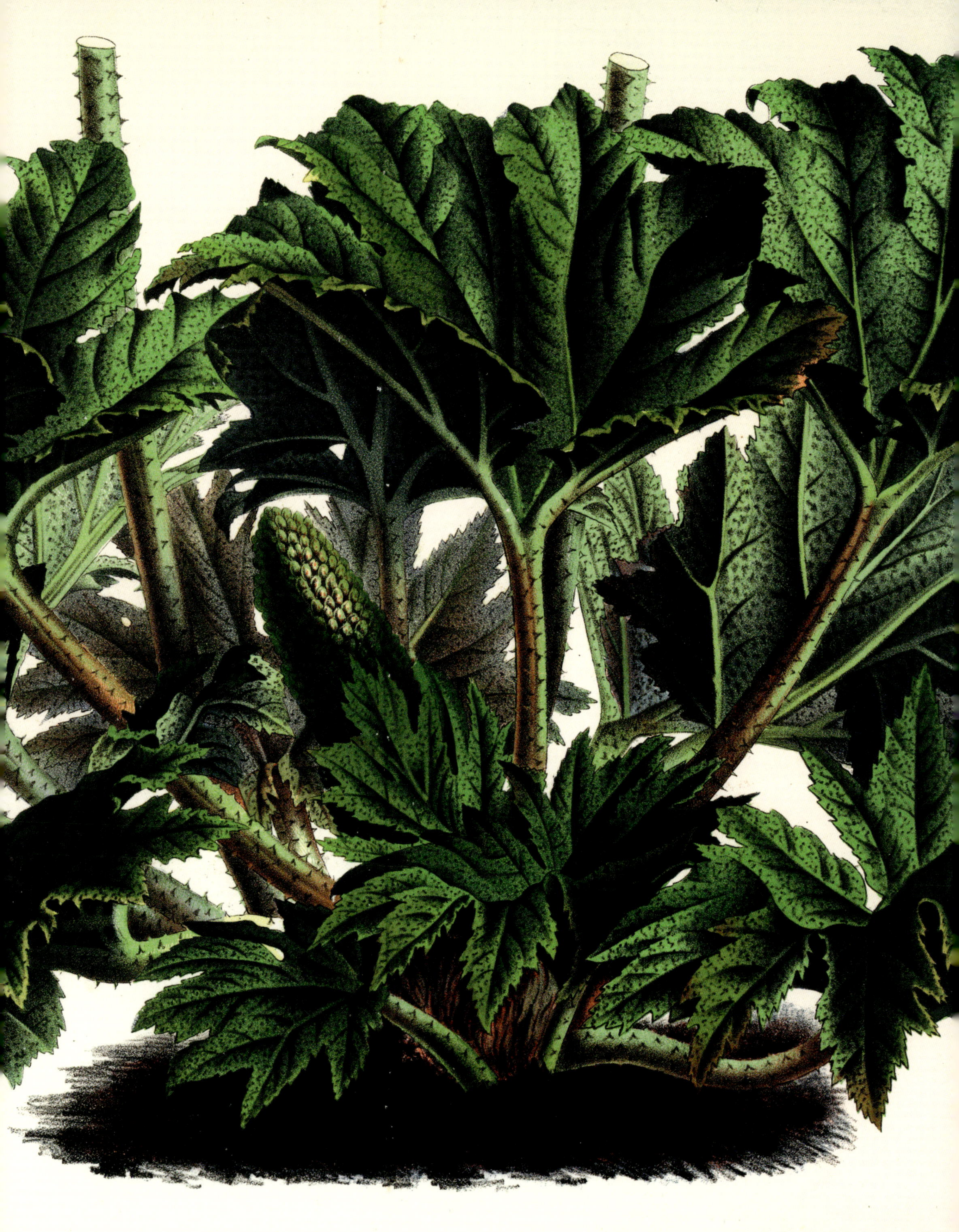

Recipes from Giant Rhubarb

The *Gunnera* genus is a relatively new arrival in Europe, and no Western historical manuscripts suggest it can be used to make a colorant. However, an absence of historical evidence does not diminish the potential of the plants to produce dyes. Today, local craftspeople and artists like myself view the various *Gunnera* species as a valuable source for creating colorants, and this is borne out by historical uses in the North and South America.

In the spring of 1782, the Spanish botanist Hipólito Ruiz López returned from Culenco to Concepción in Chile, following a ten-year South American expedition on which he collected many plants from around Fuerte de Nacimiento (Fort Nativity). He noted that the Indigenous Mapuche people used the thick roots and large leaves of a plant called *Panke acaulis* Molina (now *Gunnera tinctoria* var. *tinctoria*) for tanning and dyeing leather black.

Decades later, at the end of 1834, Charles Darwin reported that *Panke tinctoria* (syn. *Gunnera scabra*), the name of which was changed to *Gunnera tinctoria* in 1805 by Charles-François Brisseau de Mirbel, was used by the Chono, one of the Indigenous peoples of Chile, to tan leather and to make a black dye from the roots.

Due to mid-19th-century campaigns led by the Chilean government, Mapuche independence diminished, which caused discord and violence between them and the settlers. This reduced the knowledge of dyeing among the Mapuche because younger generations could not benefit from the skills passed down from their elders, which are so vital in heritage craft. Expert Mapuche dyers and weavers are known as *düwkafe*, and today they re-weave the forgotten threads of their ancestors' dyeing knowledge. Unfortunately, forest access to retrieve plants is restricted due to historical and social-ecological changes, including displacement and deforestation that affects the water resources on which plants like the giant rhubarb depend. Local communities in the Pehuen settlement in the city of Lebu, in the Arauco Province, Chile, collect *nalca*, edible *Gunnera* stems that the Mapuche call *pëñal fillkün* (little lizard attached to the ravine), mainly as a food source.

The Indigenous Chilean population also used wirevine (*Muehlenbeckia hastulata*), producing a black dye from the root and leaves. I myself made a black colorant from *Gunnera manicata* roots and leaves collected from Cambridge University Botanic Garden (see *Recipe 1*).

Giant rhubarb (*Gunnera tinctoria*). Illustration by Abraham Jacobus Wendel from Heinrich Witte's *Flora* (1868).

Recipe 1: Making *Gunnera* black

Dig up the root of *Gunnera manicata*, or any *Gunnera* species, and let it dry. Heat the dried root in water—which will become acidic—add some iron sulfate, and then simmer for 30 minutes. Due to various chemical reactions, the solution will oxidize to black. Strain off the waste and reheat to reduce the water to a third and let it cool. Keep the dye-ink in a sealed jar.

The large leaves can also be cut into small pieces and heated in water with iron sulfate to produce a black dye or ink, and a yellow color when an alum mordant is used.

—Nabil's modern recipe

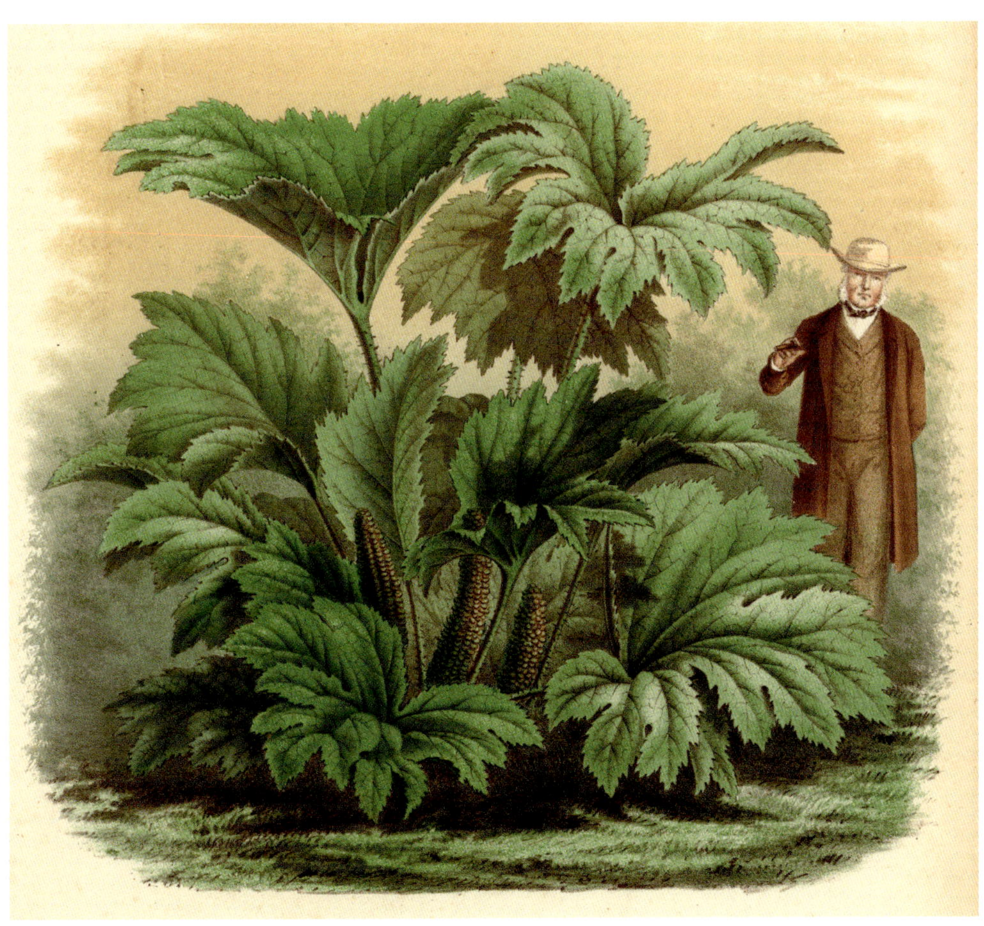

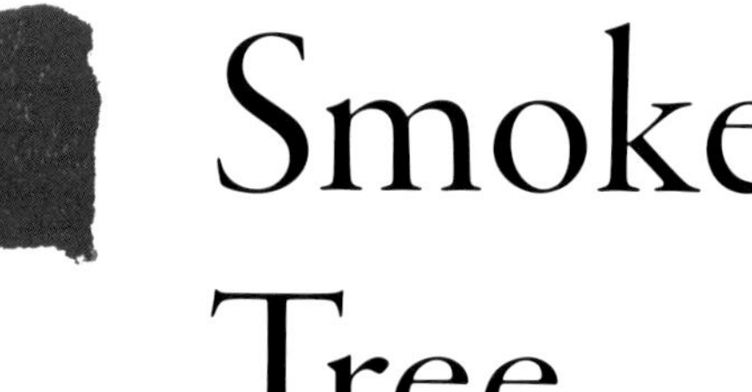

Smoke Tree

Latin name: *Cotinus coggygria*

Other names: Smoke bush, Venetian sumac, dyer's sumac, young fustic, wig tree

Family: *Anacardiaceae*

Genus: *Cotinus*

Native: South Central Europe to Central and Southern China

Type: Deciduous shrub

The smoke tree (*Cotinus coggygria*), a member of the sumac family (Anacardiaceae), is often grown as an ornamental plant. One of its common names, fustic, comes from the Arabic word *fustug*, meaning "bush." According to Chapter 41 in Book 13 of John Bostock's 1855 translation of Pliny the Elder's *Naturalis Historia*, Pliny refers to a plant called "coccygia" and notes that *Cotinus* is famous for dyeing cloth purple. Although Pliny does not provide details, he is correct in saying that *Cotinus* can produce a dye.

Ancient Japanese and Chinese gardens often included smoke tree and other plant species for contemplation. In the 3rd century BCE, Emperor Ying Zheng of the Qin Dynasty reunited China and built the legendary E'pang Palace in the Shanglin Garden on the southern bank of the Wei River, where it is said that Chinese plum (*Prunus mume*), *Citrus* species, and smoke tree grew with other plants, including the Chinese persimmon (*Diospyros kaki*), the fruits of which can be used to make a black dye.

Since the Middle Ages, the smoke tree has been used as a dye plant for tanning leather and was gathered into small bundles in the 14th century for export from Italy. It was also an important wood for carpentry, and the leaves were a good source of the tannins needed for 16th-century tapestries and Ottoman carpets and rugs, which analysis by conservation scientists reveals were dyed with *Cotinus coggygria*.

The smoke tree was introduced to Britain around the 17th century and became important economically for hundreds of years. It was described by John Gerard in his 1597 *Herball, or Generall Historie of Plantes*, in which he says that the Venetian sumac, another name for *Cotinus coggygria*, is an "excellent and most beautiful plant." I agree wholeheartedly with this, as I grow purple-leaved *Cotinus coggygria* in a pot in my garden as an ornamental shrub and to process into a black ink.

Traditionally, black dye was made with logwood (*Haematoxylum campechianum*), which is native to southern Mexico, and old fustic or dyer's mulberry (*Maclura tinctoria*). Old fustic was shipped in the 16th century from South America and the West Indies to Boston, then across the Atlantic to Europe. It was used to make khaki colors for military uniforms in the First World War.

Smoke tree (*Cotinus coggygria*). Watercolor by Francesco Peyrolery from *Iconographia Taurinensis* (Iconography of Turin; vol. 10, 1765).

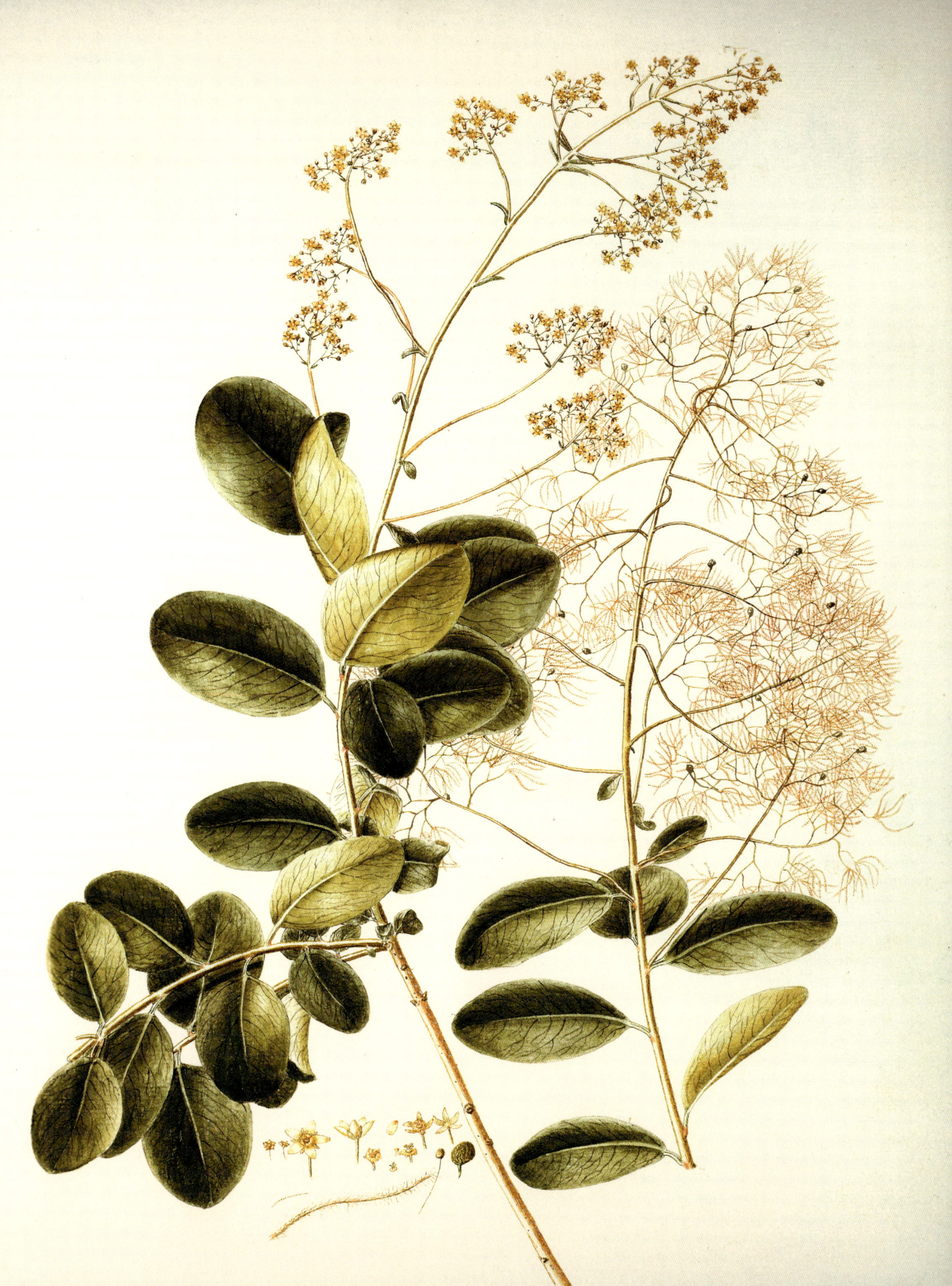

Recipes from Smoke Tree

Although the heartwood of old fustic (*Maclura tinctoria*) was used to make a yellow dye in Europe in the Middle Ages, yellow dye was primarily made from the wood of young fustic (*Cotinus coggygria*), as was a black. Using an alum mordant with the leaves produces yellow, while a potash mordant makes a green. The yellow made from the wood can be "saddened" or made darker when iron sulfate is added to produce olive-green shades, which are lightfast. The branches, berries, and leaves contain high levels of tannins, and the fresh or dried leaves were used for centuries to dye animal skins black.

William Partridge's *A Practical Treatise on Dyeing* (1823) includes numerous recipes for "Jet Black" made with multiple plants and ingredients, such as chipped logwood (*Haematoxylum campechianum*), sumac (*Rhus* species), and young fustic (see *Recipe 1*). I collected smoke tree leaves from Cambridge University Botanic Garden and successfully made black ink for brushwork.

The heartwood can be made into an orangey-yellow color with an alum mordant, and browns can be produced with a copper mordant. In the past, other dyes such as weld (*Reseda luteola*; see page 144) were added to young fustic to enhance the color. Recipes in Gioanventura Rosetti's 1548 dye book *Plictho de Larte de Tentori* (Instructions in the Art of Dyers) indicate that yellow using young fustic was the predominant color made.

A good black can be produced from smoke tree leaves, which can be added to other dyes like madder (*Rubia tinctorum*; see page 100) or brazilwood (*Biancaea sappan*; see page 84) to create a black or very dark purple. This color is similar to one outlined in Recipe 1.5.1B of the 14th-century *Liber Diversarum Arcium* (Book of Various Arts): "If you want to temper black [made from ground willow charcoal], it is tempered with gum water, and a little brazilwood is put in." Although this recipe focuses on making a paint rather than a dye, the color concept is similar. In the late 1700s to early 1800s, American physician and chemist Edward Bancroft mentioned that young fustic yellow was used as the first dye before over-dyeing with cochineal beetle and cream of tartar to produce a warmer, scarlet-red color.

Old fustic (*Maclura tinctoria*). Illustration from Michel Étienne Descourtilz and Jean Théodore Descourtilz's *Flore Médicale des Antilles, ou, Traité des Plantes Usuelles* (Medical Flora of the Antilles, or Treatise on Common Plants; 1821–29).

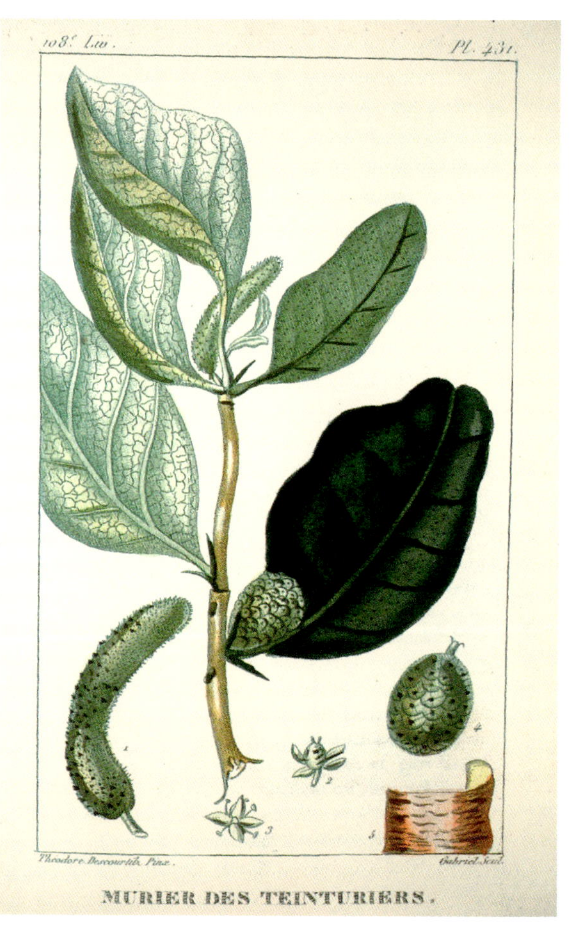

Recipe 1: Making sumac jet black

"[For boiling, use] 10 pounds of logwood, 2¼ pounds of sumac, 2¼ pounds of [young] fustic, and boil for two hours, and the cloth three, heave out, and cool as before. [Add] 5 pounds of copperas [and] ¼ pound of blue vitriol and dissolve in a bucket, and add to the liquor without boiling; the cloth to be boiled four hours, run up, heave out, and proceed as directed for the first black [that is, the cloth taken out and cooled]."

—William Partridge, *A Practical Treatise on Dyeing*, 1823

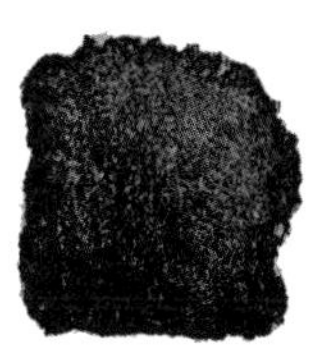

Recipe 2: Making *Cotinus*-black ink

Collect a handful of green or purple *Cotinus coggygria* leaves, then cut into small pieces and soak overnight in water or white wine. Heat the leaves with an iron mordant and reduce the water or wine by half. Add gum arabic to produce a black ink. If the ink is too watery, reheat and reduce a little more.

—Nabil's modern recipe

Oak Galls

Oak galls, or gallnuts, are growths caused by the gall wasp on oak trees. They are described as a tanning agent by Theophrastus in his 3rd-century-BCE *Historia Plantarum* (Enquiry into Plants). Like henna (*Lawsonia inermis*), leaves from the sumac tree (*Rhus* species), or fibrillium tannins from the date palm (*Phoenix dactylifera*), oak galls can also be made into a black hair dye, which is described by Pedanius Dioscorides in his 1st-century-CE *De Materia Medica*.

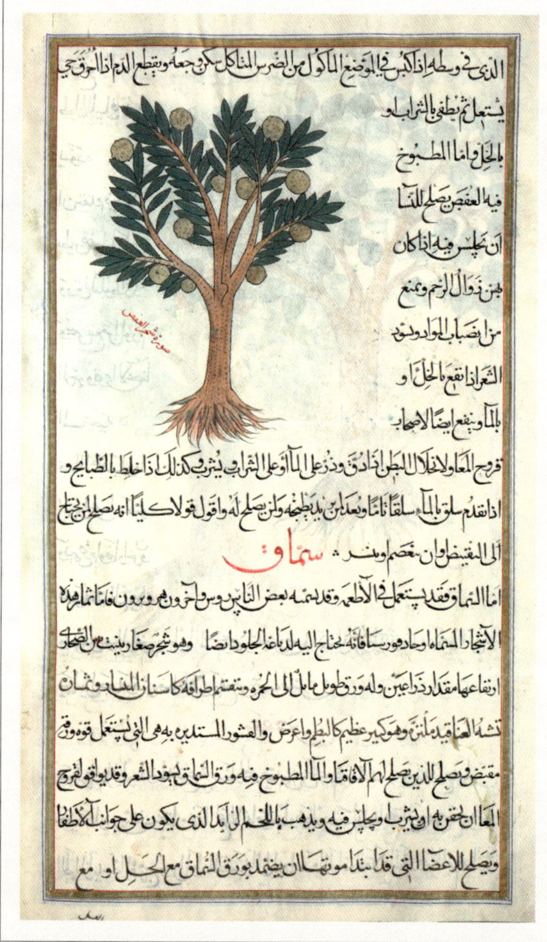

Aleppo oak (*Quercus infectoria*) with **gallnuts**. Illustration by Persian artist Mirzā Bāqir for an 1889–90 edition of a 9th-century Persian translation of Pedanius Dioscorides' *De Materia Medica* (1st century CE).

Opposite: Leaves and **galls of English oak** (*Quercus robur*) (bottom). Watercolor by Joris Hoefnagel from *Mira Calligraphiae Monumenta* (ca. 1561–62).

Oak galls were also used in a Greek military ink recipe recorded by Philo of Byzantium in the *Mechanike Syntaxis* (Compendium of Mechanics) in the 3rd century BCE. A message would be written on a person's skin with oak gall water. Once dried, it became invisible. The message would remain hidden until dabbed with iron sulfate water, which turned the letters black through oxidation.

The 4th-century-CE *Codex Sinaiticus* (Add MS 43725), in the British Library, London, and the 5th-century-CE *Papyri Graecae Magicae* (Greek Magical Papyri; PGM XII.397–400) mention a magical ink made using myrrh, vitriol (iron sulfate), oak galls, and gum arabic. Analysis shows that the *Book of Kells* (MS 58) at Trinity College, Dublin was written in iron gall ink. The ink remained a primary writing medium for technical and illuminated manuscripts until the 1970s, when it was still relied on in Europe for official government documents. The ink continues to be used today by artists, with conservation scientists replicating recipes to restore old manuscripts and artworks.

A medieval ink recipe—Recipe 1.6.1 in the 14th-century *Liber Diversarum Arcium* (Book of Various Arts)—provides a simple process for preparing iron gall ink. For example, if you collect ten oak galls weighing 3½ ounces (10 grams), you should use the same amount of powdered

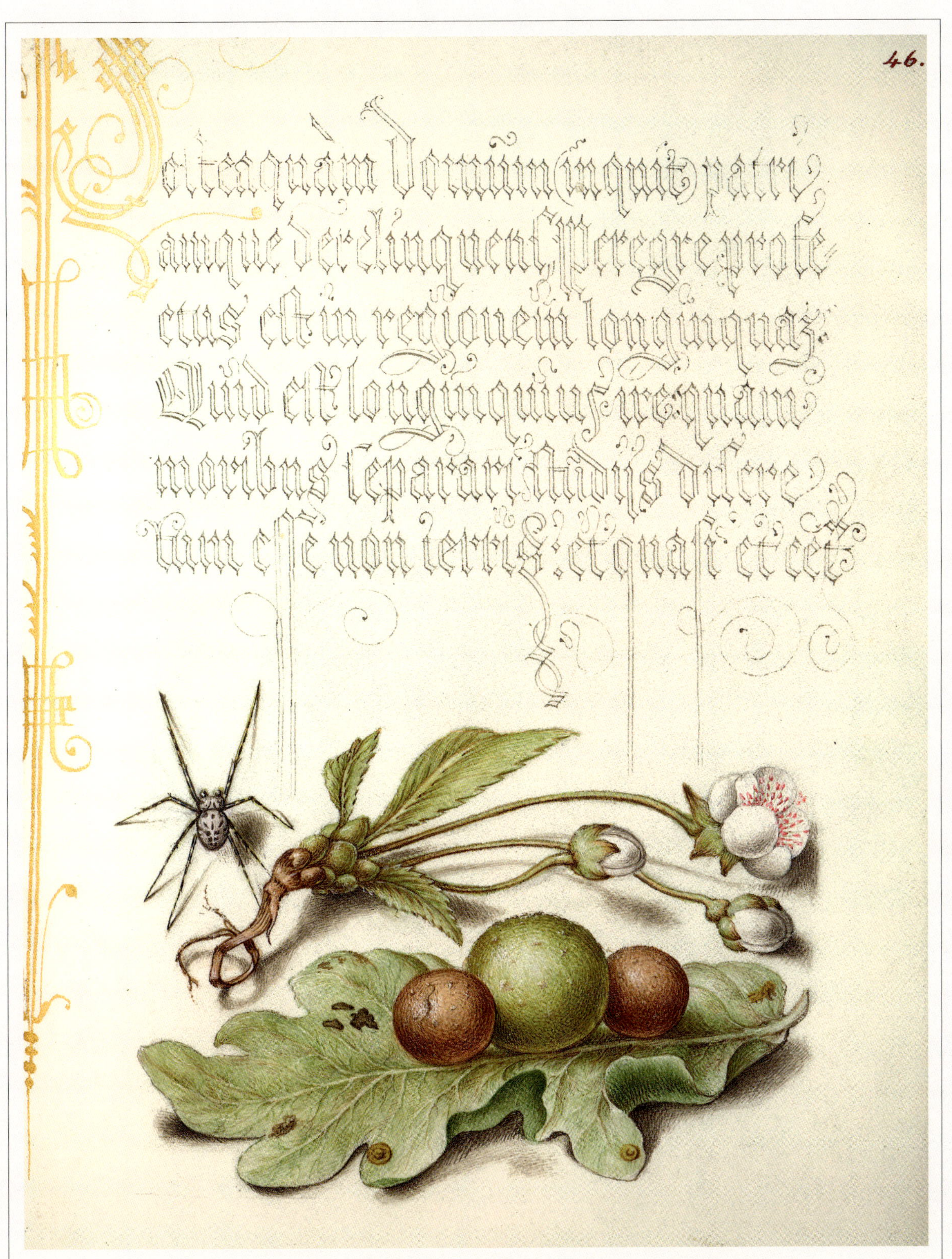
46.

gum arabic. The iron sulfate required is half the weight of the galls and is divided into two parts. Multiplying the number of oak galls by twelve gives the quantity of wine needed. To prepare the ink, heat the crushed oak galls, gum arabic, and half the iron sulfate in the wine. Then reduce the solution by half, strain out the solids, and reheat the ink. Finally, add the remaining half of the iron sulfate. Reduce slightly again, then strain once more, for a high-quality black ink.

An early 15th-century treatise on manuscript illumination called *Libro Secondo de Diversi Colori e Sise da Mettere a Oro* (Book Two on Different Colors and How to Apply Gold), held in the Biblioteca Casanatense, Rome (MS 1793), includes a recipe for a secret writing ink, similar to that of Philo of Byzantium. It suggests soaking the oak galls in water to extract the tannin, then writing the message and using vitriol (iron sulfate) to reveal it. An earlier Italian recipe for making ink, found in the flyleaf of a 14th-century Catholic breviary (Garrett MS 42), at Princeton University Library, appears to have been erased, and was rediscovered by specialists using ultraviolet light. It translates as: "Five ounces of gall and one-half of gum arabic and one-fourth of vitriol [iron sulfate] and one leaf of good white wine, and put [it] out for two nights and add two pomegranate peels to it and it is made good."

The 16th-century Italian Gioanventura Rosetti's dye book *Plictho de Larte de Tentori* (Instructions in the Art of Dyers; 1548) contains over ten recipes using oak galls to make black and gray colors, with instructions for dyeing leather and skins black, linens and wools gray, and how "to make berets black." A century or so later, Sir Isaac Newton made his own ink. In a manuscript held in Cambridge University Library (Add MS.3975 f.13r), he gives a detailed recipe for ink using ale:

"To Make Excellent Ink, ℞ *[recipe] ½lb of Galls cut in pieces or grosly beaten, ¼*℔ *of Gumm Arabick cut or broken. Put 'em into a Quart of strong beer or Ale. Let 'em stand a month stopt up, stirring them now & then. At ye. end of the moneth put in *℥1 [about 30 grams] or ℥1½ [about 45 grams] of copperas (Too much copperas makes ye ink apt to turn yellow.) Stir it & use it. Stop it up for some time with a paper prickt full of holes & let it stand in ye sunn. When you take out ink put in so much strong beer & it will endure many years. Water makes it apt to mold. Wine does not. The air also if it stand open inclines it to mold. With this Ink new made, I [Sir Isaac Newton] wrote this."*

* ℔ is an apothecary pound
* ℥ is an apothecary ounce

I reconstructed Newton's recipe and found that making the ink with wine or spirits was more pleasing due to the aroma of alcohol.

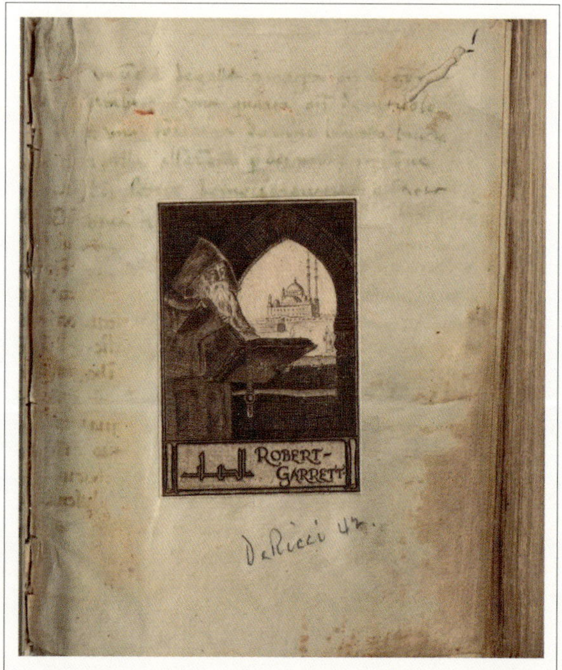

Page from a 14th-century Catholic breviary showing an **erased recipe for oak gall ink**, which was rediscovered under ultraviolet light.

Iron Gall Black Ink

This is my modern recipe for iron gall black ink, inspired by traditional techniques.

1. Gather and weigh the ingredients
If you collected ten brown oak galls (ensure there is a hole in each one which indicates the gall wasp or grub is no longer inside)—with each gall weighing approximately 1 gram—you will need 10 grams of powdered gum arabic. Divide this amount by half to find the amount of iron sulfate required. Multiply the number of galls you collected by twelve to get the quantity of good white wine you need—you can use any other alcohol, such as rum, brandy, whiskey, vodka, or clear vinegar too.

2. Heat galls in wine
Crush the galls into small pieces, then cautiously heat with gum arabic in the wine in a saucepan. When the wine reaches a high temperature, it will bubble up rapidly, so take the pan off the heat if this happens.

3. Reduce and strain
Add half the iron sulfate and reduce the wine to a third. Strain off the blackened galls (you can make a black pigment with the galls once they have dried).

4. Reheat the wine
Return the pan to the heat and add the remaining iron sulfate to make the ink. Stir the ink well until you are happy with the consistency, then finally strain again.

If you follow the equations and processes in this recipe, you will make a good black ink. Iron gall ink can also be mixed to a white substrate such as chalk or gesso to form a gray paint.

Iron gall ink

Iron gall ink mixed with gesso

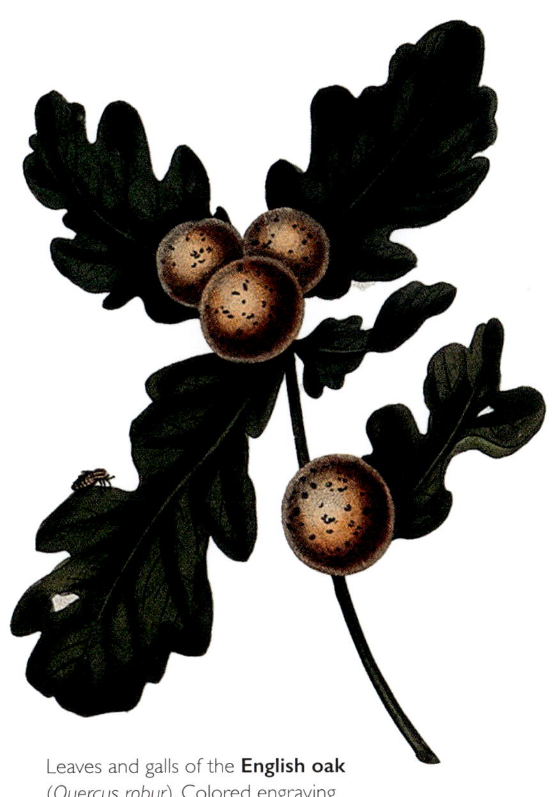

Leaves and galls of the **English oak** (*Quercus robur*). Colored engraving by J. Pass (ca. 1826).

Pomegranate

Latin name: *Punica granatum*

Other names: Carthaginian apple, Granada

Family: *Lythraceae*

Genus: *Punica*

Native: Northeasten Turkey to West and North Pakistan

Type: Deciduous shrub

Pomegranates have been used for food and medicine for thousands of years, with the 16th-century BCE *Ebers Papyrus*, held in Leipzig University Library, suggesting treating tapeworms by soaking the bark in beer or cooking the root in water for a patient to drink. The shape of the fruit inspired ceramic vessels and jewelry, and a silver and ivory pomegranate vase was found in the tomb of Tutankhamun in Egypt (ca. 1325 BCE).

Illustrations of vegetable gardens in Egypt's Old Kingdom (ca. 2649–2130 BCE) depict date palms, juniper, fig, sycamore, grapevines, and, indeed, pomegranates. They have also been found in archaeological investigations of Middle and New Kingdom tombs in Egypt, where they were given as offerings to the dead. Stone carvings of the pomegranate shrub have been excavated from the botanic garden in the festival hall of Tuthmose II (also of the 18th Dynasty).

Pomegranate fruit is associated with many deities, including Kishmo-jin, the Japanese demon and bringer of plagues. Kishmo-jin once ate children before Buddha made her more compassionate, transforming her into a mother goddess. The ancient Syrian deity Kubaba is also associated with the pomegranate, connecting its seeds to the great mother. In Ancient Greek myth, the goddess Persephone was a daughter of Zeus who was abducted by Hades, who tricked her into eating pomegranate seeds when she attempted to escape the Underworld. This compelled her to spend half the year with him and the other with her mother, Demeter, goddess of the corn.

The pomegranate represents eternal life and resurrection in the Christian tradition. The rounded fruit also symbolizes pregnancy, leading to an association with the mother goddess. The pagan worship of this goddess under various names—Astarte, Ishtar, and the Palestinian deity Ashtoreth—eventually evolved into the Christian image of the Virgin Mary as the great mother figure.

In the world of art, the 15th-century Florentine Renaissance painter Sandro Botticelli created an egg tempera painting on a wooden panel called the *Madonna and Child with Angels* (ca. 1487), also known as the *Madonna of the Pomegranate*, which is held by the Uffizi Gallery, Florence. The painting is rich in plant symbolism and features a pomegranate in the hands of the mother and child.

Pomegranate (*Punica granatum*). Illustration by Pierre-Joseph Redouté from *Choix des Plus Belles Fleurs et des Plus Beaux Fruits* (Selection of the Most Beautiful Flowers and Fruits; 1827).

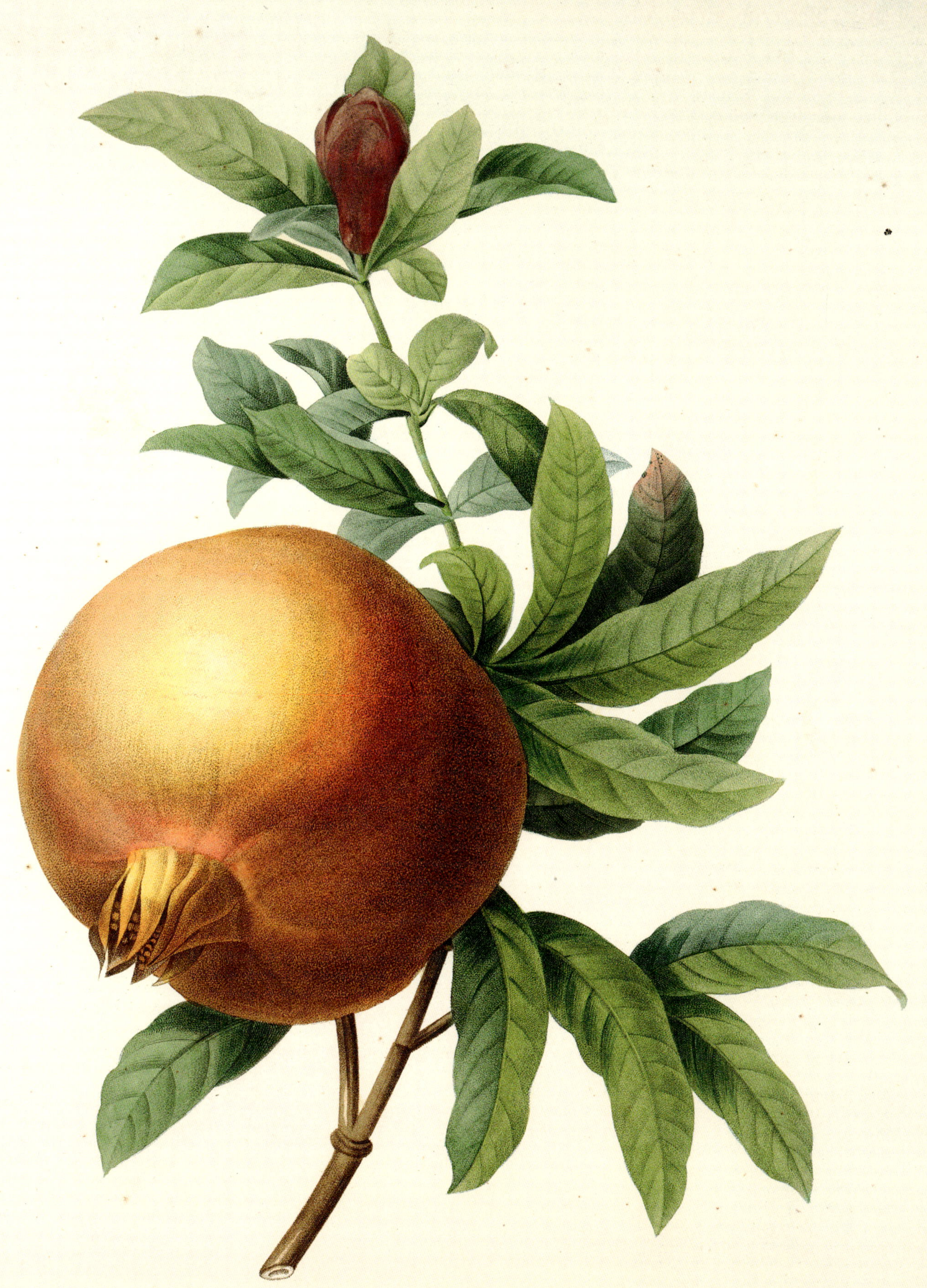

Recipes from Pomegranate

The 3rd-century CE *Leiden Papyrus* X (P.LEID.X), in the Dutch National Museum of Antiquities, and the *Papyrus Graecus Holmiensis* (ca. 300 CE), or *Stockholm Papyrus*, offer deep insight into organic colors. The *Stockholm Papyrus* contains a recipe for pomegranate added to vinegar-water and reduced by half. Iron-mordanted wool is then added to the vinegar-water to produce a "Sardian Purple" dye. The papyrus also explains how to use pomegranate blossoms and alum to produce yellows and the root with safflower (*Carthamus tinctorius*; see page 104) and a little verdigris to make a green color.

For my own experiments, I collected 30 grams of pomegranate blossoms and made a black ink using 5 grams of iron sulfate and 5 fluid ounces (150 milliliters) of water. I heated and reduced the dye to 2½ fluid ounces (75 milliliters). I added grapefruit seed extract to combat mold growth.

The 11th-century Arabic *Book of the Staff of the Scribe* by al-Muʿizz ibn Bādīs provides a recipe for making a yellow ink or glaze using pomegranate blossoms and saffron mixed with strong vinegar and gum arabic. Another recipe details a purple writing ink that can be made with sweet violet (*Viola odorata*), greater celandine (*Chelidonium majus*; see page 34), saffron (*Crocus sativa*; see page 30), and pomegranate rind in myrtle water and gum arabic. Although too much pomegranate could oxidize to a black, this would still produce a good purple color.

The Italian Gioanventura Rosetti's dye book *Plictho de Larte de Tentori* (Instructions in the Art of Dyers; 1548) includes several recipes for making "black water" by using walnut husks and pomegranate rinds (see *Recipe 1*). This combination was common when making an ink or dye, as both contain tannins, which is ideal for making a black.

A 16th-century recipe for black ink appears in a small collection of texts (MS no. 2,556) by anonymous Jewish alchemists, held in the Jewish Theological Seminary Library, New York. The "Description of the Ink of Ezra haSofer" (Ezra the Scribe, a 5th-century-BCE scribe and priest) uses myrtle and olive leaves, water, gallnuts, green walnut husks, and pomegranate peel.

Pomegranate (*Punica granatum*). Illustration from a Persian manuscript (ca. 1540–45) of a work by 13th-century author Zakarīyā ibn Muḥammad Qazwīnī.

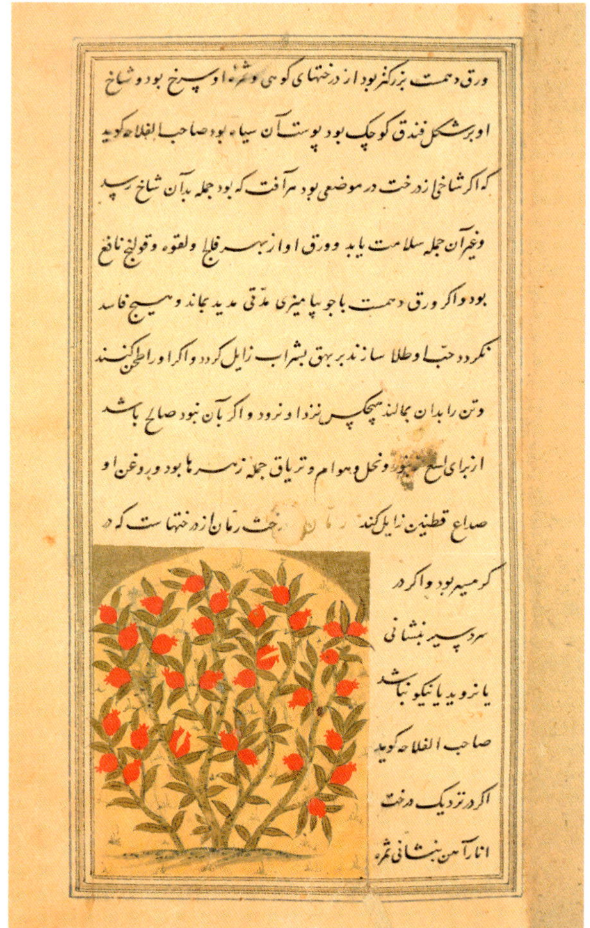

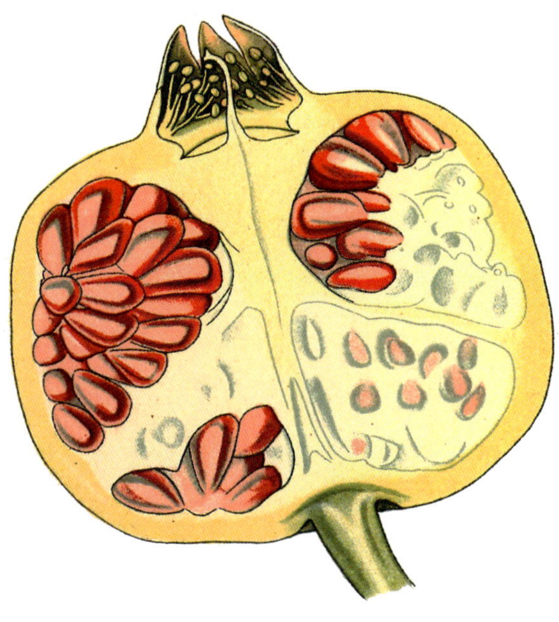
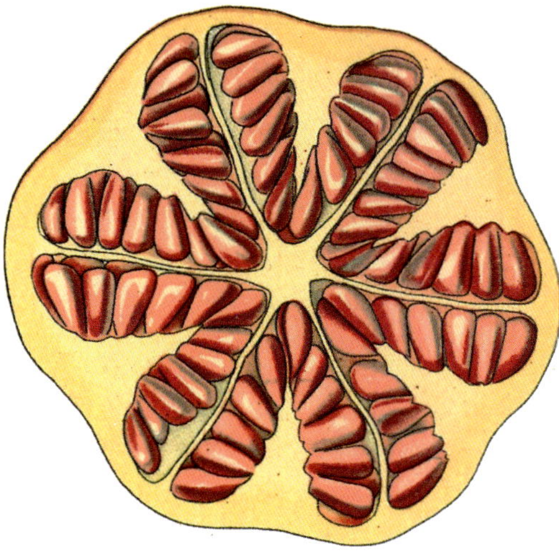

Pomegranate (*Punica granatum*). Illustration from *Köhler's Medizinal-Pflanzen* (Köhler's Medicinal Plants; vol.2, 1890).

Recipe 1: Making black water

"Measure juice of shells of fresh walnuts, juice of pomegranates, gum arabic, of all these three things, half an ounce of Roman vitriol [iron sulfate] and distill and you will make black water."

—Recipe 14, Gioanventura Rosetti, *Plictho de Larte de Tentori*, 1548

Recipe 2: Making pomegranate-black paint and ink

To make a black paint: Wash and dry 82 grams of pomegranate rinds, then heat in water and simmer for an hour. Add 14 grams of iron sulfate and reduce the water by half. Sieve the dye through a fine-mesh strainer and leave the waste to dry. Once dried, crush the blackened skins in a pestle and mortar to make a fine pigment, then add 10 grams of pigment to 5 fluid ounces (150 milliliters) of clear vinegar and leave for several days. Sieve the vinegar then let the pigment dry out before adding to gum arabic to make a black paint.

To make a black ink: Reheat the darkened vinegar with gum arabic and reduce to thicken the liquid to make a black ink.

—Nabil's modern recipe

Colors from Nabil's Practice

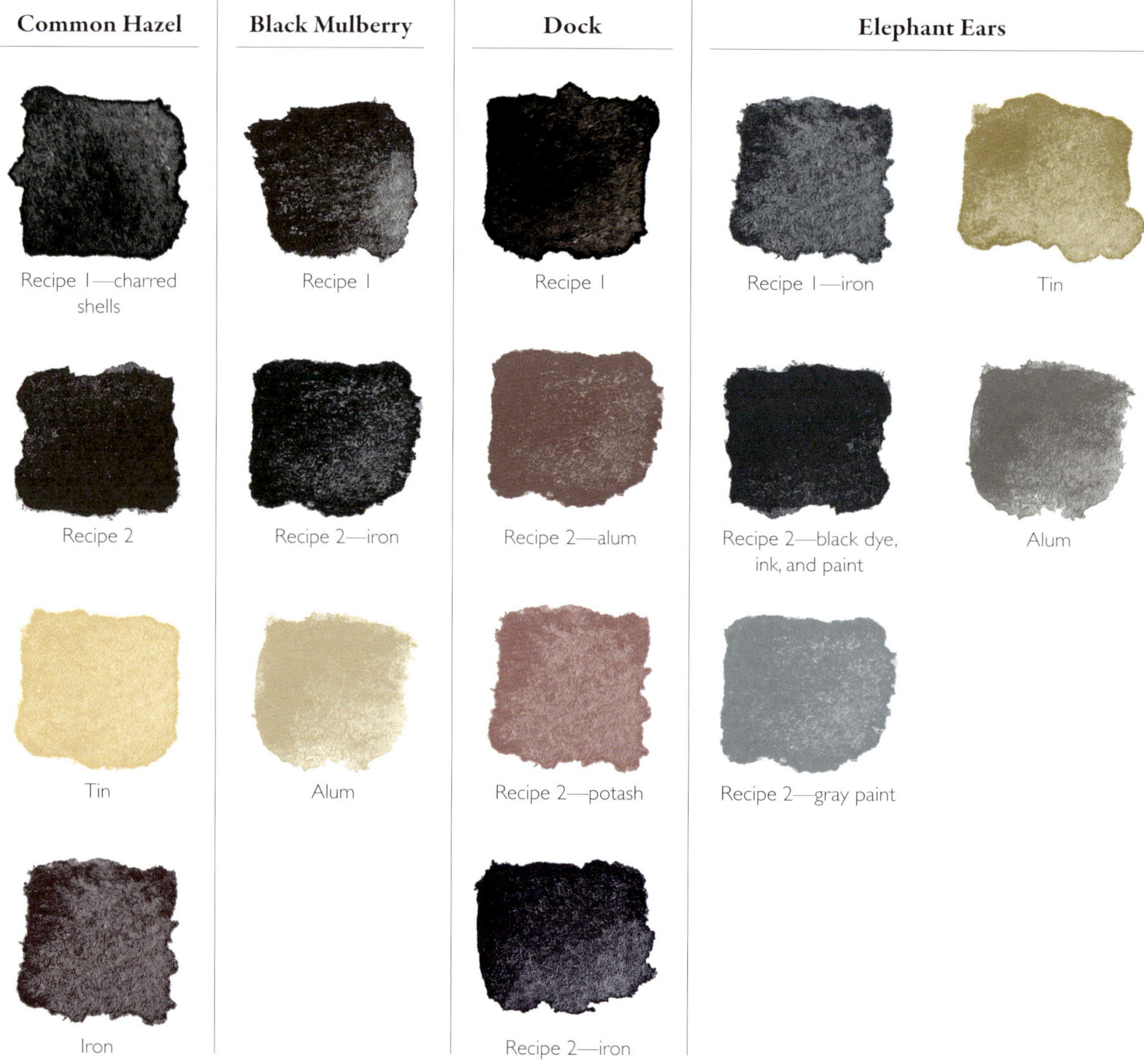

The key to making organic black using a structured framework (see pages 17–19) is based on the plant's chemical compounds. The focused ingredient is gallic tannins found in trees, fruits, and leaves, with the highest content contained within oak galls, elephant ear leaves, and the waste skins of the pomegranate fruit. Many other plants can produce a successful dark black dye ink that each possess different characteristics relating to their distinctive blackness.

Giant Rhubarb	Smoke Tree	Oak Galls	Pomegranate
Recipe 1—root and iron	Recipe 1	Iron	Recipe 1
Recipe 1—leaf and iron	Recipe 2—iron	Ink and gesso	Recipe 2—iron
Recipe 1—leaf and alum	Purple leaves—no mordant		

Public Dye Index

Making plant colors accessible through a public dye index was an initiative of the author while serving as an artist-in-residence at the Cambridge University Botanic Garden, and was supported by Arts Council England. The index is a catalog that explores site-specific plants in the garden's Living Collections and guides people toward basic recipes for making colors from non-traditional dye plants. Below is a list of the plants used, which were processed through experimentation, highlighting the mordants applied to make specific colors. The catalog can be found on the Cambridge University Botanic Garden website under "Botanic Dyes."

Common Name	**Latin Name**	**Part Used**	**Mordant = Colour**
Amazon moonflower	*Strophocactus wittii*	Flowers	Potash = Yellow
Black-grape cotoneaster	*Cotoneaster ignotus*	Berries	Alum = Purple
California allspice	*Calycanthus occidentalis*	Flowers	Alum = Green / Yellow
California poppy	*Eschscholzia californica*	Flowers	Alum or Baking Soda = Yellow / Green
Caucasian osmanthus	*Osmanthus decorus*	Fruits	Iron = Black
Chinese pokeweed	*Phytolacca polyandra*	Berries	Alum = Purple
Common alder	*Alnus glutinosa*	Leaves	Tin = Yellow; Iron = Brown
Coneflower	*Rudbeckia laciniata*	Flowers	Alum = Olive Green; Potash = Yellow-Tan
Cup and saucer plant	*Cobaea scandens*	Flowers	Alum or Tin = Purple / Green
Dahlia	*Dahlia* 'Orfeo'	Flowers	Potash = Yellow / Orange
Darwin's barberry	*Berberis darwinii*	Berries	Alum = Purple–Gray-blue
Evening primrose	*Oenothera biennis*	Flowers	Iron = Black
Fuchsia	*Fuchsia* 'Mrs Popple'	Flowers	Alum = Yellow / Brown
Gentian sage	*Salvia patens*	Flowers	Potash = Yellow / Baking Soda = Green
Great millet	*Sorghum bicolor*	Leaves	Iron = Black
Lebanese wild apple	*Malus tribolata*	Fruits	No mordant = Green with Verdigris
Ligularia	*Ligularia japonica*	Flowers	Alum = Green / Yellow
Lion's tail	*Leonotis leonurus*	Flowers	Alum = Yellow; Potash = Brown
Magnolia	*Magnolia* 'Vulcan'	Flowers	Alum or Tin = Pink / Purple; Iron = Black
Malabar ebony	*Diospyros malabarica*	Leaves	Iron = Black
May apple	*Podophyllum peltatum*	Leaves	Alum or Potash = Yellow
Monkshood	*Aconitum carmichaelii*	Flowers	Potash = Green; Alum = Yellow
Nepal mahonia	*Berberis napaulensis*	Bark	None = Yellow; Alum = Green; Potash = Pink; Iron = Orange-brown
Onion	*Allium cepa*	Skin	Alum = Yellow
Peruvian wild petunia	*Ruellia chartacea*	Flowers	Alum = Orange; Tin = Pink
Pink powderpuff	*Calliandra haematocephala*		Flowers Alum = Purple
Purple amaranth	*Amaranthus cruentus*	Flowers	Alum = Yellow
Red bistort	*Persicaria amplexicaulis*	Leaves	Alum or Potash = Yellow
Rose of Sharon	*Hibiscus syriacus*	Flowers	Green / Yellow / Brown
Rowan	*Sorbus aucuparia*	Leaves	Alum = Yellow; Potash = Reddish-salmon; Iron = Black-brown
Snow poppy	*Eomecon chionantha*	Flowers	Potash = Yellow
St. John's wort	*Hypericum perforatum*	Flowers	Alum or Potash = Green; Baking Soda = Yellow
Strawberry tree	*Arbutus unedo*	Fruits	Potash = Purple; Iron = Gray-black
Swedish whitebeam	*Sorbus intermedia*	Berries	Iron = Greenish-black
Sweet chestnut	*Castanea sativa*	Leaves	Iron = Black / Brown
Yellow deadly nightshade	*Atropa belladonna* var. *lutea*		Berries Alum = Purple–Blue; Potash = Yellow
Yellow jasmine	*Jasminum humile*	Flowers	Alum and tin = Yellow; Iron = Orange

Bibliography

Adam, F. (1977) *The Clans, Septs & Regiments of the Scottish Highlands*. London: Johnston and Bacon.

Adrosko, R. J. (2012) *Natural Dyes and Home Dyeing*. New York: Dover.

Aesop. (2016) *Aesop's Fables – Complete Collection*. Enhanced Media Publishing.

Ahmed, H. E. (2009) "Natural Dyes in North Africa 'Egypt.'" In *Handbook of Natural Colorants*, edited by T. Bechtold and R. Mussak, pp.27–37, 34. John Wiley & Sons.

Ahmed, H. E. (2019) "Natural Dyes in Historical Egyptian Textiles." *Trends in Textile & Fashion Designing* 3 (2): pp.309–10.

Albertus Magnus. (1999) *The Book of Secrets of Albertus Magnus of the Virtues: Of Herbs, Stones, and Certain Beasts, Also a Book of the Marvels of the World*. Edited by M. R. Best and F. Brightman. Boston: WeiserBooks.

Ali, N. (2018) "Colourants Made from Aphids and Ivy Gum." *Heritage Science* 6: p.38.

Ali, N. (2020) *Goat's Blood, Tablets and Sacred Ivy*. Cambridge Open Engage, Cambridge University Press. www.cambridge.org/engage.

Ali, N. (2024) *Botanic Dyes*. Index Dye-Catalogue. Cambridge University Botanic Garden. www.botanic.cam.ac.uk/collections/botanic-dyes/.

American Academy of Arts and Sciences. (1860) *Proceedings of the American Academy of Arts and Sciences* (Vol.4), 459th meeting. Boston: Welch, Bigelow & Co., p.135.

Araújo, R. et al. (2018) "Silver Paints in Medieval Manuscripts: A First Molecular Survey into Their Degradation." *Heritage Science* 6: pp.1–13.

Aretaeus and Adams, F. (1856) *The Extant Works of Aretaeus, the Cappadocian*. London: Syndenham Society.

Balfour, J. H. (1885) *The Plants of the Bible*. London: Thomas Nelson and Sons.

Balfour-Paul, J. (1998) *Indigo*. London: British Museum Press.

Bahl, B. S., and Bahl, A. (2017) *A Textbook of Organic Chemistry*. New Delhi: S. Chand Publishing.

Bancroft, E. (1794) *Experimental Researches Concerning the Philosophy of Permanent Colours: And the Best Means of Producing Them, by Dyeing, Calico Printing, &c* (Vol. 1). London: Cadell and Davies.

Barkeshli, M. (2015) "Dyes Used by Iranian Masters in Paper Dyeing Process Based on Persian Medieval Recipes." *Abstracts of Papers of the American Chemical Society* 249.

Barkeshli, M., et al. (2025) "Folium in Persian and Islamic Manuscripts (15th–19th Centuries): Historical Significance and Analytical Study." *Restaurator: International Journal for the Preservation of Library and Archival Material* 46 (1): pp.1–33.

Barrows, J. (1735) *Dictonarium Polygraphicum, or The Whole Body of Arts, Regularly Digested*. London: Hitch and Davis.

Beck, L. Y. (2011) *Pedanius Dioscorides of Anazarbus: De Materia Medica*. Hildesheim: Olms-Weidmann.

Betz, H. D. (1986) *The Greek Magical Papyri in Translation, Including the Demotic Spells*. Chicago and London: Chicago University Press.

Bicchieri, M. (2014) "The Purple Codex Rossanensis: Spectroscopic Characterisation and First Evidence of the Use of the Elderberry Lake in a Sixth Century Manuscript." *Environmental Science and Pollution Research* 21 (24): pp.14146–157.

Bioletti, S., et al. (2009) "The Examination of the Book of Kells Using Micro-Raman Spectroscopy." *Journal of Raman Spectroscopy* 40 (8): pp.1043–49.

Bishop, R. R., et al. (2014) "Seeds, Fruits and Nuts in the Scottish Mesolithic." *Proceedings of the Society of Antiquaries of Scotland* 143: pp.9–71. doi.org/10.9750/PSAS.143.9.71.

Boddy, K., Johnson, B. L., and Wickenden, A. (2023) *Saffron: Considering the History of Saffron in Cambridgeshire and Essex*. Cambridge: University of Cambridge. www.cam.ac.uk/stories/saffron.

Borradaile, R., and Borradaile, V. (eds). (1966) *The Strasbourg Manuscript: A Medieval Painters' Handbook Translated from the Old German*. London: Tiranti.

Boujard, V. (2023) *The Kew Gardener's Guide to Growing Shrubs: The Art and Science to Grow with Confidence*. London: Frances Lincoln.

Brittain, J., et al. (2005) *Plant Names Explained: Botanical Terms and Their Meaning*. David and Charles.

Britten, J. (1886) *A Dictionary of English Plant-Names*. London: For the English Dialect Society, Trübner & Co. See p.261.

Broecke, L. (2015) *Cennino Cennini's "Il libro dell'arte": A New English Translation and Commentary with Italian Transcription*. London: Archetype Publications.

Brown, S. (1995) *Stained Glass: An Illustrated History*. London: Bracken Books.

Bunney, S. (ed.). (1992) *The Illustrated Encyclopedia of Herbs: Their Medicinal and Culinary Uses*. London: Chancellor Press.

Caley, E. R. (trans.), and Jensen, W. B. (ed.). (2008) *The Leyden and Stockholm Papyri: Greco-Egyptian Chemical Documents From the Early 4th Century AD*. Cincinnati: University of Cincinnati.

Campbell, J. M., and Thompson, R. A. (1949) *A Dictionary of Assyrian Botany*. London: British Academy.

Campbell-Culver, M. (2004) *The Origin of Plants: The People and Plants That Have Shaped Britain's Garden History*. London: Eden Project Books.

Cardon, D. (2007) *Natural Dyes: Sources. Tradition, Technology and Science*. London: Archetype Publications.

Carter, H. (2014) *The Tomb of Tutankhamun, Discovered by the Late Earl of Carnarvon and Howard Carter: Vol. 2*. London: Bloomsbury Academic.

Casson, L. (2012) *The Periplus Maris Erythraei: Text with Introduction, Translation, and Commentary*. Princeton: Princeton University Press.

Chancier, R. (2003) *Madder Red: A History of Luxury and Trade*. London: Routledge.

Charles, S. J. (2024) *The Medieval Scriptorium: Making Books in the Middle Ages*. London: Reaktion Books.

Clarke, M. (2001) *The Art of All Colours: Mediaeval Recipe Books for Painters and Illuminators*. London: Archetype Publications.

Clarke, M. (2011) *Mediaeval Painters' Materials and Techniques: The Montpellier Liber Diversarum Arcium*. London: Archetype Publications.

Clarke, M. (2012) "A Unique 12th-Century Illuminator's Treatise: An Original Composition Incorporated in the Brussels Compendium Artis Picturae." In *The Artist's Process: Technology and Interpretation*, edited by S. Eyb-Green et al., pp.54–59. London: Archetype Publications.

Clarke, M. (2013) "Recovering the Medieval Palette." In *Colour in the Making: From Old Wisdom to New Brilliance*, edited by P. Ball, M. Clarke, and C. Parraman, pp.44–57. London: Black Dog Publishing.

Clarke, M. (2016) *The Crafte of Lymmyng and the Maner of Steynyng: Middle English Recipes for Painters, Stainers, Scribes, and Illuminators*. Oxford: Oxford University Press.

Clarke, M. (2018) *Tricks of the Medieval Trades: The Trinity Encyclopedia; A Collection of Fourteenth Century English Craft Recipes*. London: Archetype Publications.

Clarke, M. (2026) "Middle English Artisanal Recipes." In *The Oxford Handbook of Middle English Prose*, edited by S. Sobecki and E. Steiner. Oxford: Oxford University Press.

Collins, M. (2000) *Medieval Herbals: The Illustrative Traditions*. London: British Library and University of Toronto Press.

Cruz, Martín de la, et al. (1940) *The Badianus Manuscript, Codex Barberini, Latin 241, Vatican Library: An Aztec Herbal of 1552*. Johns Hopkins Press.

Cuneo, Pia F. (2002) *Artful Armies, Beautiful Battles: Art and Warfare in Early Modern Europe*. Leiden: Brill. See p.94.

De Cleene, M., and Lejeune, M. C. (2002) *Compendium of Symbolic and Ritual Plants in Europe, Vol. II: Herbs*. Ghent: Man and Culture.

De Rosso, V. V., and Mercadante, A. Z. (2009) *Dyes in South America: Handbook of Natural Colorants*. Chichester: John Wiley & Sons. See pp.53–64.

Dean, J. (1999) *Wild Colour*. New York: Watson-Guptil Publications.

Densmore, F. (1928) *Uses of Plants by the Chippewa Indians*. Washington, DC: US Government Printing Office.

Derraik, J. G., and Rademaker, M. (2007) "Phytophotodermatitis Caused by Contact with a Fig Tree (*Ficus carica*)." *The New Zealand Medical Journal* 120 (1261).

Dodge, C. J. (2018) "A Forgotten Century of Brazilwood: The Brazilwood Trade from the Mid-Sixteenth to Mid-Seventeenth Century." *E-Journal of Portuguese History* 16 (1): pp.1–27.

Eastaugh, N., et al. (2007) *Pigment Compendium: A Dictionary of Historical Pigments*. London: Routledge.

Ebeid, H., et al. (2013) "A Study of Dyed Endpapers during Islamic Mediaeval Times in Egypt: Purpose, Materials and Techniques." In *Paper Conservation: Decisions and Compromises*, edited by L. Watteeuw and C. Hofmann, pp.61–65. ICOM-CC Graphic Document Working Group interim meeting, Vienna, 17–19 April 2013. Vienna: Austrian National Library.

Egerton, F. N. (2008) "A History of the Ecological Sciences, Part 27: Naturalists Explore Russia and the North Pacific during the 1700s." *The Bulletin of the Ecological Society of America* 89: pp.39–60.

Eggli, U., et al. (2024) "Flowers of *Aloe vera* from Medieval Manuscripts to Renaissance Printed Books. *Aldrovandiana: Historical Studies in Natural History* 3 (1): pp.151–85.

Egmond, F. (2015) *The World of Carolus Clusius: Natural History in the Making, 1550–1610*. London: Routledge. See p.166.

El Mahdy, C. (1989) *Mummies, Myth and Magic in Ancient Egypt*. London: Thames & Hudson.

Evelyn, J. (1706) *Sylva, or A Discourse of Forest-trees, and the Propagation of Timber in His Majesty's Dominions*. London: Robert Scott.

Fairchild, T. (1722) *The City Gardener. Containing the Most Experienced Method of Cultivating and Ordering Such Ever-Greens, Fruit-Trees, Flowering Shrubs, Flowers, Exotick Plants, &c. as Will be Ornamental, and Thrive Best in the London Gardens*. London: T. Woodward.

Faraone, C. A. (2010) "A Socratic Leaf Charm For Headache (Charmides 155b–157c), Orphic Gold Leaves, and the Ancient Greek Tradition of Leaf Amulets." In *Myths, Martyrs, and Modernity: Studies in the History of Religions in Honour of Jan N. Bremmer*, edited by J. Dijkstra, J. Kroesen, and Y. Kuiper, pp.145–66. Leiden: Brill.

Flowers, T. H., et al. (2019) "Colouring of Pacific Barkcloths: Identification of the Brown, Red and Yellow Colourants Used in the Decoration of Historic Pacific Barkcloths." *Heritage Science* 7 (2). doi.org/10.1186/s40494-018-0243-9

Folkard, R. (1884) *Plant Lore, Legends and Lyrics*. London: Sampson, Low.

Gettens, R. J., and Stout, G. L. (1966) *Painting Materials: A Short Encyclopaedia*. New York: Dover.

Gibbs, L. (2002) *The Complete Fables*. New York: Oxford University Press.

Ginovyan, M., et al. (2023) "Anti-Cancer Effect of *Rumex obtusifolius* in Combination with Arginase/Nitric Oxide Synthase Inhibitors via Downregulation of Oxidative Stress, Inflammation, and Polyamine Synthesis." *The International Journal of Biochemistry and Cell Biology* 158: p.106396.

Gneuss, H., and Lapidge, M. (2014) *Anglo-Saxon Manuscripts: A Bibliographical Handlist of Manuscripts and Manuscript*. Toronto: University of Toronto Press.

Goodwin, J. (1982) *A Dyer's Manual*. London: Pelham Books.

Gordon, L. (1980) *A Country Herbal*. Exeter: Web & Bower.

Grierson, S. (1986) *The Colour Cauldron*. Massachusetts: Charles T. Branford.

Guiley, R. (2010) *The Encyclopedia of Witches, Witchcraft and Wicca*. Infobase Publishing.

Haig, E. (1913) *The Floral Symbolism of the Great Masters*. London: Kegan Paul, Trench, Trübner & Co.

Hall, J. (1974) *Dictionary of Subjects and Symbols in Art*. London: John Murray.

Harris, E., et al. (1981) *Field Guide to the Trees*

and Shrubs of Britain. London: Reader's Digest.

Havkin-Frenkel, D., et al. (2006) *Proceedings of the First International Symposium on Natural Preservatives in Food Systems* (No. 709). Belgium: International Society for Horticultural Science. See pp.16, 32, 101.

Henslow, G. (1899) *Medical Works of the Fourteenth Century Together with a List of Plants Recorded in Contemporary Writings, with Their Identifications by George Henslow*. London: Chapman & Hall.

Höfer, R. (2022) *Renewable Resources for Surface Coatings, Inks and Adhesives*. London: Royal Society of Chemistry.

Hong, D. Y., and Blackmore, S. (eds.). (2015) *The Plants of China*. Cambridge: Cambridge University Press.

Hulton, P. H., and Smith, L. (1979) *Flowers in Art from East and West*. London: British Museum Publications.

Hunt, T. (1995) "Early Anglo-Norman Receipts for Colours." *Journal of the Warburg and Courtauld Institutes* 58: pp.203–9. doi.org/10.2307/751511.

Hurry, J. B. (1930) *The Woad Plant and Its Dye*. London: Oxford University Press. Reprinted 1973, Clifton, NJ: Augustus M. Kelly.

Huxley, L. (ed.) (2011) *Life and Letters of Sir Joseph Dalton Hooker, Vol. 2*. New York: Cambridge University Press.

Islam, A. (2018) "Hazelnut Culture in Turkey." *Akademik Ziraat Dergisi* 7 (2): pp.251–58. doi.org/10.29278/azd.476665.

Jashemski, W. F., and Meyer, F. G. (eds.), (2002) *The Natural History of Pompeii*. Cambridge: Cambridge University Press.

Joyce, P. W. (1887) *The Origin and History of Irish Names and Places*. Dublin: M. H. Gill & Son.

Kalke, C. M. (1985) "The Making of a Thyrsus: The Transformation of Pentheus in Euripides' *Bacchae*." *The American Journal of Philology* 106 (4): pp.409–26.

Kandeler, R., and Ullrich, W. R. (2009) "Symbolism of Plants: Examples from European-Mediterranean Culture Presented with Biology and History of Art." *Journal of Experimental Botany* 60 (9): pp.2461–464.

Kearney, T. H., and Peebles, R.H. (2023) *Arizona Flora: Identifies 3,438 Species of Flowering Plants, Ferns, and Fern-Allies Growing Uncultivated in Arizona*. Berkeley: University of California Press.

Kelly, D. W. (1986) *Charcoal and Charcoal Burning*. Oxford: Shire Publications.

Kingsbury, N. (2016) *Garden Flora: The Natural and Cultural History of the Plants in Your Garden*. Portland: Timber Press.

Kirby, J. (1977) "A Spectrophotometric Method for the Identification of Lake Pigment Dyestuffs." *National Gallery Technical Bulletin* 1: pp.35–45.

Kirby, J., and White, R. (1996) "The Identification of Red Lake Pigment Dyestuffs and a Discussion of Their Use." *National Gallery Technical Bulletin* 17: pp.56–80.

Kirby, J. et al. (2010) *Trade in Artists' Materials: Markets and Commerce in Europe to 1700*. London: Archetype Publications.

Krekel, C., et al. (2017) "De edera et lacca: Identification of a Medieval Colorant Made from Ivy." In *The Diversity of Dyes in History and Archaeology*, edited by J. Kirby, pp.139–45. London: Archetype Publications.

Lawrence, E., & Větvička, V. (1992) *The Illustrated Encyclopedia of Trees and Shrubs*. London: Chancellor Press.

Leapman, M. (2012) *The ingenious Mr Fairchild: the forgotten father of the flower garden*. London: Faber & Faber.

Leeming, D. (2005) *The Oxford companion to world mythology*. Oxford: Oxford University Press.

Lehner, E., and Lehner, J. (1960) *Folklore and Symbolism of Flowers, Plants and Trees*. New York: Tudor Publishing Company.

Lethaby, W. R. (1916) "English Primitives: Master Walter of Colchester, The Incomparable Painter, c. 1180–1248, and the Master of the Chichester Roundel." *The Burlington Magazine for Connoisseurs* 29 (161): pp.189–96.

Levey, M. (1962) "Mediaeval Arabic Bookmaking and Its Relation to Early Chemistry and Pharmacology." *Transactions of the American Philosophical Society* 52 (4): pp.1–79.

Lewington, A. (2003) *Plants for People*. London: Transworld.

Liddell, H. G., et al. (1996) *A Greek-English Lexicon*. 9th ed. Oxford: Clarendon Press. See p.2012.

Liles, J. N. (1990) *The Art and Craft of Natural Dyeing: Traditional Recipes for Modern Use*. Tennessee: University of Tennessee Press. See pp.33–34.

Lindley, J. (1821) *Collectanea botanic, or Figures and Botanical Illustrations of Rare and Curious Exotic Plants*. London: Richard and Arthur Taylor.

Lindley, J., and Moore, T. (eds.). (1876) *The Treasury of Botany: A Popular Dictionary of the Vegetable Kingdom; with Which Is Incorporated a Glossary of Botanical Terms*. London: Longmans, Green & Co.

Longo, L., et al. (2025) "Direct Evidence for Processing *Isatis tinctoria*, a Non-Nutritional Plant, 32–34,000 Years Ago." *PLOS One*. doi.org/10.1371/journal.pone.0321262

Lurker, M. (2004) *The Routledge Dictionary of Gods and Goddesses, Devils and Demons*, London: Routledge.

Lyford, C. A. (1943) *The Crafts of the Ojibwa (Chippewa)*. A publication of the Education Division, United States Indian Affairs. Phoenix, AZ: Phoenix Indian School. See pp.152–53.

Mackay, J. G. (1924) *The Romantic Story of the Highland Garb*. Stirling: E. Mackay.

Mackenzie, C. (1821) *One Thousand Experiments in Chemistry, with illustrations of natural phenomena*. London: Sir R. Phillips and Co.

Maranhão, R., et al. (2016) *Brazilwood: The Colour and the Sound; 500 Years of History of the Wood the County was Named After*. São Paulo: Terceiro Nome.

Marcall, L. (1583) *A Profitable Booke Declaring Dyvers Approoved Remedies, to Take Out Spottes and Staines, in Silkes, Velvets, Linnnen* [sic] *and Woollen Clothes. With Divers Colours How to Die Velvets and Silkes, Linnen and Woollen, Fustian and Threade. Also to Dresse Leather, and to Colour Felles*. London: Purfoote and Pounsonbie. Available online in the digital collection Early English Books Online 2, University of Michigan Library, quod.lib.umich.edu/e/eebo2/B00420.0001.001.

Matteo, M. (2013) *The Four Books of Pseudo-Democritus*. Leeds: Maney Publishing.

Mawa, S., et al. (2013) "*Ficus carica* L.(Moraceae): Phytochemistry, Traditional Uses and Biological Activities. *Evidence-Based Complementary and Alternative Medicine*: p.974256. doi.org/10.1155/2013/974256.

Mayer, R., and Smith, E. (eds.). (1951) *The Artist's Handbook of Materials and Techniques.* London: Faber & Faber.

Mayo, C. A., and Keenan, T. J. (eds.). (1899) *American Druggist and Pharmaceutical Record, Vol. 34.* New York: American Druggist Publishing Co. See p.260.

McLean, T. (1981) *Medieval English Gardens.* London: Collins.

McRae, B. A. (1993) *Colors from Nature: Growing, Collecting, and Using Natural Dyes.* Vermont: Storey Communications.

Melo, M. J. (2009) "History of Natural Dyes in the Ancient Mediterranean World." In *Handbook of Natural Colorants*, edited by T. Bechtold and R. Mussak, pp.3–20. Chichester: Wiley.

Melo, M. J., et al. (2007) "Identification of 7,4′-Dihydroxy-5-methoxyflavylium in 'Dragon's Blood': To Be or Not To Be an Anthocyanin." *Chemistry–A European Journal* 13 (5): pp.1417–22.

Merrifield, M. P. (1849) *Original Treatises Dating from the XIIth to XVIIIth Centuries on the Arts of Painting.* 2 vols. London: John Murray.

Morrison, E., and Grollemond, L. (2019) *Book of Beasts: The Bestiary in the Medieval World.* Los Angeles: Getty Publications.

Mozley, J. (1776) *The School of Wisdom, or Repository of the Most Valuable Curiosities of Art & nature.* London: J. F. & C. Rivington.

Neven, S., and Sanyova, J. (2014) "An Elusive Colorant: Availability, Preparation and Use of Anthocyanin Colorants in European Medieval Illuminators' Workshop." In *Making and Transforming Art: Technology and Interpretation*, edited by H. Dubois et al., pp.16–22. London: Archetype Publications.

Neven, S. (2016) *The Strasbourg Manuscript: A Medieval Tradition of Artists' Recipe Collections (1400–1570).* London: Archetype Publications.

Newman, W. R. (2019) *Newton the Alchemist: Science, Enigma, and the Quest for Nature's "Secret Fire".* Princeton: Princeton University Press.

Newton, Isaac, Sir. (ca.1669–93) *Newton Papers: Laboratory Notebook* (MS Add.3975 f.13r). Cambridge University Library. Available at: https://cudl.lib.cam.ac.uk/view/MS-ADD-03975/29.

Nicholson, P. T., and Shaw, I. (2000) *Ancient Egyptian Materials and Technology.* Cambridge: University Press Cambridge.

Oman, C. (1932) "The Goldsmiths at St. Albans Abbey during the 12th and 13th Centuries." *St Albans and Hertfordshire Architectural and Archaeological Society Transactions.*

Panayotova, S., Jackson, D., and Ricciardi, P. (eds.). (2016) *Colour: The Art & Science of Illuminated Manuscripts.* London: Harvey Miller Publishers.

Patel, R. (2014) *The Jewish Alchemists: A History and Source Book.* Princeton: Princeton University Press.

Paton, W. R. (1918) *The Greek Anthology.* Cambridge, MA: Harvard University Press.

Peterson, J. F., and Terraciano, K. (eds.). (2019) *The Florentine Codex: An Encyclopedia of the Nahua World in Sixteenth-Century Mexico.* Austin: University of Texas Press.

Phipps, E. (2010) *Cochineal Red: The Art History of a Color.* New York: Metropolitan Museum of Art. See pp.28, 36.

Pliny the Elder. (1855) *The Natural History.* Translated by B. Bostock and H. T. Riley. London: Taylor and Francis. See book 21, chap. 24: "The Cyanos: The Holochrysus."

Porter, Y. (2007) *Painters, Paintings and Books: An Essay on Indo-Persian Technical Literature, 12–19th Centuries.* New Delhi: Manohar Publishers & Distributors. See p.38.

Poulain, M., Speecke, M., Ervynck, A., et al. (2025) "At the Apothecary: Life in an International District in 15th-Century Bruges." *Medieval Archaeology* 69 (1): pp.94–139. doi.org/10.1080/00766097.2025.2504289

Prance, G., and Nesbitt, M. (eds.). (2012) *The Cultural History of Plants.* New York: Routledge.

Raggio, O., and Wilmering, A. M. (1999) *The Gubbio Studiolo and Its Conservation: Italian Renaissance Intarsia and the Conservation of the Gubbio Studiolo.* New York: Metropolitan Museum of Art.

Ramos, D. E. (ed.). (1998) *Walnut Production Manual.* Oakland: University of California.

Rehak, P. (2002) "Imag(in)ing a Women's World in Bronze Age Greece: The Frescoes from Xeste 3 at Akrotiri, Thera." In *Among Women: From the Homosocial to the Homoerotic in the Ancient World*, edited by N. S. Rabinowitz and L. Auanger, pp.34–59. Austin: University of Texas Press.

Renouf, P. Le Page. (1893) *The Egyptian Book of the Dead.* London: Society of Biblical Archaeology.

Reynolds, T. (ed.). (2004) *Aloes: The Genus Aloe.* Boca Raton: CRC Press.

Rice, G. (2011) *Planting the Dry Shade Garden: The Best Plants for the Toughest Spot in Your Garden.* Portland: Timber Press.

Richmond, A. (2024) *A Short History of Flowers: The Stories That Make Our Gardens.* London: Frances Lincoln.

Robins, G. (2008) *The Art of Ancient Egypt.* Cambridge, MA: Harvard University Press.

Rohde, E. S. (1932) *The Story of the Garden.* London: Medici Society.

Rohde, E. S. (1969) *A Garden of Herbs.* New York: Dover.

Rohde, E. S. (1972) *The Old English Herbals.* London: Minerva Press.

Rosetti, G. (1969) *The Plictho: Instructions in the Art of the Dyers which Teaches the Dyeing of Woolen Cloths, Linens, Cottons, and Silk by the Great Art as Well as by the Common.* Translation of the 1st ed. of 1548 by S. M. Edelstein and H. C. Borghetty. Cambridge, MA: MIT Press.

Roux, J. (2020) *Floriography: An Illustrated Guide to the Victorian Language of Flowers.* Vol. 1. Andrews McMeel Publishing.

Rowland, I. D., and Howe, T. N. (1999) *Vitruvius' Ten Books on Architecture.* Cambridge: Cambridge University Press.

Ruiz, H. (1998) *The Journals of Hipólito Ruiz, Spanish Botanist in Peru and Chile, 1777–1788.* Portland: Timber Press.

Ruscelli, G. (1595) *The Secrets of the Reverend Maister Alexis of Piemont: Containing excellent remedies against diverse diseases, wounds, and other accidents, with the maner to make distillations, parfumes, confitures, dyings, colours, fusions, and meltings*. Translated by W. Ward. London: Peter Short for Thomas Wight.

Rydén, M. (1984) *The English Plant Names in the Grete Herball (1526): A Contribution to the Historical Study of English Plant-Name Usage*. Stockholm: Almqvist & Wiksell.

Saras, T. (2023) *Aloe Vera Unveiled: Nature's Healing Wonder*. Semarang: Tiram Media.

Savage, F. G. (1923) *The Flora and Folk Lore of Shakespeare*. Stratford-on-Avon: Shakespeare Press.

Schweppe, H. (1986) *Practical Hints on Dyeing with Natural Dyes: Production of Comparative Dyeings for the Identification of Dyes on Historic Textile Materials*. Suitland, MD: Smithsonian Institution, Conservation Analytical Laboratory.

Shibayama, N., et al. (2015) "Analysis of Natural Dyes and Metal Threads Used in 16th–18th Century Persian/Safavid and Indian/Mughal Velvets by HPLC-PDA and SEM-EDS to Investigate the System to Differentiate Velvets of These Two Cultures." *Heritage Science* 3.

Skelly, C. J. (1994) *Dictionary of Herbs, Spices, Seasonings and Natural Flavorings*. New York: Routledge.

Skemer, D. C., et al. (2013) *Medieval and Renaissance Manuscripts in the Princeton University Library*. Princeton: Princeton University Press. See p.64.

Simons, G. L. (1973) *Sex and Superstition*. London: Abelard-Schuman.

Smith, C. S., and Hawthorne, J. G. (1974) "*Mappae Clavicula*: A Little Key to the World of Medieval Techniques." *Transactions of the American Philosophical Society* 64 (4): pp.1–128.

Smith, H. H. (1932) *Ethnobotany of the Ojibwe Indians*. Milwaukee: Order of the Board of Trustees.

Smith, J. R. (1996) *Safflower*. Champaign, IL: AOCS Publishing.

Theophilus. (1979) *On Divers Arts: The Foremost Medieval Treatise on Painting, Glassmaking and Metalwork*. Translated with introduction and notes by J. G. Hawthorne and C. S. Smith. New York: Dover.

Theophrastus and Hort, A. (1916) *Theophrastus: Enquiry into Plants and Minor Works on Odours and Weather Signs*. London: Heinemann.

Thompson, D. V. (1926) "*Liber de coloribus illuminatorum siue pictorum* from Sloane Ms. No. 1754." *Speculum* 1 (3): pp.280–307.

Thompson D. V. (trans.).(1933) *Cennino d'Andrea Cennini da Colle di Val d'Elsa: Il libro dell'arte*. 2 vols. New Haven: Yale University Press. Vol. 2 (translation) also published separately as *The Craftsman's Handbook* by Yale University Press, New Haven, 1933; reprinted 1960, Dover, New York.

Thompson, D. V. (1936) *The Materials and Techniques of Medieval Painting*. London: George Allen & Unwin.

Thompson, D. V. (1956) *The Materials and Techniques of Medieval Painting*. New York: Dover.

Thompson, D. V., and Hamilton, G. H. (trans.). (1933) *An Anonymous Fourteenth-Century Treatise, "De Arte Illuminandi," The Technique of Manuscript Illumination*. Translated from the Latin of Naples MS. XII. E. 27. New Haven: Yale University Press.

Thompson, R. C. (1934) "An Assyrian Chemist's Vade-Mecum." *Journal of the Royal Asiatic Society* 66 (4): pp.771–85.

Totelin, L. M. (ed.). (2009) *Hippocratic Recipes: Oral and Written Transmission of Pharmacological Knowledge in Fifth- and Fourth-Century Greece*. Leiden: Brill.

Treveris, P. (1526) *The Grete Herball*. London.

Turner, N. K., and Oltrogge, D. (2016) "Pigment Recipes and Model Books: Mechanisms for Knowledge Transmission and the Training of Manuscript Illuminators." In *Colour: The Art & Science of Illuminated Manuscripts*, edited by S. Panayotova, D. Jackson, and P. Ricciardi, pp.89–107. London: Harvey Miller.

Turner, W. (1881) *The Names of Herbes, by William Turner. A.D. 1548. Edited with an Introduction, an Index of English Names and an Identification of the Plants, by James Britten*. London: Trübner & Co. See pp.50, 105.

Ungerer, J. T. (2024) *Grow, Gather, Heal: Goldenrod—Nature's Unsung Hero; A Deep Dive Into Goldenrod's History, Folk and Traditional Remedies, Medicinal Benefits, Recipes, Herbal Uses, Growing and Harvesting*. John T. Ungerer.

Vandivere, A., van Loon, A., Callewaert, T., et al. (2019) "Fading into the Background: The Dark Space Surrounding Vermeer's *Girl with a Pearl Earring*." *Heritage Science* 7.

Vejar, K., et al. (2020) *Journeys in Natural Dyeing: Techniques for Creating Color at Home*. New York: Abrams.

Vest, M., and Wouters, J. (1999) "Dyestuffs and Pigments in 12th to 18th Century Alum Tawed Bookbinding Leather in European Collections." In *ICOM Committee for Conservation 12th Triennial Meeting, Lyon, 29 August–3 September 1999*, Vol.2, pp.714–20.

Wallert, A. (1995) "*Libro Secondo de Diversi Colori e Sise da Mettere a Oro*: A 15th-Century Technical Treatise on Manuscript Illumination." In *Historical Painting Techniques, Materials, and Studio Practice*, edited by A. Wallert, E. Hermens, and M. F. J. Peek, pp. 38–47. Los Angeles: Getty Conservation Institute.

Watt, M., and Sellar, W. (2012) *Frankincense & Myrrh: Through the Ages, and a Complete Guide to Their Use in Herbalism and Aromatherapy Today*. New York: Random House.

Watts, D. (2007) *Dictionary of Plant Lore*. London: Academic Press.

Whigham, P., and Jay, P. (1975) *The Poems of Meleager*. Berkeley: University of California Press.

White, R. (1986) "Brown and Black Organic Glazes, Pigments and Paints." *National Gallery Technical Bulletin* 10: pp.58–71.

Wilkinson, A. (1998) *The Garden in Ancient Egypt*. London: Rubicon Press.

Wilson, E. H. (1913) *A Naturalist in Western China with Vasculum, Camera, and Gun*. London: Methuen & Co.

Withering, W. , et al. (1787) *A Botanical Arrangement of British Plants; Including the Uses of Each Species ... With an Easy Introduction to the Study of Botany ... Illustrated by Copper Plates*. 2nd ed. London: M. Swinney.

Yoshida, K., et al. (2009) "Blue Flower Color Development by Anthocyanins: From Chemical Structure to Cell Physiology." *Natural Product Reports* 26 (7): pp.884–915.

Manuscript Resources

Bern (Switzerland), City and University Library
Colmar Art Book (150 MS.Helv.Hist.XII 45) – 15th century (Neven, unpublished)

Bologna, University Library
Arabic manuscript (Cod. Arab. 2954) – 13th century (Eggli et al., 2024)

Segreti per Colori (Secrets of Color) (MS. 2861) – 15th century (Merrifield, 1849)

Brussels, Royal Library
Compendium Artis Picturae (Compendium of the Art of Painting) (MS. 10 152) – 12th century (Clarke, 2012)

Cambridge, Trinity College
Trinity Encyclopedia (MS. O.9.39) – 14th century (Clarke, 2018)

Cambridge, University Library
(MS. Dd.v.76) – late 14th century/ early 15th century (Clarke, 2018)

Philosophiæ Naturalis Principia Mathematica (Mathematical Principles of Natural Philosophy) (Adv.b.39.1) – 17th century (Newton, 1687)

Sir Isaac Newton's Iron Gall ink recipe (Add MS. 3975 f.13r) – 17th century (Newton, ca.1669–93)

Dioscorides, Pedanius
De Materia Medica – 1st century (Beck, 2011)

Dublin, Trinity College
The Book of Kells (MS. 58) – 8th–9th century (Bioletti et al., 2009)

Florence, Medici Laurentian Library
Il Libro dell'Arte (The Craftsman's Handbook) (MS. XXIII, Plut.78–23 cc.44–87v) – 15th century (Broecke, 2015)

Bernardino de Sahagún, *Il Codice Fiorentino* (The Florentine Codex) (Med. Palat. 218–220) – 16th century (Peterson and Trerraciano, 2019)

Heildelberg, Heidelberg University Library
Farb und Tintenrezepte (Paint and Ink Recipes) (Cod. Pal. Germ.489) – 16th century. Available at: https://digi.ub.uni-heidelberg.de/diglit/cpg489/0059

Periplus Maris Erythraei (Periplus of the Erythraean Sea) (Cod. Pal. Grace.398) – Late 9th century (Casson, 2012)

Göttingen (Germany), Lower Saxony State and University Library
Göttingen Musterbuch (Göttingen Model Book) (Cod. MS. 8° Uff.51 Cim) – 16th century (Turner and Oltrogge, 2016)

Leiden, National Museum of Antiquities
Papyri Demoticae Magicae (Demotic Magical Papyri) (PDM.XIV.563–74) – 5th century

Papyri Graecae Magicae (Greek Magical Papyri) (PGM.CXXIII) (PGM XII.397–400) – 5th century (Betz, 1986)

Leipzig (Germany), University Library
Ebers Medical Papyrus – 16th century (Bryan, 1930)

London, British Library
Bald's Leechbook (Royal MS. 12 D XVII) – 9th century (Gneuss and Lapidge, 2014)

Codex Sinaiticus (Book of Sinai) (Add MS. 43725) – 4th-century (Charles, 2024)

De Diversis Artibus (On Various Arts) (MS. Harley 3915) (Royal MS.7.F VIII) – 13th century (Hawthorne and Smith, 1979)

Roger Bacon, *Epistola de Secretis Operibus Artis et Naturae, et de Nullitate Magiae* (Letter on the Secret Works of Art and Nature, and the Nullity of Magic) – 13th century (Gray et al., 1982)

(Sloane MS 1975, f.38) – 12th century (Collins, 2000)

The Art of Making Colours for to Lymme for Books (MS. Sloane 122) – 16th century (Clarke, 2016)

(MS. Sloane 4) – 16th century (Clarke, 2016)

(MS. Sloane 73) – 14th century (Clarke, 2018)

(MS. Sloane 122) – 16th century (Clarke, 2016)

(MS. Sloane 3548) – 15th century (Clarke, 2016)

Jawar-i Simi (MS. Or. 7465) – 18th century (Barkeshli, 2015)

Plictho de Larte de Tentori, Che Insegna Tenger Pañi Telle Banbasi et Sede si per Larthe Magiore Come per la Commune (Instructions in the Art of Dyers, Which Teaches the Dyeing of Woolen Cloths, Linens, Cotton, and Silk by the Great Art as well as by the Common) (1044.i.19) – 16th century (Rosetti, trans. Edelstein and Borghetty, 1969)

London, British Museum
Papyri Demoticae Magicae (Demotic Magical Papyri) (PDM.XIV.563–74) (P. Lond. 10070.1) (PGM VII.222–49) (P. Lond.121) – 5th century (Betz, 1986)

Sloane Manuscript (MS Sloane 416) – 15th century (Thompson and Hamilton, 1933)

De Coloribus et Mixtionibus (Of Colors and Mixtures) (MS. Cotton Titus D.XXIV) – 12th century (Hunt, 1995)

(MS. Egerton 2852) –14th century (Henslow, 1899)

London, Royal Academy
Dictonarium Polygraphicum or *The Whole Body of Arts, Regularly Digested* (15/1253) – 18th century (Barrow, 1735)

Lucca, Feliniana Chapter Library
Lucca Manuscript (Codex 490) – 8th century (Smith and Hawthorne, 1974)

Massachusetts, Harvard University Houghton Library
MS. Lat.235 – 14th–15th century (Clarke, 2016)

Massachusetts, Harvard University Herbaria and Libraries
De Historia Stirpium Commentarii Insignes (Notable commentaries on the history of plants) – 16th century (Fuchs, 1542)

Mexico City, Library of the National Institute of Anthropology and History
Libellus de Medicinalibus Indorum Herbis (A Book on the Medicinal Herbs of the Indians) (Cruz-Badianus Codex), formally known as (Vatican Library, Codex Barberini, Latin 241) – 16th century (Cruz et al., 1940)

Milan (Italy), State University Library
Papyri Graecae Magicae (Greek Magical

Papyri) (PGM.CXXIII) (P.Mil.Vogl.inv.1245) – 5th century (previously kept at Pisa University Library, Papyrus Cazzaniga no. 1) (Betz, 1986)

Montpellier, Library of the Faculty of Medicine
Liber Diversarum Arcium (Book of Various Arts) (MS. H 277) – 14th century (Clarke, 2011)

Munich, Bavarian State Library
Liber Illuministarum, pro fundamentis auri et coloribus ac consimilibus collectus ex diversis (The Book of the Illuminator) (MS. BSB Cgm. 821) – early 16th century (Clarke, unpublished)

Naples, National Library of Vittorio Emanuele II.
De Arte Illuminandi (On the Art of Illumination) (MS. Latin XII.E.27) – 14th century (Thompson and Hamilton, 1933)

New Jersey, Princeton University Library
Catholic prayer book (MS. Garrett 42) – 14th century (Skemer and Bennett, 2013). Available at: https://catalog.princeton.edu/catalog/9940764823506421

New York, Corning Museum of Glass
Phillipps-Corning Manuscript – late 12th century (Smith and Hawthorne, 1974)

New York, Metropolitan Museum of Art
The Grete Herball (no. 44.7.44) – 16th century (Ryden, 1984), (Treveris, 1529)

New York, Jewish Theological Seminary Library
Treatise on the work of the kinds of colors of ink (No.2556, fol.13a) – 16th century (Patel, 2014)

Nottingham, University Library
Wollaton Antiphonal (Latin MS. 250) – 15th century (Porter, unpublished)

Oxford, Bodleian Library
(MS. Rawlinson C.506) –15th century (Clarke, 2018)

Liber de Coloribus Illuminatorum Sive Pictorum (Book of Colors for Illuminators or Painters) (MS. Sloane 1754) – 15th century (Thompson, 1926), (Clarke, 2001)

(MS Douce 45) – 16th century (Clarke, 2016)

Jawhar-i Simi (Simi's Jewel) (Ms. Or. 7465, fols. 38a–49b) – 15th century (Barkeshli, 2015)

Oxford, Pembroke College
(MS. 21, fol. 272r) – 15th century (Clarke, 2016)

Parma, Palatine Library
O Libro de Komo Se Fazen As Kores (The Book on How to Make Colors) (Parma MS. 1959) – 15th century (Strolovitch, 2010)

Paris, National Library of France
Greek manuscript (BNF Gr.2179) – 9th-century (Eggli et al., 2024)

Experimenta de Coloribus (Experiments upon Colors) (MS. Latin 6741) – 15th century (Merrifield, 1849)

Alī Şeyrafi, *Golzār-e Safā* (Garden of Joy) – 15th century (Barkeshli, 2015)

Liber Magistri Petri de Sancto Audemaro de Coloribus Faciendis (Book of Master Peter of St. Audemar—On Making Colors) (MS. 6741) – 15th century (Merrifield, 1849)

Philo of Byzantium (3rd century BCE), *Mechanike Syntaxis* (Compendium of Mechanics): *Peri Epistolon* (On Secret Letters) (Parisinus Graecus 2522) – 14th century copy (Murray, 2012), (Clarke, unpublished)

Pliny the Elder, *Naturalis Historia* (Natural History) (Ms. lat. 9325) – 1st century (Pliny, trans. Bostock, 1855), (Rackham, 1938–62)

Rome, Library of Casanatense
Libro Secondo de Diversi Colori e Sise da Mettere a Oro (The Second Book of the Different Color from Sis da Mettere a Oro) (MS. 1793) – early 15th century (Wallert, 1995)

Plictho de Larte de Tentori (Instructions in the Art of Dyers) – 16th century (Rosetti, trans. Edelstein and Borghetty, 1969)

Rome, National Central Library
Albertus Magnus, *Liber Aggregationis, seu Liberetta Secretorum de Virtutibus Herbarum Lapidum et Animalium Quorundam. De Mirabilibus Mundi* (The Book of Secrets, of the Virtues of Herbs, Stones, and Certain Beasts, a Book of the Wonders of the World) (InC. 20) – 15th century copy (Albertus Magnus, ed. Best and Brightman, 1973)

Rossano (Italy), Diocesan Museum of Sacred Art
Codex Purpureus Rossanensis (Purple Rossano Gospels) (042) – 6th century (Bicchieri, 2014)

Sélestat (France), Humanist Library
(MS. 17) – 10th century (Smith and Hawthorne, 1974)

St Gallen (Switzerland), Cantonal Library Vadiana
(MS. Vad.429, ff.68v–70r) – 15th century (Neven, unpublished)

Stockholm, National Library of Sweden
Papyrus Graecus Holmiensis (Greek Papyrus of Stockholm) (Acc 2013/75) —5th century (Caley and Jensen, 2008)

Strasbourg (France), National University Library
Strasbourg Manuscript (MS. A VI 19) – 15th century; original manuscript lost to a fire in 1870 (Neven, 2016)

Tehran (Iran), Parliament Library
Resāleh dar Bayān-e Kāğad̲ Morakkab va Ḥall-e Alvān (A Treatise about Paper, Ink and Dissolving Dyes) (No.1 and No.4767) – 15th century (Barkeshli, 2015)

Tehran (Iran), Malek National Library
Resāleh dar Bayān-e Ṭariqeh-ye Sāk̲tan-e Morakkab va Kāğad̲-e Alvān (A Treatise about the Technique of Preparation of Colored Paper and Inks) (No.2870) – 16th century (Barkeshli, 2015)

Vatican City, Vatican Apostolic Library
Theophrasti, Plantarum Historia (Enquiry Into Plants) (Vat.gr.1759) – 3rd–4th century (Theophrastus, Hort, 1916)

Vienna, National Library
Juliana Anicia Codex or *Vienna Dioscurides* (Codex Vindobonensis med. gr.1) – 6th–early 7th century

Vienna, University Library
Innsbruck Manuscript (MS. Cod. 355) – 14th century (Ploss, 1962)

Venice, Marciana National Library
Pseudo-Democritus, *Physika Kai Mystika* (Natural and Mystical Matters) (Marc. Graec. 299) – 1st century (Martelli, 2013)

Vitruvius, *De Architectura Libri Decem* (The Ten Books on Architecture) (Ms It. IV, 152) – 1st century BCE (Rowland and Howe, 1999)

Washington, DC, Library of Congress
The Vertuose Boke of Distyllacyon of the Waters of All Maner of Herbes – 16th century (Brunshwig, 1528)

Index

Credits

Picture Credits
The publisher would like to thank the following for permission to reproduce material. All reasonable efforts have been made to contact copyright holders to obtain permission for the use of copyright material. The publisher apologizes for any errors or omissions and will gratefully incorporate any corrections in future reprints if notified.

Alamy Stock Photo/Album 56, 105, 137, 193; /bilwissedition Ltd. & Co. KG 29, 59; / Florilegius 59, 68, 71, 75, 91, 92; /Historical image collection by Bildagentur-online 86; /Interfoto 7; /Les Archives Digitales 55; /Penta Springs Limited 27, 103, 165; /Quagga Media 61; /The Picture Art Collection 146; /The Reading Room 127

Biodiversity Heritage Library 15, 69, 121, 122, 130, 131, 202

Bridgeman Images/© Florilegius 197; / From the British Library archive 60, 174; /© Look and Learn 194; /© NPL - DeA Picture Library / G. Cigolini 201; /© NPL - DeA Picture Library 72

Cambridge University Botanic Garden 21

Flickr/Bergianska Stiftelsen, Stockholm 139; /Biodiversity Heritage Library 85

Internet Archive 31, 39, 101, 113, 125, 133, 138, 153, 154, 161, 169, 189, 209, 211

J. Paul Getty Museum, Los Angeles (83.ML.102) 11; (86.MV.527) 205

Metropolitan Museum of Art New York, Purchase, The Cloisters Collection, Lila Acheson Wallace Gift, and Rogers Fund, 2019 (2019.197) 90

Missouri Botanical Garden, St. Louis, U.S.A. 64

New York Botanical Garden, LuEsther T. Mertz Library 135

New York Public Library/General Research Division 47; /George Arents Collection 185; /Rare Book Division 27, 181, 187; /Spencer Collection (Persian MS. 39) 40, 48, 102, 142, 162, 171r, 190, 204 / Spencer Collection (Persian MS. 49) 210; /The Miriam and Ira D. Wallach Division of Art, Prints and Photographs: Picture Collection 186

Österreichische Nationalbibliothek, Vienna (Cod. Med. gr. 1 facsimile, fol.198v) 106; /(Cod. Med. gr. 1, fol.147v) 166

Princeton University Library (Garrett MS. 42. Gift of Robert Garrett, 1942.) 206

Rawpixel 22, 28, 35, 43, 67, 73, 76, 79, 89, 97, 115, 123, 143, 157, 158, 163, 167, 173, 183

RHS Lindley Collections 126

Wellcome Collection 12, 16, 32, 44, 207

Wikimedia Commons 87, 98, 99, 129, 155, 170, 182; /British Library 94; / Hortus Botanicus Leiden 63, 199; /Loyola University New Orleans 8; /Missouri Botanical Garden 65; /Naturalis Biodiversity Center/Japanese drawings / Cock Blomhoff Collection 191; /New York Botanical Garden, LuEsther T. Mertz Library 118; /The Norwegian National Library 114, 134, 141, 171l; /Universitätsbibliothek, Leiden(MS. Q 9) 36; /www.BioLib.de 57, 117, 119, 145

Acknowledgments
I am deeply gratefully to the team at UniPress for their direction and expertise, which enabled me to develop and write this book—Daniel Mills, Ruth Patrick, Alex Coco, Caroline West, Luke Herriott, Julia Ruxton, Jason Hook, and Jenny Manstead—and also to Princeton University Press.

I would especially like to thank Mark Clarke for his tremendous support on my journey of exploration and his solid authority on the subject of manuscripts, historical painters, and illuminators' craft practices, along with his comments and professional knowledge.

I also want to express my gratitude to Sam Brockington, Curator at Cambridge University Botanic Garden, for providing me with the opportunity to explore the garden's Living Collections. This experience contributed significantly to the development of some of the colors mentioned in this book. I am sincerely thankful to all the horticultural staff, the Learning and admin team, and the garden staff at Cambridge University Botanic Garden.

Additionally, I would like to thank the following individuals and organizations for their support: Cambridge University Herbarium; Cory and Herbarium Libraries; Arts Council England; Isaac Newton Trust; The Fitzwilliam Museum; London's Natural History Museum; Louise Walsh; Jo Kirby; Sylvie Neven; Maria João Melo; Leslie Carlyle; Jenny Balfour-Paul; Ljiljana Fruk and the FrukLab team at CEB; former Head Gardener Richard Gant at Madingley Hall, Cambridge; Jenny Goodwin; Association of Guilds of Weavers, Spinners and Dyers; Chad Harris, Master Judge with the American Iris Society; and the specialists and experts from around the world with whom I corresponded.

Finally, I owe an enormous debt of gratitude to my parents, family, and friends for their patience and support over the past twenty-five years of my plant journey through art.

Nabil Ali is a visual artist who conducts research working from translated manuscripts containing painters' and illuminators' recipes for plant-based dyes and paints. His work has led him to develop the Botanic Dyes public index dye catalogue, utilising site-specific plants growing in Cambridge University Botanic Garden. Nabil is a visiting tutor at the University of Cambridge and has taught workshops at sites including the Cambridge University Botanic Garden; Ferens Art Gallery; Fitzwilliam Museum; and Radcliffe Observatory Quarter, University of Oxford.

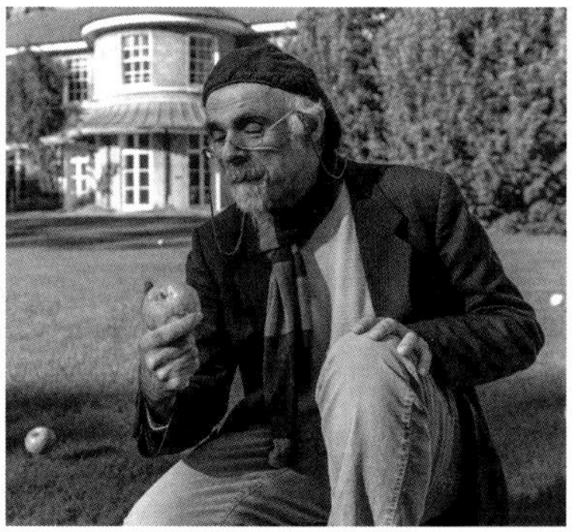